Proceedings of the 3rd RILEM Spring Convention and Conference (RSCC 2020)

RILEM BOOKSERIES

Volume 35

RILEM, The International Union of Laboratories and Experts in Construction Materials, Systems and Structures, founded in 1947, is a non-governmental scientific association whose goal is to contribute to progress in the construction sciences, techniques and industries, essentially by means of the communication it fosters between research and practice. RILEM's focus is on construction materials and their use in building and civil engineering structures, covering all phases of the building process from manufacture to use and recycling of materials. More information on RILEM and its previous publications can be found on www.RILEM.net.

Indexed in SCOPUS, Google Scholar and SpringerLink.

More information about this series at http://www.springer.com/series/8781

Vítor M. C. F. Cunha · Mohammadali Rezazadeh ·
Chandan Gowda
Editors

Proceedings of the 3rd RILEM Spring Convention and Conference (RSCC 2020)

Volume 4: Shift to a Circular Economy

Springer

Editors
Vítor M. C. F. Cunha
Department of Civil Engineering
Institute for Sustainability and Innovation
in Structural Engineering (ISISE)
University of Minho
Guimarães, Portugal

Mohammadali Rezazadeh
Department of Civil Engineering
Institute for Sustainability and Innovation
in Structural Engineering (ISISE)
University of Minho
Guimarães, Portugal

Chandan Gowda
Department of Civil Engineering
Institute for Sustainability and Innovation
in Structural Engineering (ISISE)
University of Minho
Guimarães, Portugal

ISSN 2211-0844 ISSN 2211-0852 (electronic)
RILEM Bookseries
ISBN 978-3-030-76542-2 ISBN 978-3-030-76543-9 (eBook)
https://doi.org/10.1007/978-3-030-76543-9

This Springer imprint is published by the registered company Springer Nature Switzerland AG
The registered company address is: Gewerbestrasse 11, 6330 Cham, Switzerland

Committee

Executive Chair

Eduardo B. Pereira, ISISE—University of Minho, Portugal

Vice-chair

Fábio Figueiredo

Honorary Chair

Joaquim A. O. Barros

Topic 4: Shift to a Circular Economy

Topic Lead

Enzo Martinelli, Università degli Studi di Salerno, Italy

Topic Scientific Committee

Alessandro Pasquale Fantilli, Politecnico di Torino, Italy
Belén González Fonteboa, University of A Coruña, Spain
Carlo Pellegrino, Università di Padova, Italy
Carlos Thomas, Universidad de Cantabria, Spain
Eddie Koenders, Technische Universität Darmstadt, Germany
Jian-Guo Dai, The Hong Kong Polytechnic University, Hong Kong

Jorge de Brito, University of Lisbon, Portugal
Liberato Ferrara, Politecnico di Milano, Italy
Maria Antonietta Aiello, Università del Salento, Italy
Miguel José Pereira, University of Algarve, Portugal
Holmer Savastano Junior, University of São Paulo, Brazil
Asif Hussain Shah, Shaqra University, Saudi Arabia
Carlos Rebelo, University of Coimbra, Portugal
Gintaris kaklauskas, University of Vilnius, Lithuania
Nicolas Roussel, IFSTTAR, France
Nuno Cristelo, University of Trás-os-Montes and Alto Douro, Portugal
Pietro Crespi, Politecnico di Milano, Italy
Žiga Turk, University of Ljubljana, Slovenia
Tiago Miranda, University of Minho, Portugal
Domenico Asprone, University of Naples "Federico II", Italy
Mário Pimentel, University of Porto, Portugal

Preface

The topic of *RSCC 2020 Shift to a Circular Economy* promoted the discussion on a wide range of sustainability-related themes, embracing research and technology on the use and development of sustainable materials and structural systems, recycling and reusing, implementation of industrial processes leading to waste mitigation, digital fabrication and deconstruction, as well as integrative approaches towards achieving a circular economy. Within this topic, twenty-four attendees (from seventeen countries) presented twenty-seven papers.

The construction industry has a crucial role on addressing the worldwide challenges of climate change, urbanization and sustainability. Consequently, it is imperative to develop innovative products, technologies and solutions that can propel the enhancement of the material-, structural-, production- efficiency, as well as the increase of resilience, durability and sustainability of the built environment. Nonetheless, construction industry has the dubious distinction of being the first consumer of global raw materials, as well as producing about half of the world's solid waste. As an illustration, concrete is not only the most used construction material, but also the most widely consumed substance on Earth after water. Moreover, many of the concrete components are consumed faster than they can be replenished (resource depletion). This has an impact on rivers' ecosystems (e.g. fisheries), on hydrological functions (e.g. change in water flows), on infrastructures (e.g. damage to bridges and river embankments), on landscape (e.g. costal erosion) and on extreme events (e.g. decline of protection against extreme events like floods). The previously enunciated environmental impacts just highlight the importance of the sector in contributing to a sustainable development.

Accordingly, the awareness has to be increased and appropriate innovative technical solutions have to be found, in order to pave the way for a more sustainable world for the forthcoming generations. Sustainable construction as a multidimensional concept should aim to promote the increase of structural service life, diverting waste materials (resulting from structure demolition and retrofitting) from disposal into landfills, reducing the need for resource extraction, reduction of non-renewable energy, reducing materials and water consumption, as well as contributing to the sharp decrease of emissions, residues and pollutants. Therefore, the upcoming years are awaited with great expectation regarding eco-innovation within the construction sector.

To close, the editors would like to acknowledge the authors for their contribution, as well as their participation, under unprecedented circumstances, during the outbreak of such a ruthless pandemic. To you, our sincere thanks!

Guimarães, Portugal

Vítor M. C. F. Cunha
Mohammadali Rezazadeh
Chandan Gowda

Contents

RILEM Publications

The following list is presenting the global offer of RILEM Publications, sorted by series. Each publication is available in printed version and/or in online version.

RILEM Proceedings (PRO)

PRO 1: Durability of High Performance Concrete (ISBN: 2-912143-03-9; e-ISBN: 2-351580-12-5; e-ISBN: 2351580125); *Ed. H. Sommer*

PRO 2: Chloride Penetration into Concrete (ISBN: 2-912143-00-04; e-ISBN: 2912143454); *Eds. L.-O. Nilsson and J.-P. Ollivier*

PRO 3: Evaluation and Strengthening of Existing Masonry Structures (ISBN: 2-912143-02-0; e-ISBN: 2351580141); *Eds. L. Binda and C. Modena*

PRO 4: Concrete: From Material to Structure (ISBN: 2-912143-04-7; e-ISBN: 2351580206); *Eds. J.-P. Bournazel and Y. Malier*

PRO 5: The Role of Admixtures in High Performance Concrete (ISBN: 2-912143-05-5; e-ISBN: 2351580214); *Eds. J. G. Cabrera and R. Rivera-Villarreal*

PRO 6: High Performance Fiber Reinforced Cement Composites—HPFRCC 3 (ISBN: 2-912143-06-3; e-ISBN: 2351580222); *Eds. H. W. Reinhardt and A. E. Naaman*

PRO 7: 1st International RILEM Symposium on Self-Compacting Concrete (ISBN: 2-912143-09-8; e-ISBN: 2912143721); *Eds. Å. Skarendahl and Ö. Petersson*

PRO 8: International RILEM Symposium on Timber Engineering (ISBN: 2-912143-10-1; e-ISBN: 2351580230); *Ed. L. Boström*

PRO 9: 2nd International RILEM Symposium on Adhesion between Polymers and Concrete ISAP '99 (ISBN: 2-912143-11-X; e-ISBN: 2351580249); *Eds. Y. Ohama and M. Puterman*

PRO 10: 3rd International RILEM Symposium on Durability of Building and Construction Sealants (ISBN: 2-912143-13-6; e-ISBN: 2351580257); *Ed. A. T. Wolf*

PRO 11: 4th International RILEM Conference on Reflective Cracking in Pavements (ISBN: 2-912143-14-4; e-ISBN: 2351580265); *Eds. A. O. Abd El Halim, D. A. Taylor and El H. H. Mohamed*

PRO 12: International RILEM Workshop on Historic Mortars: Characteristics and Tests (ISBN: 2-912143-15-2; e-ISBN: 2351580273); *Eds. P. Bartos, C. Groot and J. J. Hughes*

PRO 13: 2nd International RILEM Symposium on Hydration and Setting (ISBN: 2-912143-16-0; e-ISBN: 2351580281); *Ed. A. Nonat*

PRO 14: Integrated Life-Cycle Design of Materials and Structures—ILCDES 2000 (ISBN: 951-758-408-3; e-ISBN: 235158029X); (ISSN: 0356-9403); *Ed. S. Sarja*

PRO 15: Fifth RILEM Symposium on Fibre-Reinforced Concretes (FRC)—BEFIB'2000 (ISBN: 2-912143-18-7; e-ISBN: 291214373X); *Eds. P. Rossi and G. Chanvillard*

PRO 16: Life Prediction and Management of Concrete Structures (ISBN: 2-912143-19-5; e-ISBN: 2351580303); *Ed. D. Naus*

PRO 17: Shrinkage of Concrete—Shrinkage 2000 (ISBN: 2-912143-20-9; e-ISBN: 2351580311); *Eds. V. Baroghel-Bouny and P.-C. Aïtcin*

PRO 18: Measurement and Interpretation of the On-Site Corrosion Rate (ISBN: 2-912143-21-7; e-ISBN: 235158032X); *Eds. C. Andrade, C. Alonso, J. Fullea, J. Polimon and J. Rodriguez*

PRO 19: Testing and Modelling the Chloride Ingress into Concrete (ISBN: 2-912143-22-5; e-ISBN: 2351580338); *Eds. C. Andrade and J. Kropp*

PRO 20: 1st International RILEM Workshop on Microbial Impacts on Building Materials (CD 02) (e-ISBN 978-2-35158-013-4); *Ed. M. Ribas Silva*

PRO 21: International RILEM Symposium on Connections between Steel and Concrete (ISBN: 2-912143-25-X; e-ISBN: 2351580346); *Ed. R. Eligehausen*

PRO 22: International RILEM Symposium on Joints in Timber Structures (ISBN: 2-912143-28-4; e-ISBN: 2351580354); *Eds. S. Aicher and H.-W. Reinhardt*

PRO 23: International RILEM Conference on Early Age Cracking in Cementitious Systems (ISBN: 2-912143-29-2; e-ISBN: 2351580362); *Eds. K. Kovler and A. Bentur*

PRO 24: 2nd International RILEM Workshop on Frost Resistance of Concrete (ISBN: 2-912143-30-6; e-ISBN: 2351580370); *Eds. M. J. Setzer, R. Auberg and H.-J. Keck*

PRO 25: International RILEM Workshop on Frost Damage in Concrete (ISBN: 2-912143-31-4; e-ISBN: 2351580389); *Eds. D. J. Janssen, M. J. Setzer and M. B. Snyder*

PRO 26: International RILEM Workshop on On-Site Control and Evaluation of Masonry Structures (ISBN: 2-912143-34-9; e-ISBN: 2351580141); *Eds. L. Binda and R. C. de Vekey*

PRO 27: International RILEM Symposium on Building Joint Sealants (CD03; e-ISBN: 235158015X); *Ed. A. T. Wolf*

PRO 28: 6th International RILEM Symposium on Performance Testing and Evaluation of Bituminous Materials—PTEBM'03 (ISBN: 2-912143-35-7; e-ISBN: 978-2-912143-77-8); *Ed. M. N. Partl*

PRO 29: 2nd International RILEM Workshop on Life Prediction and Ageing Management of Concrete Structures (ISBN: 2-912143-36-5; e-ISBN: 2912143780); *Ed. D. J. Naus*

PRO 30: 4th International RILEM Workshop on High Performance Fiber Reinforced Cement Composites—HPFRCC 4 (ISBN: 2-912143-37-3; e-ISBN: 2912143799); *Eds. A. E. Naaman and H. W. Reinhardt*

PRO 31: International RILEM Workshop on Test and Design Methods for Steel Fibre Reinforced Concrete: Background and Experiences (ISBN: 2-912143-38-1; e-ISBN: 2351580168); *Eds. B. Schnütgen and L. Vandewalle*

PRO 32: International Conference on Advances in Concrete and Structures 2 vol. (ISBN (set): 2-912143-41-1; e-ISBN: 2351580176); *Eds. Ying-shu Yuan, Surendra P. Shah and Heng-lin Lü*

PRO 33: 3rd International Symposium on Self-Compacting Concrete (ISBN: 2-912143-42-X; e-ISBN: 2912143713); *Eds. Ó. Wallevik and I. Níelsson*

PRO 34: International RILEM Conference on Microbial Impact on Building Materials (ISBN: 2-912143-43-8; e-ISBN: 2351580184); *Ed. M. Ribas Silva*

PRO 35: International RILEM TC 186-ISA on Internal Sulfate Attack and Delayed Ettringite Formation (ISBN: 2-912143-44-6; e-ISBN: 2912143802); *Eds. K. Scrivener and J. Skalny*

PRO 36: International RILEM Symposium on Concrete Science and Engineering —A Tribute to Arnon Bentur (ISBN: 2-912143-46-2; e-ISBN: 2912143586); *Eds. K. Kovler, J. Marchand, S. Mindess and J. Weiss*

PRO 37: 5th International RILEM Conference on Cracking in Pavements—Mitigation, Risk Assessment and Prevention (ISBN: 2-912143-47-0; e-ISBN: 2912143764); *Eds. C. Petit, I. Al-Qadi and A. Millien*

PRO 38: 3rd International RILEM Workshop on Testing and Modelling the Chloride Ingress into Concrete (ISBN: 2-912143-48-9; e-ISBN: 2912143578); *Eds. C. Andrade and J. Kropp*

PRO 39: 6th International RILEM Symposium on Fibre-Reinforced Concretes—BEFIB 2004 (ISBN: 2-912143-51-9; e-ISBN: 2912143748); *Eds. M. Di Prisco, R. Felicetti and G. A. Plizzari*

PRO 40: International RILEM Conference on the Use of Recycled Materials in Buildings and Structures (ISBN: 2-912143-52-7; e-ISBN: 2912143756); *Eds. E. Vázquez, Ch. F. Hendriks and G. M. T. Janssen*

PRO 41: RILEM International Symposium on Environment-Conscious Materials and Systems for Sustainable Development (ISBN: 2-912143-55-1; e-ISBN: 2912143640); *Eds. N. Kashino and Y. Ohama*

PRO 42: SCC'2005—China: 1st International Symposium on Design, Performance and Use of Self-Consolidating Concrete (ISBN: 2-912143-61-6; e-ISBN: 2912143624); *Eds. Zhiwu Yu, Caijun Shi, Kamal Henri Khayat and Youjun Xie*

PRO 43: International RILEM Workshop on Bonded Concrete Overlays (e-ISBN: 2-912143-83-7); *Eds. J. L. Granju and J. Silfwerbrand*

PRO 44: 2nd International RILEM Workshop on Microbial Impacts on Building Materials (CD11) (e-ISBN: 2-912143-84-5); *Ed. M. Ribas Silva*

PRO 45: 2nd International Symposium on Nanotechnology in Construction, Bilbao (ISBN: 2-912143-87-X; e-ISBN: 2912143888); *Eds. Peter J. M. Bartos, Yolanda de Miguel and Antonio Porro*

PRO 46: Concrete Life'06—International RILEM-JCI Seminar on Concrete Durability and Service Life Planning: Curing, Crack Control, Performance in Harsh Environments (ISBN: 2-912143-89-6; e-ISBN: 291214390X); *Ed. K. Kovler*

PRO 47: International RILEM Workshop on Performance Based Evaluation and Indicators for Concrete Durability (ISBN: 978-2-912143-95-2; e-ISBN: 9782912143969); *Eds. V. Baroghel-Bouny, C. Andrade, R. Torrent and K. Scrivener*

PRO 48: 1st International RILEM Symposium on Advances in Concrete through Science and Engineering (e-ISBN: 2-912143-92-6); *Eds. J. Weiss, K. Kovler, J. Marchand, and S. Mindess*

PRO 49: International RILEM Workshop on High Performance Fiber Reinforced Cementitious Composites in Structural Applications (ISBN: 2-912143-93-4; e-ISBN: 2912143942); *Eds. G. Fischer and V. C. Li*

PRO 50: 1st International RILEM Symposium on Textile Reinforced Concrete (ISBN: 2-912143-97-7; e-ISBN: 2351580087); *Eds. Josef Hegger, Wolfgang Brameshuber and Norbert Will*

PRO 51: 2nd International Symposium on Advances in Concrete through Science and Engineering (ISBN: 2-35158-003-6; e-ISBN: 2-35158-002-8); *Eds. J. Marchand, B. Bissonnette, R. Gagné, M. Jolin and F. Paradis*

PRO 52: Volume Changes of Hardening Concrete: Testing and Mitigation (ISBN: 2-35158-004-4; e-ISBN: 2-35158-005-2); *Eds. O. M. Jensen, P. Lura and K. Kovler*

PRO 53: High Performance Fiber Reinforced Cement Composites—HPFRCC5 (ISBN: 978-2-35158-046-2; e-ISBN: 978-2-35158-089-9); *Eds. H. W. Reinhardt and A. E. Naaman*

PRO 54: 5th International RILEM Symposium on Self-Compacting Concrete (ISBN: 978-2-35158-047-9; e-ISBN: 978-2-35158-088-2); *Eds. G. De Schutter and V. Boel*

PRO 55: International RILEM Symposium Photocatalysis, Environment and Construction Materials (ISBN: 978-2-35158-056-1; e-ISBN: 978-2-35158-057-8); *Eds. P. Baglioni and L. Cassar*

PRO 56: International RILEM Workshop on Integral Service Life Modelling of Concrete Structures (ISBN 978-2-35158-058-5; e-ISBN: 978-2-35158-090-5); *Eds. R. M. Ferreira, J. Gulikers and C. Andrade*

PRO 57: RILEM Workshop on Performance of cement-based materials in aggressive aqueous environments (e-ISBN: 978-2-35158-059-2); *Ed. N. De Belie*

PRO 58: International RILEM Symposium on Concrete Modelling—CONMOD'08 (ISBN: 978-2-35158-060-8; e-ISBN: 978-2-35158-076-9); *Eds. E. Schlangen and G. De Schutter*

PRO 59: International RILEM Conference on On Site Assessment of Concrete, Masonry and Timber Structures—SACoMaTiS 2008 (ISBN set: 978-2-35158-061-5; e-ISBN: 978-2-35158-075-2); *Eds. L. Binda, M. di Prisco and R. Felicetti*

PRO 60: Seventh RILEM International Symposium on Fibre Reinforced Concrete: Design and Applications—BEFIB 2008 (ISBN: 978-2-35158-064-6; e-ISBN: 978-2-35158-086-8); *Ed. R. Gettu*

PRO 61: 1st International Conference on Microstructure Related Durability of Cementitious Composites 2 vol., (ISBN: 978-2-35158-065-3; e-ISBN: 978-2-35158-084-4); *Eds. W. Sun, K. van Breugel, C. Miao, G. Ye and H. Chen*

PRO 62: NSF/ RILEM Workshop: In-situ Evaluation of Historic Wood and Masonry Structures (e-ISBN: 978-2-35158-068-4); *Eds. B. Kasal, R. Anthony and M. Drdácký*

PRO 63: Concrete in Aggressive Aqueous Environments: Performance, Testing and Modelling, 2 vol., (ISBN: 978-2-35158-071-4; e-ISBN: 978-2-35158-082-0); *Eds. M. G. Alexander and A. Bertron*

PRO 64: Long Term Performance of Cementitious Barriers and Reinforced Concrete in Nuclear Power Plants and Waste Management—NUCPERF 2009 (ISBN: 978-2-35158-072-1; e-ISBN: 978-2-35158-087-5); *Eds. V. L'Hostis, R. Gens and C. Gallé*

PRO 65: Design Performance and Use of Self-consolidating Concrete—SCC'2009 (ISBN: 978-2-35158-073-8; e-ISBN: 978-2-35158-093-6); *Eds. C. Shi, Z. Yu, K. H. Khayat and P. Yan*

PRO 66: 2nd International RILEM Workshop on Concrete Durability and Service Life Planning—ConcreteLife'09 (ISBN: 978-2-35158-074-5; ISBN: 978-2-35158-074-5); *Ed. K. Kovler*

PRO 67: Repairs Mortars for Historic Masonry (e-ISBN: 978-2-35158-083-7); *Ed. C. Groot*

PRO 68: Proceedings of the 3rd International RILEM Symposium on 'Rheology of Cement Suspensions such as Fresh Concrete (ISBN 978-2-35158-091-2; e-ISBN: 978-2-35158-092-9); *Eds. O. H. Wallevik, S. Kubens and S. Oesterheld*

PRO 69: 3rd International PhD Student Workshop on 'Modelling the Durability of Reinforced Concrete (ISBN: 978-2-35158-095-0); *Eds. R. M. Ferreira, J. Gulikers and C. Andrade*

PRO 70: 2nd International Conference on 'Service Life Design for Infrastructure' (ISBN set: 978-2-35158-096-7, e-ISBN: 978-2-35158-097-4); *Eds. K. van Breugel, G. Ye and Y. Yuan*

PRO 71: Advances in Civil Engineering Materials—The 50-year Teaching Anniversary of Prof. Sun Wei' (ISBN: 978-2-35158-098-1; e-ISBN: 978-2-35158-099-8); *Eds. C. Miao, G. Ye and H. Chen*

PRO 72: First International Conference on 'Advances in Chemically-Activated Materials—CAM'2010' (2010), 264 pp., ISBN: 978-2-35158-101-8; e-ISBN: 978-2-35158-115-5; *Eds. Caijun Shi and Xiaodong Shen*

PRO 73: 2nd International Conference on 'Waste Engineering and Management—ICWEM 2010' (2010), 894 pp., ISBN: 978-2-35158-102-5; e-ISBN: 978-2-35158-103-2, *Eds. J. Zh. Xiao, Y. Zhang, M. S. Cheung and R. Chu*

PRO 74: International RILEM Conference on 'Use of Superabsorbent Polymers and Other New Additives in Concrete' (2010) 374 pp., ISBN: 978-2-35158-104-9; e-ISBN: 978-2-35158-105-6; *Eds. O.M. Jensen, M.T. Hasholt, and S. Laustsen*

PRO 75: International Conference on 'Material Science—2nd ICTRC—Textile Reinforced Concrete—Theme 1' (2010) 436 pp., ISBN: 978-2-35158-106-3; e-ISBN: 978-2-35158-107-0; *Ed. W. Brameshuber*

PRO 76: International Conference on 'Material Science—HetMat—Modelling of Heterogeneous Materials—Theme 2' (2010) 255 pp., ISBN: 978-2-35158-108-7; e-ISBN: 978-2-35158-109-4; *Ed. W. Brameshuber*

PRO 77: International Conference on 'Material Science—AdIPoC—Additions Improving Properties of Concrete—Theme 3' (2010) 459 pp., ISBN: 978-2-35158-110-0; e-ISBN: 978-2-35158-111-7; *Ed. W. Brameshuber*

PRO 78: 2nd Historic Mortars Conference and RILEM TC 203-RHM Final Workshop—HMC2010 (2010) 1416 pp., e-ISBN: 978-2-35158-112-4; *Eds. J. Válek, C. Groot and J. J. Hughes*

PRO 79: International RILEM Conference on Advances in Construction Materials Through Science and Engineering (2011) 213 pp., ISBN: 978-2-35158-116-2, e-ISBN: 978-2-35158-117-9; *Eds. Christopher Leung and K.T. Wan*

PRO 80: 2nd International RILEM Conference on Concrete Spalling due to Fire Exposure (2011) 453 pp., ISBN: 978-2-35158-118-6; e-ISBN: 978-2-35158-119-3; *Eds. E.A.B. Koenders and F. Dehn*

PRO 81: 2nd International RILEM Conference on Strain Hardening Cementitious Composites (SHCC2-Rio) (2011) 451 pp., ISBN: 978-2-35158-120-9; e-ISBN: 978-2-35158-121-6; *Eds. R.D. Toledo Filho, F.A. Silva, E.A.B. Koenders and E.M. R. Fairbairn*

PRO 82: 2nd International RILEM Conference on Progress of Recycling in the Built Environment (2011) 507 pp., e-ISBN: 978-2-35158-122-3; *Eds. V.M. John, E. Vazquez, S.C. Angulo and C. Ulsen*

PRO 83: 2nd International Conference on Microstructural-related Durability of Cementitious Composites (2012) 250 pp., ISBN: 978-2-35158-129-2; e-ISBN: 978-2-35158-123-0; *Eds. G. Ye, K. van Breugel, W. Sun and C. Miao*

PRO 84: CONSEC13—Seventh International Conference on Concrete under Severe Conditions—Environment and Loading (2013) 1930 pp., ISBN: 978-2-35158-124-7; e-ISBN: 978-2- 35158-134-6; *Eds. Z.J. Li, W. Sun, C.W. Miao, K. Sakai, O.E. Gjorv and N. Banthia*

PRO 85: RILEM-JCI International Workshop on Crack Control of Mass Concrete and Related issues concerning Early-Age of Concrete Structures—ConCrack 3—Control of Cracking in Concrete Structures 3 (2012) 237 pp., ISBN: 978-2-35158-125-4; e-ISBN: 978-2-35158-126-1; *Eds. F. Toutlemonde and J.-M. Torrenti*

PRO 86: International Symposium on Life Cycle Assessment and Construction (2012) 414 pp., ISBN: 978-2-35158-127-8, e-ISBN: 978-2-35158-128-5; *Eds. A. Ventura and C. de la Roche*

PRO 87: UHPFRC 2013—RILEM-fib-AFGC International Symposium on Ultra-High Performance Fibre-Reinforced Concrete (2013), ISBN: 978-2-35158-130-8, e-ISBN: 978-2-35158-131-5; *Eds. F. Toutlemonde*

PRO 88: 8th RILEM International Symposium on Fibre Reinforced Concrete (2012) 344 pp., ISBN: 978-2-35158-132-2; e-ISBN: 978-2-35158-133-9; *Eds. Joaquim A.O. Barros*

PRO 89: RILEM International workshop on performance-based specification and control of concrete durability (2014) 678 pp., ISBN: 978-2-35158-135-3; e-ISBN: 978-2-35158-136-0; *Eds. D. Bjegović, H. Beushausen and M. Serdar*

PRO 90: 7th RILEM International Conference on Self-Compacting Concrete and of the 1st RILEM International Conference on Rheology and Processing of Construction Materials (2013) 396 pp., ISBN: 978-2-35158-137-7; e-ISBN: 978-2-35158-138-4; *Eds. Nicolas Roussel and Hela Bessaies-Bey*

PRO 91: CONMOD 2014—RILEM International Symposium on Concrete Modelling (2014), ISBN: 978-2-35158-139-1; e-ISBN: 978-2-35158-140-7; *Eds. Kefei Li, Peiyu Yan and Rongwei Yang*

PRO 92: CAM 2014—2nd International Conference on advances in chemically-activated materials (2014) 392 pp., ISBN: 978-2-35158-141-4; e-ISBN: 978-2-35158-142-1; *Eds. Caijun Shi and Xiadong Shen*

PRO 93: SCC 2014—3rd International Symposium on Design, Performance and Use of Self-Consolidating Concrete (2014) 438 pp., ISBN: 978-2-35158-143-8; e-ISBN: 978-2-35158-144-5; *Eds. Caijun Shi, Zhihua Ou and Kamal H. Khayat*

PRO 94 (online version): HPFRCC-7—7th RILEM conference on High performance fiber reinforced cement composites (2015), e-ISBN: 978-2-35158-146-9; *Eds. H.W. Reinhardt, G.J. Parra-Montesinos and H. Garrecht*

PRO 95: International RILEM Conference on Application of superabsorbent polymers and other new admixtures in concrete construction (2014), ISBN: 978-2-35158-147-6; e-ISBN: 978-2-35158-148-3; *Eds. Viktor Mechtcherine and Christof Schroefl*

PRO 96 (online version): XIII DBMC: XIII International Conference on Durability of Building Materials and Components (2015), e-ISBN: 978-2-35158-149-0; *Eds. M. Quattrone and V.M. John*

PRO 97: SHCC3—3rd International RILEM Conference on Strain Hardening Cementitious Composites (2014), ISBN: 978-2-35158-150-6; e-ISBN: 978-2-35158-151-3; *Eds. E. Schlangen, M.G. Sierra Beltran, M. Lukovic and G. Ye*

PRO 98: FERRO-11—11th International Symposium on Ferrocement and 3rd ICTRC—International Conference on Textile Reinforced Concrete (2015), ISBN: 978-2-35158-152-0; e-ISBN: 978-2-35158-153-7; *Ed. W. Brameshuber*

PRO 99 (online version): ICBBM 2015—1st International Conference on Bio-Based Building Materials (2015), e-ISBN: 978-2-35158-154-4; *Eds. S. Amziane and M. Sonebi*

PRO 100: SCC16—RILEM Self-Consolidating Concrete Conference (2016), ISBN: 978-2-35158-156-8; e-ISBN: 978-2-35158-157-5; *Ed. Kamal H. Kayat*

PRO 101 (online version): III Progress of Recycling in the Built Environment (2015), e-ISBN: 978-2-35158-158-2; *Eds I. Martins, C. Ulsen and S. C. Angulo*

PRO 102 (online version): RILEM Conference on Microorganisms-Cementitious Materials Interactions (2016), e-ISBN: 978-2-35158-160-5; *Eds. Alexandra Bertron, Henk Jonkers and Virginie Wiktor*

PRO 103 (online version): ACESC'16—Advances in Civil Engineering and Sustainable Construction (2016), e-ISBN: 978-2-35158-161-2; *Eds. T.Ch. Madhavi, G. Prabhakar, Santhosh Ram and P.M. Rameshwaran*

PRO 104 (online version): SSCS'2015—Numerical Modeling—Strategies for Sustainable Concrete Structures (2015), e-ISBN: 978-2-35158-162-9

PRO 105: 1st International Conference on UHPC Materials and Structures (2016), ISBN: 978-2-35158-164-3; e-ISBN: 978-2-35158-165-0

PRO 106: AFGC-ACI-fib-RILEM International Conference on Ultra-High-Performance Fibre-Reinforced Concrete—UHPFRC 2017 (2017), ISBN: 978-2-35158-166-7; e-ISBN: 978-2-35158-167-4; *Eds. François Toutlemonde and Jacques Resplendino*

PRO 107 (online version): XIV DBMC—14th International Conference on Durability of Building Materials and Components (2017), e-ISBN: 978-2-35158-159-9; *Eds. Geert De Schutter, Nele De Belie, Arnold Janssens and Nathan Van Den Bossche*

PRO 108: MSSCE 2016—Innovation of Teaching in Materials and Structures (2016), ISBN: 978-2-35158-178-0; e-ISBN: 978-2-35158-179-7; *Ed. Per Goltermann*

PRO 109 (2 volumes): MSSCE 2016—Service Life of Cement-Based Materials and Structures (2016), ISBN Vol. 1: 978-2-35158-170-4; Vol. 2: 978-2-35158-171-4; Set Vol. 1&2: 978-2-35158-172-8; e-ISBN : 978-2-35158-173-5; *Eds. Miguel Azenha, Ivan Gabrijel, Dirk Schlicke, Terje Kanstad and Ole Mejlhede Jensen*

PRO 110: MSSCE 2016—Historical Masonry (2016), ISBN: 978-2-35158-178-0; e-ISBN: 978-2-35158-179-7; *Eds. Inge Rörig-Dalgaard and Ioannis Ioannou*

PRO 111: MSSCE 2016—Electrochemistry in Civil Engineering (2016); ISBN: 978-2-35158-176-6; e-ISBN: 978-2-35158-177-3; *Ed. Lisbeth M. Ottosen*

PRO 112: MSSCE 2016—Moisture in Materials and Structures (2016), ISBN: 978-2-35158-178-0; e-ISBN: 978-2-35158-179-7; *Eds. Kurt Kielsgaard Hansen, Carsten Rode and Lars-Olof Nilsson*

PRO 113: MSSCE 2016—Concrete with Supplementary Cementitious Materials (2016), ISBN: 978-2-35158-178-0; e-ISBN: 978-2-35158-179-7; *Eds. Ole Mejlhede Jensen, Konstantin Kovler and Nele De Belie*

PRO 114: MSSCE 2016—Frost Action in Concrete (2016), ISBN: 978-2-35158-182-7; e-ISBN: 978-2-35158-183-4; *Eds. Marianne Tange Hasholt, Katja Fridh and R. Doug Hooton*

PRO 115: MSSCE 2016—Fresh Concrete (2016), ISBN: 978-2-35158-184-1; e-ISBN: 978-2-35158-185-8; *Eds. Lars N. Thrane, Claus Pade, Oldrich Svec and Nicolas Roussel*

PRO 116: BEFIB 2016—9th RILEM International Symposium on Fiber Reinforced Concrete (2016), ISBN: 978-2-35158-187-2; e-ISBN: 978-2-35158-186-5; *Eds. N. Banthia, M. di Prisco and S. Soleimani-Dashtaki*

PRO 117: 3rd International RILEM Conference on Microstructure Related Durability of Cementitious Composites (2016), ISBN: 978-2-35158-188-9; e-ISBN: 978-2-35158-189-6; *Eds. Changwen Miao, Wei Sun, Jiaping Liu, Huisu Chen, Guang Ye and Klaas van Breugel*

PRO 118 (4 volumes): International Conference on Advances in Construction Materials and Systems (2017), ISBN Set: 978-2-35158-190-2; Vol. 1: 978-2-35158-193-3; Vol. 2: 978-2-35158-194-0; Vol. 3: ISBN:978-2-35158-195-7; Vol. 4: ISBN:978-2-35158-196-4; e-ISBN: 978-2-35158-191-9; *Ed. Manu Santhanam*

PRO 119 (online version): ICBBM 2017—Second International RILEM Conference on Bio-based Building Materials, (2017), e-ISBN: 978-2-35158-192-6; *Ed. Sofiane Amziane*

PRO 120 (2 volumes): EAC-02—2nd International RILEM/COST Conference on Early Age Cracking and Serviceability in Cement-based Materials and Structures, (2017), Vol. 1: 978-2-35158-199-5, Vol. 2: 978-2-35158-200-8, Set: 978-2-35158-197-1, e-ISBN: 978-2-35158-198-8; *Eds. Stéphanie Staquet and Dimitrios Aggelis*

PRO 121 (2 volumes): SynerCrete18: Interdisciplinary Approaches for Cement-based Materials and Structural Concrete: Synergizing Expertise and Bridging Scales of Space and Time, (2018), Set: 978-2-35158-202-2, Vol.1: 978-2-35158-211-4, Vol.2: 978-2-35158-212-1, e-ISBN: 978-2-35158-203-9; *Eds. Miguel Azenha, Dirk Schlicke, Farid Benboudjema, Agnieszka Knoppik*

PRO 122: SCC'2018 China—Fourth International Symposium on Design, Performance and Use of Self-Consolidating Concrete, (2018), ISBN: 978-2-35158-204-6, e-ISBN: 978-2-35158-205-3; *Eds. C. Shi, Z. Zhang, K. H. Khayat*

PRO 123: Final Conference of RILEM TC 253-MCI: Microorganisms-Cementitious Materials Interactions (2018), Set: 978-2-35158-207-7, Vol.1: 978-2-35158-209-1, Vol.2: 978-2-35158-210-7, e-ISBN: 978-2-35158-206-0; *Ed. Alexandra Bertron*

PRO 124 (online version): Fourth International Conference Progress of Recycling in the Built Environment (2018), e-ISBN: 978-2-35158-208-4; *Eds. Isabel M. Martins, Carina Ulsen, Yury Villagran*

PRO 125 (online version): SLD4—4th International Conference on Service Life Design for Infrastructures (2018), e-ISBN: 978-2-35158-213-8; *Eds. Guang Ye, Yong Yuan, Claudia Romero Rodriguez, Hongzhi Zhang, Branko Savija*

PRO 126: Workshop on Concrete Modelling and Material Behaviour in honor of Professor Klaas van Breugel (2018), ISBN: 978-2-35158-214-5, e-ISBN: 978-2-35158-215-2; *Ed. Guang Ye*

PRO 127 (online version): CONMOD2018—Symposium on Concrete Modelling (2018), e-ISBN: 978-2-35158-216-9; *Eds. Erik Schlangen, Geert de Schutter, Branko Savija, Hongzhi Zhang, Claudia Romero Rodriguez*

PRO 128: SMSS2019—International Conference on Sustainable Materials, Systems and Structures (2019), ISBN: 978-2-35158-217-6, e-ISBN: 978-2-35158-218-3

PRO 129: 2nd International Conference on UHPC Materials and Structures (UHPC2018-China), ISBN: 978-2-35158-219-0, e-ISBN: 978-2-35158-220-6

PRO 130: 5th Historic Mortars Conference (2019), ISBN: 978-2-35158-221-3, e-ISBN: 978-2-35158-222-0; *Eds. José Ignacio Álvarez, José María Fernández, Íñigo Navarro, Adrián Durán, Rafael Sirera*

PRO 131 (online version): 3rd International Conference on Bio-Based Building Materials (ICBBM2019), e-ISBN: 978-2-35158-229-9; *Eds. Mohammed Sonebi, Sofiane Amziane, Jonathan Page*

PRO 132: IRWRMC'18—International RILEM Workshop on Rheological Measurements of Cement-based Materials (2018), ISBN: 978-2-35158-230-5, e-ISBN: 978-2-35158-231-2; *Eds. Chafika Djelal, Yannick Vanhove*

PRO 133 (online version): CO2STO2019—International Workshop CO2 Storage in Concrete (2019), e-ISBN: 978-2-35158-232-9; *Eds. Assia Djerbi, Othman Omikrine-Metalssi, Teddy Fen-Chong*

PRO 134: 3rd ACF/HNU Conference on UHPC Materials and Structures—UHP'2020, ISBN: 978-2-35158-233-6, e-ISBN: 978-2-35158-234-3; *Eds. Caijun Shi, Jiaping Liu*

RILEM REPORTS (REP)

Report 19: Considerations for Use in Managing the Aging of Nuclear Power Plant Concrete Structures (ISBN: 2-912143-07-1); *Ed. D. J. Naus*

Report 20: Engineering and Transport Properties of the Interfacial Transition Zone in Cementitious Composites (ISBN: 2-912143-08-X); *Eds. M. G. Alexander, G. Arliguie, G. Ballivy, A. Bentur and J. Marchand*

Report 21: Durability of Building Sealants (ISBN: 2-912143-12-8); *Ed. A. T. Wolf*

Report 22: Sustainable Raw Materials—Construction and Demolition Waste (ISBN: 2-912143-17-9); *Eds. C. F. Hendriks and H. S. Pietersen*

Report 23: Self-Compacting Concrete state-of-the-art report (ISBN: 2-912143-23-3); *Eds. Å. Skarendahl and Ö. Petersson*

Report 24: Workability and Rheology of Fresh Concrete: Compendium of Tests (ISBN: 2-912143-32-2); *Eds. P. J. M. Bartos, M. Sonebi and A. K. Tamimi*

Report 25: Early Age Cracking in Cementitious Systems (ISBN: 2-912143-33-0); *Ed. A. Bentur*

Report 26: Towards Sustainable Roofing (Joint Committee CIB/RILEM) (CD 07) (e-ISBN 978-2-912143-65-5); *Eds. Thomas W. Hutchinson and Keith Roberts*

Report 27: Condition Assessment of Roofs (Joint Committee CIB/RILEM) (CD 08) (e-ISBN 978-2-912143-66-2); *Ed. CIB W 83/RILEM TC166-RMS*

Report 28: Final report of RILEM TC 167-COM 'Characterisation of Old Mortars with Respect to Their Repair (ISBN: 978-2-912143-56-3); *Eds. C. Groot, G. Ashall and J. Hughes*

Report 29: Pavement Performance Prediction and Evaluation (PPPE): Interlaboratory Tests (e-ISBN: 2-912143-68-3); *Eds. M. Partl and H. Piber*

Report 30: Final Report of RILEM TC 198-URM 'Use of Recycled Materials' (ISBN: 2-912143-82-9; e-ISBN: 2-912143-69-1); *Eds. Ch. F. Hendriks, G. M. T. Janssen and E. Vázquez*

Report 31: Final Report of RILEM TC 185-ATC 'Advanced testing of cement-based materials during setting and hardening' (ISBN: 2-912143-81-0; e-ISBN: 2-912143-70-5); *Eds. H. W. Reinhardt and C. U. Grosse*

Report 32: Probabilistic Assessment of Existing Structures. A JCSS publication (ISBN 2-912143-24-1); *Ed. D. Diamantidis*

Report 33: State-of-the-Art Report of RILEM Technical Committee TC 184-IFE 'Industrial Floors' (ISBN 2-35158-006-0); *Ed. P. Seidler*

Report 34: Report of RILEM Technical Committee TC 147-FMB 'Fracture mechanics applications to anchorage and bond' Tension of Reinforced Concrete Prisms—Round Robin Analysis and Tests on Bond (e-ISBN 2-912143-91-8); *Eds. L. Elfgren and K. Noghabai*

Report 35: Final Report of RILEM Technical Committee TC 188-CSC 'Casting of Self Compacting Concrete' (ISBN 2-35158-001-X; e-ISBN: 2-912143-98-5); *Eds. Å. Skarendahl and P. Billberg*

Report 36: State-of-the-Art Report of RILEM Technical Committee TC 201-TRC 'Textile Reinforced Concrete' (ISBN 2-912143-99-3); *Ed. W. Brameshuber*

Report 37: State-of-the-Art Report of RILEM Technical Committee TC 192-ECM 'Environment-conscious construction materials and systems' (ISBN: 978-2-35158-053-0); *Eds. N. Kashino, D. Van Gemert and K. Imamoto*

Report 38: State-of-the-Art Report of RILEM Technical Committee TC 205-DSC 'Durability of Self-Compacting Concrete' (ISBN: 978-2-35158-048-6); *Eds. G. De Schutter and K. Audenaert*

Report 39: Final Report of RILEM Technical Committee TC 187-SOC 'Experimental determination of the stress-crack opening curve for concrete in tension' (ISBN 978-2-35158-049-3); *Ed. J. Planas*

Report 40: State-of-the-Art Report of RILEM Technical Committee TC 189-NEC 'Non-Destructive Evaluation of the Penetrability and Thickness of the Concrete Cover' (ISBN 978-2-35158-054-7); *Eds. R. Torrent and L. Fernández Luco*

Report 41: State-of-the-Art Report of RILEM Technical Committee TC 196-ICC 'Internal Curing of Concrete' (ISBN 978-2-35158-009-7); *Eds. K. Kovler and O. M. Jensen*

Report 42: 'Acoustic Emission and Related Non-destructive Evaluation Techniques for Crack Detection and Damage Evaluation in Concrete'—Final Report of RILEM Technical Committee 212-ACD (e-ISBN: 978-2-35158-100-1); *Ed. M. Ohtsu*

Report 45: Repair Mortars for Historic Masonry—State-of-the-Art Report of RILEM Technical Committee TC 203-RHM (e-ISBN: 978-2-35158-163-6); *Eds. Paul Maurenbrecher and Caspar Groot*

Report 46: Surface delamination of concrete industrial ffioors and other durability related aspects guide—Report of RILEM Technical Committee TC 268-SIF (e-ISBN: 978-2-35158-201-5); *Ed. Valerie Pollet*

Circular CO_2 Utilization Strategies for More Sustainable Concrete

Sean Monkman and Mike Thomas

Abstract The sustainability goals of the cement and concrete industry cannot be met by simple improvements to process efficiencies but instead demand innovative solutions. New processes have been developed to reduce the carbon footprint of ready mixed concrete through CO_2 utilization strategies concerning three components: binder, water and aggregates. The injection of an optimized dose of waste carbon dioxide into concrete reacts with the cement binder to form in-situ nanoscale calcium carbonate particles that can improve the compressive strength of the mix. The increased cement efficiency allows the concrete to be produced with less cement thereby realizing a GHG benefit through both the mineralized CO_2 and the avoided cement CO_2 emissions. Concrete wash water, a by-product of concrete production that is typically a waste stream and a challenge to reuse, can be treated with carbon dioxide. The CO_2 is mineralized in a reaction with the waste cement suspended in the slurry. The treated slurry can more readily be used as mix water in a new concrete batch. The performance benefit of using the recycled slurry, in particular the cementitious nature of the treated wash water solids, allows for a cement reduction. Finally, the performance of recycled concrete aggregate can be improved through a carbon dioxide treatment. The carbon dioxide reacts with the hydrated cement paste component of the crushed concrete to form $CaCO_3$. A combination of the three strategies can realize a net carbon benefit of around 76.9 kg per m^3 of concrete including recycling of 50.3 kg of CO_2.

Keywords Concrete · CO_2 utilization · Carbon upcycling · Sustainability

---

S. Monkman (✉)
CarbonCure Technologies, Dartmouth, NS B3B 1R6, Canada
e-mail: smonkman@carboncure.com

M. Thomas
University of New Brunswick, Fredericton, NB E3B 5A3, Canada

V. M. C. F. Cunha et al. (eds.), *Proceedings of the 3rd RILEM Spring Convention and Conference (RSCC 2020)*, RILEM Bookseries 35,
https://doi.org/10.1007/978-3-030-76543-9_1

1 Introduction

The modern world depends on concrete. It is a building material that is versatile, durable, low cost, resilient, locally sourced and sustainable. However, there are concerns about the environmental impacts of concrete production; 5% of anthropogenic carbon dioxide emissions are associated with the production of cement (the primary binder component of concrete representing about 10–15% of its mass) [1]. The environmental impacts are most accurately considered in relation to the scale of concrete production. The material itself is not inherently worse for the environment than other building materials [2]; the cumulative impacts are primarily due to the amount of concrete produced per capita. The annual output of concrete is estimated to be around 20 times greater than the amount of steel and around 40 times greater than the amount of wood [3].

Cement production results in CO_2 emissions due to the high temperature calcination of limestone and the energy required to drive the calcination forward. In turn, the cement is used, along with water and aggregates, as a component in concrete. The cement and concrete industries have identified approaches to reduce the environmental impact of cement and concrete production [4–6]. Business-as-usual approaches (improved cement kiln efficiencies, use of lower carbon fuels, increasing use of conventional secondary cementitious materials) are projected to be insufficient to reach carbon reduction goals. Sustainable concrete production will require contributions from novel and emerging technologies. One such approach is the concept of CO_2 utilization for concrete production [7, 8]. The use of carbon dioxide as a process feedstock has the potential to both mineralize CO_2 permanently and to reduce the carbon footprint of the concrete so produced.

2 CO_2 Utilization Approaches

2.1 Addition to Ready Mix Concrete

Ready mixed concrete is a mixture of aggregates (graded stones and sand), binder (Portland cement possibly along with one or more of fly ash, blast furnace slag, silica fume, and limestone), and water. Chemical admixtures are often included to provide specific performance benefits such as workability, set control, air entrainment, etc.

Carbon dioxide can be added to ready-mixed concrete, like an admixture, to provide an improvement the concrete properties. The injection of an optimized dose of waste carbon dioxide into concrete while batching promotes in-situ nanoscale calcium carbonate seeding that can improve the compressive strength of the mix [9]. The increased cement efficiency allows the concrete to be produced with less cement thereby realizing a greenhouse gas benefit through both the mineralization of waste CO_2 and the avoided cement CO_2 emissions. Cement reductions have typically been around 5% [10]. The concrete so-produced maintains the required durability [11]. The approach is being realized at more than producers around the world [12].

2.2 Treatment of Concrete Wash Water for Recycling

Concrete wash water is a byproduct of the concrete industry. This water, which may contain suspended solids in the form of sand, aggregate and/or cementitious materials, is generated by the washing-out of concrete mixers and trucks following the production and delivery of concrete. This water is alkaline in nature and requires specialized treatment, handling and disposal.

While this water can be suitable for reuse in the production of concrete, it has been documented that negative impacts to the concrete properties can arise. Wash water is mainly a mixture of cement and, in many cases, supplementary cementitious materials (SCMs) in water. The cement hydrates with time and changes the chemistry of the water. The evolution of the water, along with the hydration products, can cause a host of issues when the water is used as mix water including set acceleration, increased water demand, and reduced 7-day strength [13, 14]. These issues generally worsen as the amount of cement in the water increases, and/or the water ages.

A carbon dioxide treatment of the wash water slurry can allow it to be beneficially reused as concrete mix water. The CO_2 is mineralized in a reaction with the waste cement suspended in the slurry to create calcium carbonate [15]. The carbonate mineral makes the slurry solids more stable with age allowing the performance outcomes of the concrete produced with the slurry to be more predictable. Set acceleration can be reduced or eliminated. The treated slurry can more readily be used as mix water in a new concrete batch. The performance benefit of using the recycled slurry, in particular the cementitious nature of the treated wash water solids, allows for a cement replacement. The approach has been the subject of laboratory study.

2.3 Treatment of Recycled Concrete Aggregate

Recycled concrete can be processed and reused as aggregates in new concrete production. Concrete made with recycled concrete aggregates (RCA) typically is weaker than concrete made with natural (virgin) aggregates. The RCA can be less dense, have a higher water absorption and a lower crushing value than natural aggregates [16]. The concrete so produced can have a significantly increased drying shrinkage and chloride ion diffusion coefficient thereby creating durability concerns.

Treatment of the recycled concrete aggregates with carbon dioxide can serve to improve the properties of the aggregates and the concrete produced with the aggregates while also serving to mineralize CO_2. The carbon dioxide reacts with the hydrated cement paste component of the crushed concrete to form $CaCO_3$. The approach is entering the first rounds of industrial-scale testing [17].

2.4 *Source of CO_2*

Carbon dioxide for industrial uses is typically sourced as a byproduct of another industrial process (commonly ethanol refining, hydrogen production, ammonia production [18]). Its origin as a byproduct means that it can be characterized as a waste that is being repurposed, recycled or upcycled. The industrial gas supply chain is mature and widely distributed.

The use of carbon dioxide as a feedstock means that the energy consumption related to its capture and compression must be considered. The industrial gas market seeks integration with CO_2 point sources that represent the best economic cases whether it be aligned with starting purity or location or both. CO_2 is liquefied for transport. The energy to create liquid CO_2 from an input flue gas stream is related to the mole fraction of carbon dioxide and is around 200 kWh/tonne for typical high purity streams appropriate for the merchant supply chain [19].

A circular economy synergy may be realized if the source of carbon dioxide is cement production. Cement plant CO_2 capture is in the early stages of development and implementation [20–25] but is not a primary source for the merchant market given the low fraction of CO_2 in the raw flue gas. It is anticipated that the concept will not be ready for wide deployment in cement kilns for at least to 10 years [26]. In contrast to merchant the CO_2 supply chain, CO_2 from cement plants would be favourably located with respect to concrete production.

3 Environmental Benefits

The potential to reduce the carbon dioxide impacts of concrete production can be assessed through the consideration of a generic mix design. United States industry averages can be taken from a 2019 survey from the National Ready Mixed Concrete Association [27]. Given that the results are an average of many mix designs it may not be representative of any viable mix design but the intent of the present consideration is to base analysis on the average cement, water and aggregate loadings for different strength classes. Mix designs for three strength classes are reported in Table 1.

According to the 2016 NRMCA survey, 94% of concrete shipped had a specified strength of 34 MPa (5000 psi) or less with 49% being 24 MPa (3500 psi) or less. A generic mix for this analysis is taken as a 28 MPa mix with the other extremes (17 and 55 MPa) representing the range of expected generic mixes.

Table 1 Generic concrete mix designs after NRMCA survey data [27]

Design strength* (MPa)	17	28	55
Portland cement (kg/m^3 concrete)	210	282	427
Fly ash (kg/m^3 concrete)	37	49	75
Slag (kg/m^3 concrete)	10	14	21
Total cementitious (kg/m^3 concrete)	257	345	522
Coarse aggregate (kg/m^3 concrete)	996	958	900
Fine aggregate (kg/m^3 concrete)	861	828	778
Mix water (L/m^3 concrete)	181	181	202
CO_2 footprint (kg CO_2/m^3 concrete)	288	393	624

*The design strengths are reported as 2500, 4000 and 8000 psi in the NRMCA survey

3.1 *Ready Mixed Concrete*

A dose of 0.1–0.2% CO_2 by weight of cement typically provides a strength improvement that supports a 5% reduction in Portland cement usage. A model 0.15% CO_2 dose supplies a quantity of carbon dioxide in proportion to the amount of cement and is in the range of 0.32–0.64 kg CO_2/m^3 concrete (0.42 kg for the generic mix design). The CO_2 is mineralized at a rate of about 90% of the dose for 0.38 kg CO_2 converted per m^3 concrete.

The environmental impact of cement production varies according to location but can be modeled as 0.863 units of CO_2 per units of finished cement [28]. The benefit realized from the avoided cement ranges from 9.1 to 18.4 kg CO_2/m^3 concrete (12.2 kg for the generic mix). In the generic case the combined environmental impact of the mineralized CO_2 and avoided cement is 12.5 kg CO_2/m^3 concrete.

3.2 *Washwater*

A generic wash water slurry is taken to have a specific gravity of 1.10 and the suspended solids are a mixture of cement (82%), fly ash (14%) and slag (4%) as per the binder proportions of the generic 28 MPa mix design. According to the blended densities of the binder components the suspended solids have a representative density of 2.98 g/ml.

The mix water used in the generic mix is 181 L/m^3 concrete. In the present analysis it is completely replaced by the treated wash water slurry. The use of 199 kg slurry/m^3 concrete comprises 181 L of water and 18.1 kg of solids of which 14.8 kg are cement. The CO_2 treatment mineralizes carbon dioxide at a proportion of 30% by weight of the cement (and no reaction with the fly ash and slag) for a total of 4.4 kg CO_2/m^3 concrete. The mass of the wash water solids has thereby increased to 22.5 kg and has the capacity to substitute for virgin cement at a ratio of 1:1. The quantity of avoided cement has an environmental impact of 19.4 kg CO_2/m^3 for the generic and low strength mix designs. In the high strength mix design there is a greater mixing water usage (202 L/m^3 concrete) and the carbon impact of the cement reduction is increased to 21.7 kg CO_2/m^3. In the generic case the combined environmental impact of the mineralized CO_2 and avoided cement is 23.9 kg CO_2/m^3 concrete (and up to 26.7 in the 55 MPa mix).

3.3 *Carbonated Recycled Concrete Aggregate*

Researchers have established that a fine aggregate can mineralize more CO_2 than a coarse aggregate. A survey of the literature [29–32] suggests that a fine aggregate (nominally of size less than 1 mm) can mineralize CO_2 at about 4.7% by final aggregate mass while coarse aggregate can mineralize CO_2 on the order of 2.2%.

The amount of CO_2 mineralized depends upon the amount of aggregate used and the amount of virgin aggregate substituted for recycled aggregate. The generic mix contains 958 kg of coarse aggregate and 828 kg of fine aggregate. At substitutions of 100% and 50%, respectively, the amount of mineralized CO_2 is 21.2 and 19.4 kg CO_2/m^3 concrete respectively. The aggregate replacement goals are ambitious but are forward-looking for purposes of modeling potential impacts.

3.4 *Overall Carbon Dioxide Benefit*

The three CO_2 utilization strategies can be combined to estimate the potential benefit. The benefit comes from the mineralization of carbon dioxide and the avoided emissions from reduced cement usage. A summary of the combined effects is presented in Table 2.

Table 2 Combined carbon impact (gross) per unit concrete for three combined CO_2 utilization technologies at three strength classes

Design strength (MPa)	17	28	55
Ready mix—mineralized CO_2 (kg/m^3 concrete)	0.2	0.3	0.4
Ready mix—avoided CO_2 (kg/m^3 concrete)	9.1	12.2	18.4
Wash water—mineralized CO_2 (kg/m^3 concrete)	4.4	4.4	5.0
Wash water—avoided CO_2 (kg/m^3 concrete)	19.3	19.4	21.7
CRCA coarse—mineralized CO_2 (kg/m^3 concrete)	22.0	21.2	19.9
CRCA fine—mineralized CO_2 (kg/m^3 concrete)	20.2	19.4	18.2
Total mineralized CO_2 (kg/m^3 concrete)	46.9	45.3	43.6
Total avoided CO_2 (kg/m^3 concrete)	28.5	31.6	40.1
Total CO_2 impact (kg CO_2/m^3 concrete)	75.4	76.9	83.7

With respect to the generic (28 MPa) case, 16% of the carbon impact reduction is from the ready mix approach, 31% is from the wash water, and 53% is from the recycled aggregate (28% from the coarse and 23% from fine). 59% of the carbon impact is associated with CO_2 mineralization while 41% is attributable to avoided CO_2 from cement reductions.

3.5 *Net Environmental Benefit*

The potential implementation of the technologies is accompanied by net new power generation and transport. Carbon dioxide supplied by the merchant market would have an energy consumption of about 200 kWh/tonne liquid CO_2 produced [19]. The CO_2 emissions rate from industrial power consumption is 51.8 g CO_2/MJ or 186 g CO_2/kWh [33]. The energy emissions associated with the capture, compression, and liquefaction of one tonne of liquid carbon dioxide results in about 0.037 tonnes CO_2 emitted. Additionally, the operation of the injection equipment is associated with an energy consumption (estimated at 0.037 kWh/kg CO_2 injected) [34]. The transport of the carbon dioxide to the locations of utilization will be associated with a carbon emission. A representative rate of CO_2 emissions is 62 g CO_2/tonne-km of freight for chemical transport [35]. Moving one tonne of liquid CO_2 200 km from the point source emitter to the utilization site would result in emissions of 0.012 tonnes CO_2.

The other factors related to CO_2 injection (e.g. the production and transport of the injection equipment), are minimal compared to the gas processing, gas transport and equipment operation [34]. The mineralization rate is not assumed to be perfect; in the present analysis 80% of the carbon dioxide dosed is mineralized as carbonate. The overall impacts of the capture, transport and injection of CO_2 are in proportion to the amount of carbon dioxide utilized per unit of concrete. The compiled results are summarized in Table 3 and show that the process emissions are about 10% of the total benefit. The baseline carbon footprint is reported as per the NRMCA industry

Table 3 Net environmental benefit estimate including process emissions

Design strength (MPa)	17	28	55
Overall CO_2 benefit (kg CO_2/m^3 concrete)	75.4	76.9	83.7
Total CO_2 utilized (kg/m^3 concrete)	52.1	50.3	48.4
Power consumption (kWh/m^3 concrete)	1.9	1.9	1.8
Operation emissions (kg CO_2/m^3 concrete)	0.4	0.3	0.3
Freight emissions (kg CO_2/m^3 concrete)	2.6	2.5	2.4
Total emissions (kg CO_2/m^3 concrete)	2.9	2.8	2.7
Net benefit (kg CO_2/m^3 concrete)	72.4	74.1	81.0
Process emissions rate versus overall benefit (%)	3.9%	3.7%	3.3%
Relative carbon footprint improvement (%)	25%	19%	13%

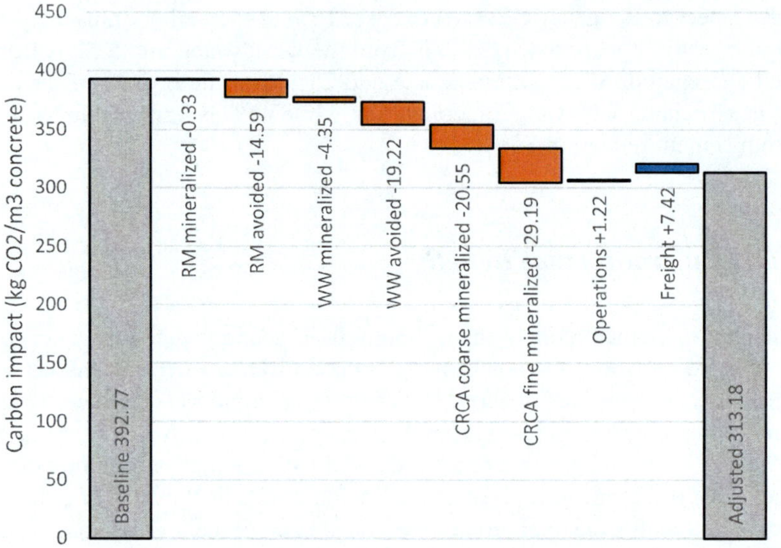

Fig. 1 Cumulative model net impacts of three CO_2 utilization approaches—ready mix admixture (RM), recycled concrete wash water slurry (WW) and carbonated recycled concrete aggregate (CRCA)—with respect to a generic 28 MPa US concrete mix

average report [27]. A visualization of the combined (net) impacts for the generic 28 MPa mix design is provided in Fig. 1.

The gross environmental benefit is around an order of magnitude larger than the process impacts. The net benefits are projected to be 96% of the gross benefit. The CO_2 utilization approaches are viable pathways to reduce the carbon footprint of concrete with reductions of 13–25% depending on the mix design.

4 Conclusions

The future will likely see these approaches combined. They can provide a combined GHG benefit of around 80 kg CO_2/m^3 concrete through simple modifications to the concrete production process. When the CO_2 utilization approaches use carbon dioxide sourced from a cement kiln, the concept of the circular economy in cement/concrete production is achieved. Not only is CO_2 from cement production upcycled, but, in parallel, the waste streams of concrete wash water and demolished concrete are repurposed into new concrete. Saving of virgin inputs of water and aggregates are realized. CO_2 utilization is a platform for creating a more sustainable built environment.

References

1. Uwasu, M., Hara, K., Yabar, H.: World cement production and environmental implications. Environ. Dev. **10**, 36–47 (2014)
2. Barcelo, L., Kline, J., Walenta, G., Gartner, E.: Cement and carbon emissions. Mater. Struct. **47**(6), 1055–1065 (2014)
3. Monteiro, P.J.M., Miller, S.A., Horvath, A.: Towards sustainable concrete. Nat. Mater. **16**(7), 698–699 (2017)
4. IEA and WBCSD: Technology Roadmap: Low-Carbon Transition in the Cement Industry. OECD/IEA and WBCSD, Paris and Geneva (2018)
5. De Wolf, C., Scrivener, K., Habert, G., Favier, A.: A Sustainable Future for the European Cement and Concrete Industry: Technology Assessment for Full Decarbonisation of the Industry by 2050. ETH Zurich (2018)
6. Lehne, J., Preston, F.: Making Concrete Change. Innovation in Low-Carbon Cement and Concrete. The Royal Institute of International Affairs, Chatham House, Cambridge (2018)
7. Ashraf, W.: Carbonation of cement-based materials: challenges and opportunities. Constr. Build. Mater. **120**, 558–570 (2016)
8. Zhang, D., Ghouleh, Z., Shao, Y.: Review on carbonation curing of cement-based materials. J. CO_2 Utiliz. **21**, 119–131 (2017)
9. Monkman, S., Grandfield, K., Langelier, B.: On the mechanism of using carbon dioxide as a beneficial concrete admixture. In: SP 329 Proceedings Twelfth International Conference, pp. 415–428. American Concrete Institute, Beijing, China (2018)
10. Monkman, S.: Waste CO_2 upcycling as a means to improve ready mixed concrete sustainability. In: Papers and Posters Proceedings. Research Institute of Binding Materials Prague, Prague (2019)
11. Monkman, S., MacDonald, M., Hooton, R.D., Sandberg, P.: Properties and durability of concrete produced using CO_2 as an accelerating admixture. Cement Concr. Compos. **74**, 218–224 (2016)
12. Find a Producer—CarbonCure Technologies. https://www.carboncure.com/producers. Last accessed 2019/12/17
13. Meininger, R.: Recycling Mixer Wash Water—Its Effect on Ready Mixed Concrete. National Ready Mixed Concrete Association, Silver Spring, MD (1973)
14. Lobo, C., Mullings, G.: Recycled Water in Ready Mixed Concrete Operations. http://www.nrmca.org/research/33CIF03-1washwater.pdf (2003)
15. Monkman, G., Sandberg, P., Cail, K., Forgeron, D., MacDonald, M.: Methods and Compositions for Treatment of Concrete Wash Water. Patent application WO2018232507A1 (2018)
16. Guo, H., Shi, C., Guan, X., Zhu, J., Ding, Y., Ling, T.-C., Zhang, H., Wang, Y.: Durability of recycled aggregate concrete—a review. Cement Concr. Compos. **89**, 251–259 (2018)
17. Picaud, E.: Le béton recyclé en route vers sa recarbonatation. https://www.lemoniteur.fr/article/le-beton-recycle-en-route-vers-sa-recarbonatation.2062769 (2019)
18. Garvey, M.D.: Report on Shifts in Merchant CO_2: New production sources on-stream now, those coming in 2017. Gasworld Mag. **55**(5) (2017)
19. Häring, H.-W. (ed.): Industrial Gases Processing. Wiley-VCH Verlag GmbH & Co. KGaA, Weinheim (2008)
20. Vatopoulos, K., Tzimas, E.: Assessment of CO_2 capture technologies in cement manufacturing process. J. Clean. Prod. **32**, 251–261 (2012)
21. Li, J., Tharakan, P., Macdonald, D., Liang, X.: Technological, economic and financial prospects of carbon dioxide capture in the cement industry. Energy Policy **61**, 1377–1387 (2013)
22. Bjerge, L.-M., Brevik, P.: CO_2 capture in the cement industry, Norcem CO_2 capture project (Norway). Energy Proc. **63**, 6455–6463 (2014)
23. Nelson, T.O., Coleman, L.J.I., Mobley, P., Kataria, A., Tanthana, J., Lesemann, M., Bjerge, L.-M.: Solid sorbent CO_2 capture technology evaluation and demonstration at Norcem's Cement Plant in Brevik, Norway. Energy Proc. **63**, 6504–6516 (2014)

24. Hills, T., Florin, N., Fennell, P.S.: Decarbonising the cement sector: a bottom-up model for optimising carbon capture application in the UK. J. Clean. Prod. **139**, 1351–1361 (2016)
25. Hills, T., Sceats, M., Rennie, D., Fennell, P.: LEILAC: low cost CO_2 capture for the cement and lime industries. Energy Proc. **114**, 6166–6170 (2017)
26. Hills, T., Leeson, D., Florin, N., Fennell, P.: Carbon capture in the cement industry: technologies, progress, and retrofitting. Environ. Sci. Technol. **50**(1), 368–377 (2016)
27. Athena Sustainable Materials Institute: Appendix D: NRMCA Member National and Regional Life Cycle Assessment Benchmark (Industry Average) Report—Version 3.0. https://www.nrmca.org/sustainability/EPDProgram/Downloads/NRMCA_REGIONAL_BENCHMARK_Nov2019.pdf. Last accessed 2020/3/14
28. Miller, S.A., John, V.M., Pacca, S.A., Horvath, A.: Carbon dioxide reduction potential in the global cement industry by 2050. Cem. Concr. Res. **114**, 115–124 (2018)
29. Zhan, B., Poon, C.S., Liu, Q., Kou, S., Shi, C.: Experimental study on CO_2 curing for enhancement of recycled aggregate properties. Constr. Build. Mater. **67**, 3–7 (2014)
30. Zhang, J., Shi, C., Li, Y., Pan, X., Poon, C.-S., Xie, Z.: Performance enhancement of recycled concrete aggregates through carbonation. J. Mater. Civ. Eng. **27**(11), 04015029 (2015)
31. Xuan, D., Zhan, B., Poon, C.S.: Assessment of mechanical properties of concrete incorporating carbonated recycled concrete aggregates. Cement Concr. Compos. **65**, 67–74 (2016)
32. Fang, X., Xuan, D., Poon, C.S.: Empirical modelling of CO_2 uptake by recycled concrete aggregates under accelerated carbonation conditions. Mater. Struct. **50**(4), 200 (2017)
33. IEA: CO_2 intensity of selected countries, 1990–2017. IEA, Paris. https://www.iea.org/data-and-statistics/charts/co2-intensity-of-selected-countries-1990-2017
34. Monkman, S., MacDonald, M.: On carbon dioxide utilization as a means to improve the sustainability of ready-mixed concrete. J. Clean. Prod. **167**, 365–375 (2017)
35. McKinnon, A., Piecyk, M.: Measuring and Managing CO_2 Emissions of European Chemical Transport. CEFIC, Edinburgh (2011)

Use of Waste Calcium Carbonate in Sustainable Cement

Luca Valentini, Ludovico Mascarin, Hassan Ez-zaki, Mark Bediako, Joseph Mwiti Marangu, and Maurizio Bellotto

Abstract The quest for a new-generation concrete, designed to be compatible with the need of mitigating the effect of greenhouse gas emissions on the climate, has prompted applied research to define a broad range of low-CO_2 cement-based materials. While minimizing CO_2 emissions is a goal of the utmost importance, research into sustainable building materials must also tackle the issue of raw material depletion (including limestone, clay and aggregate deposits, as well as water resources) in favor of secondary raw materials. One possible solution is that of minimizing the impact of quarrying by a circular economy approach that envisages the reuse of waste from stone extraction and processing. It is estimated that 200 Mt waste are produced by the stone industry worldwide each year. This includes slurries obtained from the quarrying, cutting and polishing of marble, which can be used as a source of calcium carbonate, alternative to primary limestone. This contribution illustrates the use of waste calcium carbonate, obtained from marble slurry (waste marble, WM), in sustainable cement materials alternative to Portland cement. The possibility of exploiting locally available resources is explored, and the effect of WM additions up to 50% by total mass on the macroscopic properties is investigated experimentally. It is shown that binders with adequate fresh and hardened state properties can be obtained by moderate additions of WM, which greatly enhances the environmental performance by reducing the amount of primary resources used in the mix. By reducing the amount of thermally treated clay in alkali-activated blends, the use of WM also results in a net decrease of the embodied energy.

L. Valentini (✉) · L. Mascarin · H. Ez-zaki
University of Padua, 35131 Padua, Italy
e-mail: luca.valentini@unipd.it

M. Bediako
CSIR- Building and Road Research Institute, Kumasi, Ghana

J. M. Marangu
Meru University of Science and Technology, Meru, Kenya

M. Bellotto
CIRCe Centre for the Study of Cement Materials, 35131 Padua, Italy

V. M. C. F. Cunha et al. (eds.), *Proceedings of the 3rd RILEM Spring Convention and Conference (RSCC 2020)*, RILEM Bookseries 35,
https://doi.org/10.1007/978-3-030-76543-9_2

Keywords Waste calcium carbonate · Calcined clays · Alkali activation · Blended cements

1 Introduction

It has become a *must*, in recent years, to begin scientific papers into cement materials with a few sentences on the effect of buildings and the construction industry on CO_2 emissivity and its aftermath. This is reasonable, of course, and the reasons are obvious and shall not be reiterated here. However, there is a second, equally important issue, associated with the building materials industry, which is resource depletion and the impact of quarrying. Mining and quarrying activities have a huge impact on societies, the environment and landscape modification [1, 2] and their role in the cement industry deserves attention. Therefore, considering the increasing demand of building materials, it will be necessary to adopt a rational approach to sourcing raw materials for cement production.

One possible approach, explored in this contribution, is that of mitigating the utilization of primary resources (limestone, clay) by implementing a circular supply chain that converts waste from the stone industry into secondary raw materials.

It is estimated that about 200 Mt of the gross 300 Mt worldwide stone production is constituted by quarrying and processing waste [3]. Limestone and marble represent over 50% of the quarried stone [3] and their production leads to the formation of dust and slurries, during quarrying, cutting and polishing, whose disposal constitutes a severe environmental issue, due to the risk of soil and water contamination [4–6].

Using waste marble and limestone for the production of cement materials is beneficial as it contributes to mitigating the consumption of primary limestone. Moreover, the reutilization of such waste materials mitigates the risk associated with disposal and reduces net production costs.

In this contribution, some possible applications of waste marble in sustainable cements are briefly illustrated.

2 Waste Marble Properties

The waste marble (WM) used is obtained in form of a slurry, from marble quarries located in the Apuan Alps (Tuscany). A white powder (Fig. 1) is obtained by oven drying at 60 °C for 24 h and grinding using an agate mortar.

This powder consists, as per XRD analysis, of 100% calcite with Mg impurities ($MgO = 0.6$ wt%). The particle sized distribution after micronization, obtained by laser scattering, is reported in Fig. 2.

Fig. 1 Aspect of the waste marble powder used in the study

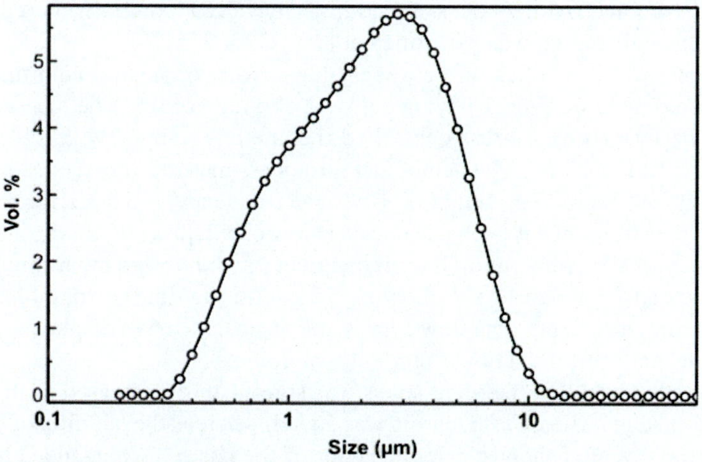

Fig. 2 Particle size distribution of the used waste marble powder

3 Blended Cements

Ternary blends of OPC, calcined clays and limestone are emerging as a possible, less impacting cement binder [7]. The environmental performance of such alternative binders can be further enhanced if waste calcium carbonate is used in the mix [8].

Several studies have investigated the hardened properties of these blended cements, but the fresh state properties have not been explored as accurately so far [9].

Here, we present the results of preliminary rheological investigations performed on a blended cement consisting of 50% OPC clinker, 30% calcined clay, 15% WM and 5% gypsum. The clay, consisting of kaolinite and minor phases that include

Table 1 XRF composition (wt%) of the Kenyan clay

SiO_2	TiO_2	Al_2O_3	Fe_2O_3	MnO	MgO	CaO	Na_2O	K_2O	P_2O_5
73.48	0.56	22.80	1.97	0.11	0.14	0.22	0.17	0.47	0.08

Table 2 Summary of the blended cements measured by rheometry

	% clinker	% calcined clay	% waste marble	% gypsum
95Clink_5G	95	–	–	5
50Clink_30Clay_15WM_5G	50	30	15	5
50Clink_45WM_5G	50	–	45	5
50Clink_45Clay_5G	50	45	–	5

quartz, feldspars and oxides, was sampled in Kenya (XRF composition reported in Table 1). Clay calcination was performed at 850 °C for 3 h.

The rheological properties of the ternary blend were compared with that of the mixes reported in Table 2, each having a water/solid ratio of 0.5. The measurements have been performed by a stress-controlled rheometer equipped with a plate-plate system and a gap of 2 mm. The actual measurements have been preceded by a pre-shear at high oscillation amplitude of 10% and frequency of 10 Hz for 1 min. A recovery of the structure has been performed at low oscillation amplitude of 5.10^{-3}% and frequency of 1 Hz for 2 min. Then, a triangular procedure and logarithmic ramps of shear rate of 10^{-2} s^{-1} to 10 s^{-1} allowing 5 s measuring time per point have been used for the measurements. The flow curves and viscosity curves displayed in Fig. 3 represent the analyses of the down-ramp measurements.

It can be observed that the yield stress was significantly increased in the binary and ternary systems when waste marble was added, whereas the addition of calcined clay alone did not alter the viscosity behavior of the pastes. When added together, WM and calcined clay apparently display a synergic effect in increasing the paste viscosity.

4 Alkali-Activated GGBFS

Clinker can be reduced in a significative amount when using ground granulated blast-furnace slag (GGBFS) to manufacture slag cements (CEM III) or composite cements (CEM V). However, in the current industrial practice these cements are limited to specific applications where their low hydration heat or their high resistance to aggressive environments is needed, because of their slow hardening kinetics.

Alkaline activation of GGBFS is needed to formulate binders of more general use, able to replace ordinary Portland cement in everyday applications. A "one part" solid alkaline activation results in a binder that sets and hardens upon mixing with water, like Portland cement, and is thus easy and safe to use.

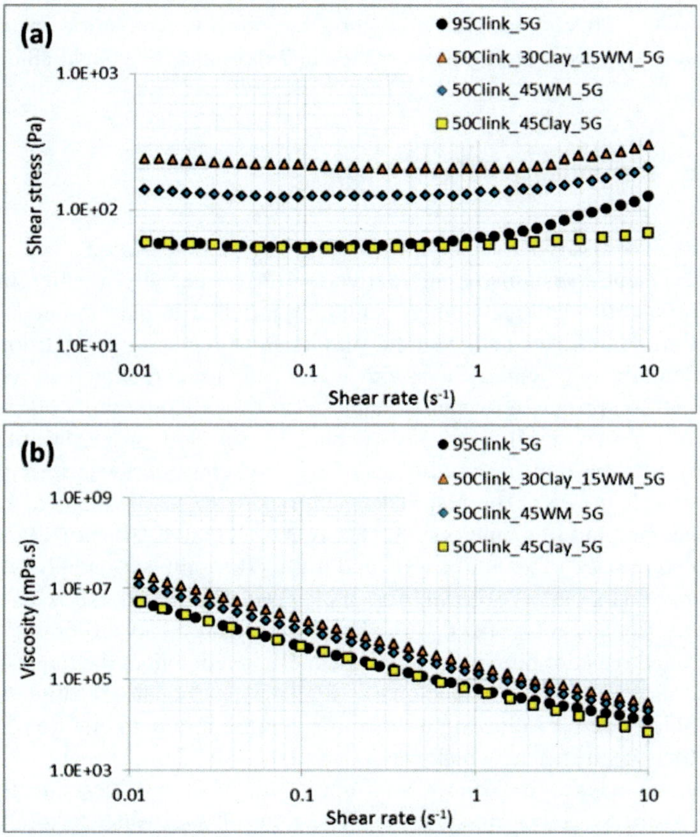

Fig. 3 Flow curves (**a**) and viscosity curves (**b**) of the studied blended cements

Activation by sodium carbonate or sodium sulfate can be used, but neither is enough by itself to provide sufficient reactivity. Sodium carbonate activation is widely practiced, with setting and hardening occurring when the carbonate ions in solution are exhausted [10]. Different approaches have been adopted to accelerate the depletion of carbonate ions in solution, the most straightforward being adding Ca^{2+} ions in solution to precipitate calcium carbonate [11]. Hydrated lime or Portland cement clinker can be used as a source of Ca^{2+} ions.

Here, as an alternative, micronized WM was used as a source of Ca^{2+} ions in solution, in combination with GGBFS (XRF composition reported in Table 3). A one-part alkali-activated binder was formulated by blending 47.5% GGBFS, 47.5% WM

Table 3 XRF composition (wt%) of the GGBFS

SiO_2	TiO_2	Al_2O_3	Fe_2O_3	MnO	MgO	CaO	Na_2O	K_2O	P_2O_5
37.12	0.70	10.96	0.74	0.28	7.59	41.92	0.23	0.46	0.01

and 5% Na_2CO_3. This binder presented good workability, even at low water/binder ratios, and a compressive strength of 26 MPa at 2 days and 49 MPa at 28 days.

5 Alkali-Activated Calcined Clays

The production of alkali-activated binders using low-grade calcined clays as precursors may represent a sustainable alternative to OPC, especially in emerging countries at subequatorial latitudes, where soils are enriched in clay minerals and the supply of Portland clinker or limestone may have to rely on import from foreign markets. Blending of calcined clays with waste calcium carbonate can reduce the environmental footprint to a further extent.

Here, we report a preliminary assessment of selected properties of binders produced by alkaline activation of blends of calcined clays of diverse compositions and waste marble powder. The materials used include three clay soils of kaolinitic composition, sampled in Cameroon (K1), Croatia (K2) and Ghana (K3), and two clays sampled in Italy, of smectitic (S1) and illitic (I1) composition. XRF compositions are reported in Table 4. The amount of clay minerals in these soils, as obtained by XRD and Rietveld refinement, varies from 28% (illitic clay) to 66% (smectitic clay), with the three kaolinitic clays having a kaolinite content in the range 40–60%. Minor mineralogical phases include quartz, feldspars, carbonates, apatite, Fe and Ti oxides. All clays were calcined in a laboratory muffle at a temperature of 800 °C, except for the smectitic clay, which was calcined at 750 °C.

The calcined clays were blended with up to 50% WM and subsequently mixed with sodium silicate activating solutions. The ratio of total water to total mass of the binders varied between 0.24 and 0.31. After mixing, the alkali-activated paste samples were placed in teflon molds and cured at ambient temperature and 95% RH. After three days, the samples were demolded and cured in air at ambient conditions.

Figure 4 reports a compilation of compressive strengths for the studied samples, plotted as a function of Na_2O_{eq}%, which is used as a qualitative indicator of the environmental impact, given that the major contributor to the ecological footprint of alkali-activated materials is represented by the alkaline activators. From this perspective, in addition to reducing: (1) the consumption of primary limestone and (2) the energy associated with clay calcination, the addition of waste marble also results in

Table 4 XRF composition (wt%) of the calcined clays used for alkaline activation

	SiO_2	TiO_2	Al_2O_3	Fe_2O_3	MnO	MgO	CaO	Na_2O	K_2O	P_2O_5
K1	47.49	1.88	34.50	15.55	0.05	0.10	0.12	0.00	0.15	0.16
K2	58.73	1.33	25.96	9.86	0.15	1.05	0.69	0.47	1.60	0.15
K3	69.11	0.11	27.47	1.10	0.00	0.01	0.13	1.36	0.70	0.00
I1	60.49	0.85	16.52	5.03	0.10	4.98	7.19	1.45	3.14	0.26
S1	54.15	3.63	17.10	9.59	0.13	5.57	7.38	0.13	0.91	1.42

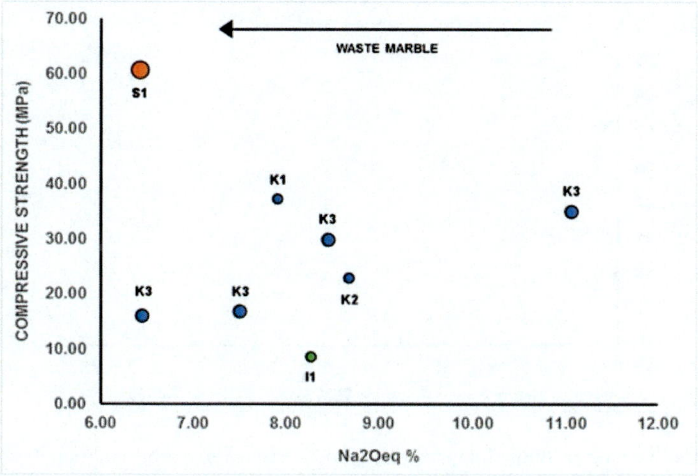

Fig. 4 Compressive strengths measured for the alkali-activated calcined clays blended with WM, as a function of equivalent Na_2O%. Blue: kaolinitic clays; green: illitic clay; orange: smectitic clay. The symbols' size is proportional to the time at which the compressive strength tests were performed (small circles: 1 week; medium circles: 2 weeks; large circle: 3 weeks)

a significant net decrease in the environmental impact by reducing the amount of activator needed.

The samples prepared with the kaolinitic clay from Ghana (K3) showed an overall positive trend in the plot, meaning that both compressive strength and Na_2O_{eq}% decreased as the amount of calcined clay is progressively diluted by WM addition. However, in general, the compressive strength does not simply depend on the amount of WM, but also on the quality of the specific clay used. For instance, clay K1 (Cameroon) has an excellent mechanical performance (37.35 MPa after 7 days) with an addition of 30% WM (7.91 Na_2O_{eq}%). Sample S1 gave a compressive strength of 60.70 MPa after 3 weeks, with an addition of 25% WM (6.42 Na_2O_{eq}%).

The addition of WM also enhanced the workability of the fresh pastes, as measured by mini-slump tests (Fig. 5). A more quantitative study of the effect of WM additions on the fresh state properties will be performed in a follow-up work, by means of systematic rheological measurements.

6 Conclusions

We reported a short preliminary account on the potential of waste calcium carbonate from the stone industry as an SCM in sustainable binders, including blended and alkali-activated cements. Retrieving accurate numbers on the world availability (which can vary locally) of calcium carbonate waste is not easy. However, considering that calcareous stones represent more than 50% of the total stone production

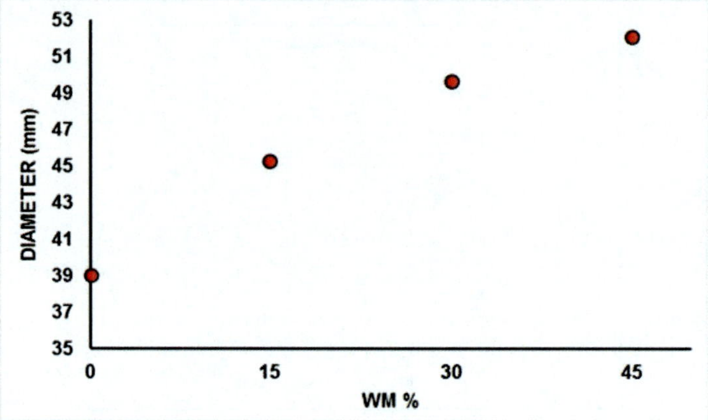

Fig. 5 Measured diameter (mini-slump test) as a function of waste marble addition, for sample K2

and that about two thirds of the global stone production is represented by waste [3], a rough estimate of 100 Mt per year of waste calcium carbonate can be inferred. It is also worth underlying the potential for local production of sustainable binders, using raw materials that are locally available in areas of emerging economic growth, such as Sub-Saharan Africa. Specifically, in this study, tests have been performed on the effect of waste marble on alternative cements formulated with clays from Kenya (ternary OPC-calcined clay-waste marble blend), Cameroon and Ghana (sodium-silicate-activated blends of calcined clays and waste marble). The waste marble used in these blends was of EU provenance, however, based on recent reports, the amount of waste from the stone industry in Africa amounts to about 7 Mt per year [3]. Specifically, marble and limestone waste have previously been used as a secondary raw material in the African continent [12, 13]. For alkali-activated calcined clays, the supply of sodium silicate solutions in the African continent may remain an issue in terms of overall cost and environmental footprint, however the possible use of carbonates as alkaline activators, even in combination with sustainable sources of soluble silica, such as rice husk ash, may potentially alleviate the economic and environmental burden, given the broad availability of alkaline carbonates of geological origin across the continent [14].

The results of this study show that moderate additions of WM are not detrimental to mechanical strength development. Excellent values of the compressive strength could be obtained for a sodium-carbonate-activated GGBFS formulated with 47.5% WM and alkali-activated smectitic and kaolinitic clays, with WM additions of 25% and 30% respectively.

The effect on the fresh properties needs to be further elucidated by systematic rheological measurements. However, it has been observed by preliminary tests that the addition of WM improves the workability of alkali-activated calcined clays. Moreover, the combined use of WM and calcined clays apparently exert a synergic effect in controlling the viscosity of ternary blends.

References

1. Burtynsky, E., Baichwal, J., De Pencier, N.: Anthropocene, 1st edn. Steidl, Göttingen (2018)
2. Subramanian, M.: Humans versus earth: the quest to define the Anthropocene. Nature **572**, 168–170 (2019)
3. Montani, C.: XXVIII World Marble and Stones Report (2017)
4. Adewole, M.B., Adesina, M.A.: Impact of marble mining on soil properties in a part of Guinea Savanna zone of Southwestern Nigeria. Ethiopian J. Environ. Stud. Manage. **4**, 1–8 (2011)
5. Careddu, N., Siotto, G.: Promoting ecological sustainable planning for natural stone quarrying: the case of the Orosei marble producing area in Eastern Sardinia. Resour. Policy **36**(4), 304–314 (2011)
6. Mulk, S., Azizullah, A., Korai, A.L., Khattak, M.N.K.: Impact of marble industry effluents on water and sediment quality of Barandu River in Buner District, Pakistan. Environ. Monitoring Assess. **187**, 8 (2015)
7. Scrivener, K., Martirena, F., Bishnoi, S., Maity, S.: Calcined clay limestone cements (LC3). Cem. Concr. Res. **114**, 49–56 (2018)
8. Krishnan, S., Kanuja, S.K., Mithia, S., Bishnoi, S.: Hydration kinetics and mechanisms of carbonates from stone wastes in ternary blends with calcined clay. Constr. Build. Mater. **164**, 265–274 (2018)
9. Scrivener, K.L., Favier, A.: Calcined Clays for Sustainable Concrete. Springer, Dordrecht (2015)
10. Bernal, S.A., Provis, J.L., Myers, R.J., San Nicolas, R., van Deventer, S.J.S.: Role of carbonates in the chemical evolution of sodium carbonate-activated slag binders. Mater. Struct. **48**, 517–529 (2015)
11. Bellotto, M., Dalconi, M.C., Contessi, S., Garbin, E., Artioli, G.: Formulation, performance, hydration and rheological behavior of 'just add water' slag-based binders. In: 1st International Conference on Innovation in Low-Carbon Cement and Concrete Technology (2019)
12. Okagbue, C.O., Onyeobi, T.U.S.: Potential of marble dust to stabilise red tropical soils for road construction. Eng. Geol. **53**, 371–390 (1999)
13. Boukhelkhal, A., Azzouz, L., Belaidi, A.S.E., Benabed, B.: Effects of marble powder as a partial replacement of cement on some engineering properties of self-compacting concrete. J. Adhesion Sci. Technol. **30**, 2405–2419 (2016)
14. US Geological Survey. Mineral Commodity Summaries 2019 (2019)

Classification of Recycled Aggregates Using Deep Learning

Jean David Lau Hiu Hoong, Jérôme Lux, Pierre-Yves Mahieux, Philippe Turcry, and Abdelkarim Aït-Mokhtar

Abstract The European Union is promoting sustainable development and circular economy by inciting its member states to recycle at least 70% of their construction and demolition waste (CDW) through the Horizon 2020 programme. CDW is crushed in order to obtain recycled aggregates (RA). The latter are a mixture of recycled concrete aggregates, natural stones, clay bricks, bituminous grains and with other materials (e.g. glass, wood and steel). The composition of RA is variable and in order to determine it, the NF EN 933-11 standard recommends manual sorting. However, it is time-consuming and it is performed only occasionally on recycling plants. Our work focuses on the development of a novel method to determine the composition of RA faster and in an automated way. It makes use of deep learning, particularly convolutional neural networks (CNN). CNNs can analyse images of RA and identify the nature of every aggregate in order to give the composition of the RA instantly. A labelled database was created for learning of the CNNs. It consists of approximately 36,000 images of individual grains classified according to their nature. The best-performing CNN is now able to identify correctly the class (i.e. the nature) of 97% of aggregates present on a picture of RA. Moreover, we proposed a method to evaluate the mass of the grains by assuming that those of a given nature have a constant form and density. We are also working on the automatic extraction of the grains from a picture of RA using Mask R-CNN.

Keywords Circular economy · Convolutional neural networks · Deep learning · Image analysis · Recycled aggregates

1 Introduction

In the context of sustainable development and circular economy, Europe is inciting its member states to recycle at least 70% of their construction and demolition waste (CDW) by 2020 through the Horizon 2020 programme. A survey [1, 2] showed that

J. D. Lau Hiu Hoong (✉) · J. Lux · P.-Y. Mahieux · P. Turcry · A. Aït-Mokhtar
Laboratoire des Sciences de l'Ingénieur Pour l'Environnement (LaSIE), UMR CNRS 7356, La Rochelle Université, Avenue Michel Crépeau, 17000 La Rochelle, France
e-mail: jean.lau_hiu_hoong1@univ-lr.fr

V. M. C. F. Cunha et al. (eds.), *Proceedings of the 3rd RILEM Spring Convention and Conference (RSCC 2020)*, RILEM Bookseries 35,
https://doi.org/10.1007/978-3-030-76543-9_3

227.5 Mt of CDW (70% of all wastes) were produced in France in 2014. 155.9 Mt (69%) were recycled mainly through reuse on road construction sites and as quarry backfill. However, 70 Mt still ended up in storage sites while the recycling of more CDW would help preserve the natural resources.

Recycled aggregates (RA) are obtained from the crushing of inert CDW. They are a mixture of recycled concrete aggregates, natural stones, clay bricks, bituminous grains, along with traces of other materials (e.g. glass, wood and steel). The composition of CDW is variable as it depends on the site from which they originate and on the crushing technique used [3, 4]. This variability in composition limits the valorisation of RA in applications such as concrete manufacturing, in which the use of raw materials of controlled quality is essential.

Many studies have been carried out in which natural aggregates (NA) have been partially or totally replaced by RA in concrete [5–7]. It has been observed that the mechanical and durability properties of concrete containing RA tend to decrease. On the other hand, in road construction, studies have shown that RA can even outperform NA [8–11]. These studies show that the composition of RA influences their properties. Due to this constraint, RA are currently used mostly for road construction, as the regulations, defined in the NF EN 206 + A1 standard, regarding concrete manufacturing are more restrictive. It should also be noted that RA can also be valorised as a mineral addition [12] and as means to uptake CO_2 by accelerated carbonation [13].

For the valorisation of RA as high-value added products, it is essential to monitor their composition closely. Currently, the NF EN 933-11 standard recommends manual sorting in order to determine that composition. However, it is time-consuming and is carried out only occasionally. This study focuses on the development of a novel method in order to determine the composition of CDW in near real time and in an automatic way. The goal of our work is to improve the characterisation of RA so that the other actors of the building sector, for instance concrete manufacturers, can rightfully use RA as value-added products.

Our method relies on deep learning which is a type of machine learning in which deep neural networks (DNN) analyse datasets of different types in order to model the data. The particular type of DNNs used for image classification is called a convolutional neural network (CNN). In order to classify objects, a CNN has to be trained on a large set of labelled images.

For our study, we created a labelled database consisting of approximately 36,000 images of RA for the learning process of CNNs. The latter analyse images of RA in order to give their composition. We have also customised a commonly used CNN to get improved performances. Moreover, in order to determine the mass of each grain, we have evaluated form factor-density constants for each type of RA by assuming that grains of a given nature have a constant form and density. Currently, we are working on the automatic extraction of grains from a photograph of RA.

2 Methods and Materials

2.1 Tested Architectures of Convolutional Neural Networks

The working principle of a convolutional neural networks (CNN) is detailed in [14]. The Residual Network, ResNet, [15] is a type of CNN used for image classification. The 34-layer Residual Network, ResNet34 (RN34) is used as a reference architecture. The 50-layer Residual Network, ResNet50 (RN50) is a variant of the RN34. It has a better accuracy than RN34 for only a slight increase in the required computing power.

In this work, we modified the architecture of RN34 in order to improve further its performance. Our custom ResNet34 (C-RN34) has full preactivation [16] and the main difference is the use of depthwise separable convolutions (DSC) [17], instead of standard convolutions. Furthermore, a squeeze-and-excitation block [18] is added at the end of each building block. During our tests, these changes have improved the validation accuracy. In fact, C-RN34 architecture gives better results than the RN50 as shown in Sect. 3.1. For this study, we used the RN50 and the C-RN34.

2.2 Creation of the Labelled Database

The labelled database of recycled aggregates (RA) used to train the convolutional neural networks (CNN) consists of images of individual grains with their corresponding sub-class. The creation of this database is detailed below.

Preparation of the samples and manual sorting. The recycled aggregates (RA) used in this work are of 0/31.5 mm fraction. They have been sampled from a recycling platform in La Rochelle, France, called Planète Recyclage, subsidiary of Charier Company. The RA are rinsed in order to remove the fines that cover them, thereby preventing their identification. Then, the RA are sorted manually. Note that the 0/4 mm fraction is not used in order to determine the composition of RA because the grains are too small to be manually sorted. The same is applied in the NF EN 933-11 standard.

This standard defines the procedures of manual sorting and the classes of constituents of RA. However, the elements in some of those classes are visually heterogeneous and so not suitable for image classification by a CNN. Hence, some classes were divided into sub-classes. Table 1 gives the classes and corresponding sub-classes of RA. Figure 1 shows examples of elements present in each sub-class. This division of the 6 classes into 16 sub-classes also allows us to obtain a composition of RA, which is more precise than that obtained by the standardised manual sorting.

Apparatus. An apparatus (Fig. 2) was developed in order to photograph the RA. It consists of a Fujifilm X-T20 camera (1) fixed on a copy stand (2), two LED lamps (3), a transparent plastic sheet (4) on which the grains are placed and a background

Table 1 EN 933-11 classes and corresponding sub-classes of recycled aggregates

EN 933-11 classes	Corresponding sub-classes	Description
Rc	Rc	Concrete grains
Ru	Ru01	White stones such as limestone
	Ru02	Grey stones such as basalt & others of similar colours
	Ru03	Light coloured grainy stones (majority of quartz and feldspar)
	Ru04	Coloured or dark coloured siliceous and rather angular stones
	Ru05	Light-coloured and rounded alluvial stones
	Ru06	Slate
Rb	Rb01	Clay bricks
	Rb02	Ceramic tiles
Ra	Ra	Bituminous grains
Rg	Rg	Glass
X	X01	Wood
	X02	Plastics
	X03	Steel
	X04	Paper and cardboard
	X05	Others

Fig. 1 Examples of elements in each sub-class

(5) of a different colour (e.g. blue) from that commonly found in RA. Furthermore, a 5 cent euro coin is used as a reference for conversion between pixels and millimetres.

Extraction of individual grains from the photographs. In order to build up the database, about 43 kg of 4/31.5 mm recycled aggregates (RA) have been sampled, manually sorted according to the sub-classes and photographed. It should be noted

Fig. 2 Photograph apparatus

that this sampling mass is more than four times the minimum value (10 kg) recommended by the NF EN 933-11 standard. After a photograph was taken, the overall mass of all the grains on the picture was measured. The grains were not weighed individually because this would have been too time-consuming.

More than 360 photographs (Fig. 3) were taken and analysed using a software based on mathematical morphology, which was developed at the laboratory in order to extract the grains individually. The geometrical data (length of minor and major axes, position and dimension of bounding box, projected area and perimeter) about each grain is also obtained after this step. This data can be used to characterise the shape of the aggregates. After extraction, 36,000 individual images of grains were obtained. Some sub-classes (Rc, Ru01, Ru02, Ru04, Ru05, Rb01, Rb02, Ra) have between 2000 and 6000 grains. However, other sub-classes (Ru03, Ru06, Rg, X01-X05) have less than 1100 particles because they are difficult to find on recycling platforms. This uneven distribution can affect the accuracy of the CNNs. This issue is mitigated through data augmentation.

Resizing and data augmentation. Both convolutional neural network (CNN) architectures tested (RN50 and C-RN34) require images of a fixed size in order to

Fig. 3 Example of a photograph of recycled aggregates

work. Therefore, the images of individual recycled aggregates (RA) extracted at the previous step have been resized to 256 × 256 pixels.

Moreover, in the labelled database, it is better if each sub-class of RA contains the same number of elements. This ensures that grains from each sub-class are classified with sufficient precision. In order to remedy the under-representation of some sub-classes, we use data augmentation techniques (random addition of salt and pepper noise, grain rotation and gamma correction). However, these augmentations are limited. Each image of the database is submitted to these modifications until the number of images per class reaches 2000 in the training set and 500 in the validation set. The training set is used to train the CNN so that it improves its ability to predict the nature of the grains. Images from the validation set test the ability of the CNN to generalize, that is to correctly classify images different from those seen during training. The parameters of the CNN are not updated during validation. Some sub-classes (Ru03, Ru06, Rg and X02-X05) were not used to create the labelled database because they do not contain enough elements (an element would have been repeated too many times using the data augmentation techniques described above).

Therefore, the labelled database used to train the CNNs consists of 22,500 images: 2500 images (training set + validation set) in each of the following 9 sub-classes: Rc, Ru01, Ru02, Ru04, Ru05, Rb01, Rb02, Ra and X01. It should be noted that the grains in the validation set are different from those in the training set. Figure 4 shows part of the labelled database of the recycled aggregates.

Fig. 4 Part of the labelled database of recycled aggregates

2.3 *Form Factor-Density Constants of the Aggregates*

As described above, a trained convolutional neural network (CNN) will be able to predict the nature of a recycled aggregate (RA) from its image. Moreover, the geometrical data (length of minor and major axes and projected area) about each grain is also available. Figure 5a shows a grain and Fig. 5b shows its geometrical data. According to the NF EN 933-11 standard, the composition of RA obtained by manual sorting is given in terms of mass proportions. Therefore, in order to compare our proposed method with manual sorting, we have to estimate the mass of the grains that undergo image analysis. Our approach is similar to that proposed by [19]. The mass m_i of a grain i can be expressed as shown in Eq. (1).

$$m_i = S_i \times L_{\min,i} \times F_k \times \rho_k \quad (1)$$

where, S_i is the projected surface area (cm^2) and $L_{\min,i}$ is the minor axis (cm). F_k and ρ_k are the form factor and density and are assumed to be constants for all the grains belonging to a sub-class k.

As it is not possible to measure the mass of each grain individually, we measured the total mass of all grains of each photograph j of a given sub-class k, denoted M_{jk}. Eventually, the product between F_k and ρ_k (form factor-density constant) can be evaluated, for each sub-class k, as shown in Eq. (2). The results are shown in Sect. 3.2.

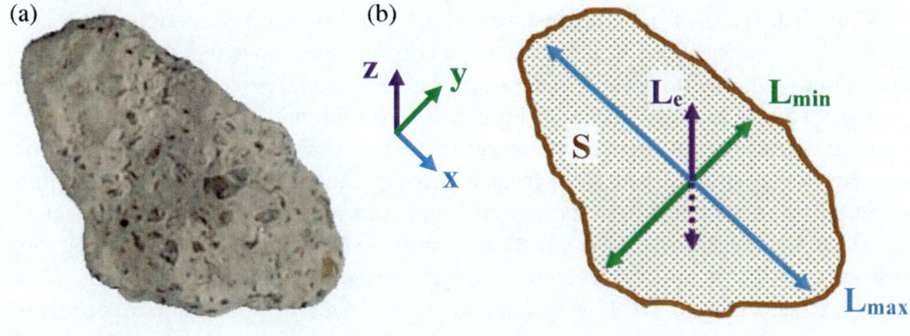

Fig. 5 **a** Original grain and **b** geometrical data of the grain

$$F_k.\rho_k = \frac{\sum_j M_{jk}}{\sum_{i,j}\left(S_{i,j} \times L_{\min,i,j}\right)} \quad (2)$$

where $\sum_{i,j} (S_{i,j} \times L_{\min,i,j})$ is the sum of the product of the surface area by the minor axis length for all grains i in all the j photographs of the sub-class k. Once determined, $F_k \rho_k$ can be used to estimate the mass of other aggregates.

3 Results and Discussion

3.1 Classification of Images of Recycled Aggregates

Comparison between the standard ResNet50 and our custom ResNet34. As explained in Sect. 2.1, a standard ResNet50 architecture (RN50) and a custom one (C-RN34) were tested. The labelled database containing 22,500 images was used to train both networks over 250 iterations.

For both architectures, the training accuracy tends to 100%. This means that both of them have been optimised and can correctly predict the sub-class of almost all the grains in the training set. The difference between them lies in the validation accuracy. When trying to predict the nature of grains that have not been used to modify its parameters (validation set), RN50 reaches a validation accuracy of only 92.0%. On the other hand, C-RN34 has a validation accuracy of 97.1%. The latter is our best-performing CNN as it has the highest validation accuracy. Therefore, it will be used for the rest of the study.

Confusion matrix. In order to analyse the predictions of the C-RN34_D40 precisely, its confusion matrix relative to the sub-classes has been represented in Fig. 6a. It compares the real sub-class of the grains (vertical axis) with the predictions (horizontal axis) of the images from the validation set. The percentage of correct predictions is given on the top-left to bottom-right diagonal and the other percentages represent wrong classifications. It can be observed that all the sub-classes have a validation accuracy between 96 and 100% except for concrete (Rc), 93%, and limestone (Ru01), 94%. The highest percentages of wrong classifications are 3 and 4% and they concern the confusion between concrete and limestone grains and vice versa. Among the natural stones, limestone has the highest percentage of confusion because it can be visually similar to light-coloured mortar.

Some confusion between concrete grains and natural stones is expected as the latter are used in order to make concrete. Moreover, most concrete grains have a part of natural stone, which is more or less visible, depending on the amount of attached mortar. On the confusion matrix concerning only the NF EN 933-11 classes, Fig. 6b, it can be seen that is mostly concrete grains that are mistaken for natural stones and not otherwise. Figure 6c, d show a concrete grain and a limestone grain respectively

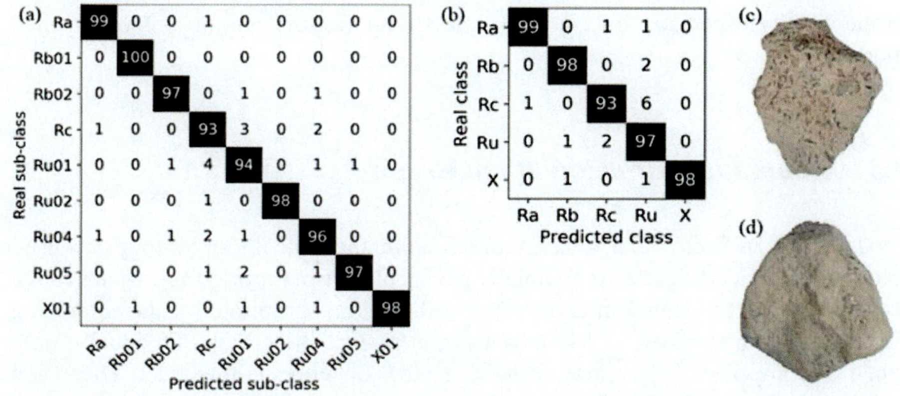

Fig. 6 **a** Confusion matrix of the sub-classes (the percentages are rounded off to the nearest integer and so the sum might be equal to 100 ± 1%); **b** confusion matrix of the NF EN 933–11 classes; **c** concrete grain and **d** limestone grain

and it can be observed that both are visually similar. It should be noted that similar confusion could be made during manual sorting.

3.2 *Form Factor-Density Constants*

In order to estimate the mass of each grain, the form factor-density constant (F.ρ) of the 9 sub-classes in the labelled database were determined as described in Sect. 2.3. The results are presented in Table 2.

However, the determination of $F.\rho$ relies on the hypothesis that in a given sub-class, all the grains have the same form and density. In reality, this might not always be the case. This is because the shape of the grains are closely related to the crushing technique used [3, 4]. Additionally, in a given sub-class, there might be variations

Table 2 Form factor-density constant (F.ρ) of the 9 sub-classes in the labelled database

Sub-class	$F.\rho$ (g/cm^3)
Rc	0.746
Ru01	0.658
Ru02	0.782
Ru04	0.624
Ru05	0.873
Rb01	0.517
Rb02	0.555
Ra	0.853
X01	0.141

in the density especially for concrete grains with varying proportions of attached mortar.

3.3 *Automatic Extraction of the Grains*

As described in Sect. 2.2, a software based on mathematical morphology was used in order to extract the grains individually from a picture of recycled aggregates (RA). This step requires manual intervention and thus is inappropriate for the automated process we aim to develop. Moreover, it can extract the grains individually only when they are placed separately from each other or with slight contact. Therefore, we are currently testing another type of convolutional neural network (CNN), known as Mask R-CNN [20], which can locate the bounding box of each grain (part of the photograph that contain the grain) and extract them individually. Figure 7 illustrates examples of bounding boxes and masks (shape and projected surface area of a grain) that have been detected using Mask R-CNN. The extracted grains can then be sent to the previously trained CNN for classification.

Fig. 7 Bounding boxes and masks extracted using mask R-CNN. Note that the colours of the masks do not represent the sub-class of the grains. They are just used to differentiate different grains

4 Conclusion and Future Work

In this study, we aimed at enhancing the value of recycled aggregates (RA) by proposing a novel method in order to obtain their composition in near real time using an automated process. Our method is based on deep learning, particularly on image analysis by convolutional neural networks (CNN).

A labelled database consisting of 36 000 images of individual RA was created for the learning process of the CNNs. A standard ResNet50 architecture was tested but it did not exceed a maximum validation accuracy of 92.0%. Therefore, we modified the ResNet34 architecture in order to improve its performance. This proved to be successful, since our custom ResNet34 architecture reached a validation accuracy of 97.1%.

Since the RA composition obtained by the standardised (NF EN 933-11) manual sorting is in terms of mass proportions, we proposed a method to evaluate the mass of our grains. By assuming that all the grains in a given sub-class have the same shape and density, we have evaluated form factor-density constants for each sub class. These constants can then be used to evaluate the mass of each grain as explained in Sect. 2.3. However, in a given sub-class not all the grains have the same shape and density. Therefore, we plan to use a separate deep neural network (DNN) in order to find a direct correlation between the image of the grains and their mass.

Another aspect that we are looking to improve is the automatic extraction of grains in real conditions, that is, when they are in contact with each other. The method based on mathematical morphology presented above is limited to grains that are only slightly in contact with each other. Therefore, we will use another type of CNN, Mask R-CNN, which localises the parts of the photograph that contain a grain (bounding box). Once the bounding boxes have been located, the grains can be extracted individually and sent to the previously trained CNN for classification.

Once the method is fully automated, it can be adapted to the conditions of a recycling platform. For instance, after the crushing process, the RA would be carried on a conveyor belt. A camera found above it would take pictures of the RA. Afterwards, these pictures would be analyzed by the deep learning method that we are developing in order to get the composition of the RA automatically in near real-time.

Acknowledgements The Region and the European Union support the project <CPER-FEDER Bâtiment durable Axis 2 MADUR Project: High-performance building materials with low environmental impact, sustainable and resilient> within the framework of the «Operational Program FEDER/FSE 2015-2020».

References

1. Ministère de l'Environnement, de l'Énergie et de la Mer—Ministry of Environment, Energy and Sea, Bilan 2014 de la production de déchets en France **4** (2017)

2. Ministère de l'Environnement, de l'Énergie et de la Mer—Ministry of Environment, Energy and Sea, Entreprises du BTP: 227.5 millions de tonnes de déchets en 2014 **4** (2017)
3. Akbarnezhad, A., Ong, K.C.G., Tam, C.T., Zhang, M.H.: Effects of the parent concrete properties and crushing procedure on the properties of coarse recycled concrete aggregates. J. Mater. Civ. Eng. **25**, 1795–1802 (2013). https://doi.org/10.1061/(ASCE)MT.1943-5533.0000789
4. Guimaraes, M.S., Valdes, J.R., Palomino, A.M., Santamarina, J.C.: Aggregate production: fines generation during rock crushing. Int. J. Miner. Process. **81**, 237–247 (2007). https://doi.org/10.1016/J.MINPRO.2006.08.004
5. Kou, S.-C., Poon, C.-S., Wan, H.-W.: Properties of concrete prepared with low-grade recycled aggregates. Constr. Build. Mater. **36**, 881–889 (2012). https://doi.org/10.1016/J.CONBUILDMAT.2012.06.060
6. Omary, S., Ghorbel, E., Wardeh, G., Nguyen, M.D.: Mix design and recycled aggregates effects on the concrete's properties. Int. J. Civ. Eng. **16**, 973–992 (2018). https://doi.org/10.1007/s40999-017-0247-y
7. Bravo, M., de Brito, J., Evangelista, L., Pacheco, J.: Durability and shrinkage of concrete with CDW as recycled aggregates: benefits from superplasticizer's incorporation and influence of CDW composition. Constr. Build. Mater. **168**, 818–830 (2018). https://doi.org/10.1016/j.conbuildmat.2018.02.176
8. Hou, Y., Mahieux, P.-Y., Lux, J., Turcry, P., Aït-Mokhtar, A.: Qualification of the residual reactivity of compacted recycled aggregates. In: ICSBM 2019, 2nd International Conference of Sustainable Building Materials, Eindhoven, Netherlands, 2019
9. Lancieri, F., Marradi, A., Mannucci, S.: C&D waste for road construction: long time performance of roads constructed using recycled aggregate for unbound pavement layers. In: Waste Management and the Environment III, WIT Press, Southampton, UK, 2006, pp. 559–569. https://doi.org/10.2495/WM060571
10. Mahieux, P.-Y., Turcry, P., Ferber, V., Chateau, L.: Caractérisation de granulats recyclés à haute valeur ajoutée, in: 35ème Rencontres Universitaires de Génie Civil, AUGC, Nantes, France, 2017
11. Turcry, P., Mahieux, P.-Y., Hamdoun, H., Chateau, L., Ferber, V.: Projet Recyment: Etude de la réactivité des bétons concassés: application aux graves non traitées et au traitement des sols—Rapport final. ADEME **100** (2016)
12. Oksri-Nelfia, L., Mahieux, P.-Y., Amiri, O., Turcry, Ph., Lux, J.: Reuse of recycled crushed concrete fines as mineral addition in cementitious materials. Mater. Struct. **49**, 3239–3251 (2016). https://doi.org/10.1617/s11527-015-0716-1
13. Zhang, N., Duan, H., Miller, T.R., Tam, V.W.Y., Liu, G., Zuo, J.: Mitigation of carbon dioxide by accelerated sequestration in concrete debris. Renew. Sustain. Energy Rev. **117**, 109495 (2020). https://doi.org/10.1016/j.rser.2019.109495
14. LeCun, Y., Bengio, Y., Hinton, G.: Deep learning. Nature **521**, 436–444 (2015). https://doi.org/10.1038/nature14539
15. He, K., Zhang, X., Ren, S., Sun, J.: Deep Residual Learning for Image Recognition (2015). http://arxiv.org/abs/1512.03385
16. He, K., Zhang, X., Ren, S., Sun, J.: Identity mappings in deep residual networks. In: Leibe, B., Matas, J., Sebe, N., Welling, M. (eds.) Computer Vision—ECCV 2016, Springer International Publishing, Cham, 2016, pp. 630–645. https://doi.org/10.1007/978-3-319-46493-0_38
17. Howard, A.G., Zhu, M., Chen, B., Kalenichenko, D., Wang, W., Weyand, T., Andreetto, M., Adam, H.: MobileNets: Efficient Convolutional Neural Networks for Mobile Vision Applications (2017)
18. Hu, J., Shen, L., Sun, G.: Squeeze-and-Excitation Networks, pp. 7132–7141 (2018)
19. Mora, C.F., Kwan, A.K.H., Chan, H.C.: Particle size distribution analysis of coarse aggregate using digital image processing. Cem. Concr. Res. **28**, 921–932 (1998). https://doi.org/10.1016/S0008-8846(98)00043-X
20. He, K., Gkioxari, G., Dollár, P., Girshick, R.: Mask R-CNN (2018). http://arxiv.org/abs/1703.06870

Performance and Ageing Evaluation of Bituminous Mixtures with High RAP Content

C. Santos, V. Antunes, J. Neves, and A. C. Freire

Abstract The implementation of a circular economy implies turning waste into resources, avoiding its disposal in landfills and minimizing the extraction of raw materials. Every year, a considerable amount of Reclaimed Asphalt Pavement (RAP) is generated, as the road network requires pavement maintenance to ensure the safety and comfort of the users. RAP contains both constituents of a bituminous mixture (bitumen and aggregates), and is 100% recyclable, therefore being able to be re-introduced back into the cycle without its use being downgraded. However, RAP is currently being employed solely as an aggregate in unbound layers or in small percentages in new bituminous mixtures, rather than taking advantage of the aged bitumen. This paper presents a study which aimed to evaluate the performance of a surface dense-graded hot bituminous mixture before and after being subjected to an ageing procedure. The bituminous mixture containing a high RAP content—75%—was treated with a commercial vegetable rejuvenator. Short Term Oven Ageing (STOA) and Long Term Oven Ageing (LTOA) procedures were used to simulate the ageing that occurs during the mixture's production and service life, respectively. The evaluation of the recycled bituminous mixture's performance was based on stiffness, fatigue resistance, and permanent deformation tests carried out in the laboratory. In comparison with a conventional bituminous mixture, it was found that, in general, the recycled bituminous mixture, before and after ageing, showed higher stiffness and resistance to permanent deformation, yet had very similar fatigue resistance.

Keywords RAP · Bituminous mixture · Circular economy · Recycling · Ageing

C. Santos (✉)
Instituto Superior Técnico, Universidade de Lisboa, Lisboa, Portugal

V. Antunes · J. Neves
CERIS, Instituto Superior Técnico, Universidade de Lisboa, Lisboa, Portugal
e-mail: vitorfsantunes@tecnico.ulisboa.pt

V. Antunes · A. C. Freire
LNEC—National Laboratory of Civil Engineering, Lisboa, Portugal

V. M. C. F. Cunha et al. (eds.), *Proceedings of the 3rd RILEM Spring Convention and Conference (RSCC 2020)*, RILEM Bookseries 35,
https://doi.org/10.1007/978-3-030-76543-9_4

1 Introduction

The road infrastructure network is indispensable for society, enabling the functioning and development of a country by providing mobility and accessibility of people and goods. The pavement maintenance actions required to ensure an adequate quality level regarding user safety and comfort generates a considerable amount of by-products.

Environmental concerns push for the transition to a circular economy model, whose aim is to decrease waste production and natural resource depletion by reintroducing products at their end-of-life stage in the cycle, rather than disposing of them. In fact, in 2015, the European Union (EU) has implemented the Action Plan for the Circular Economy, which advocates for more efficient use of natural resources and for turning waste into secondary raw materials [1].

In Europe, bituminous mixtures are the predominant construction material of pavement bound layers and they are 100% recyclable [2]. However, the Reclaimed Asphalt Pavement (RAP) is mostly recycled as unbound granular material or in small percentages (5–20%) in new bituminous mixtures. This is not an efficient use of this resource since the bitumen's properties are not being taken advantage of, thus downgrading the RAP [3].

Furthermore, the exposure of the binder to the climate conditions during its service life is an important aspect to be considered in bituminous mixtures, in general. In the case of RAP use, especially in the case of high incorporation rates (> 25%), this ageing effect that changes the binder's properties becomes more important. In fact, the ageing process makes the binder stiffer and influences the mixture's performance, therefore, when producing new bituminous mixtures with high RAP content, it is recommended to use a softer binder or rejuvenator which improves the aged binder's properties [4].

Indeed, in order to be viewed as a viable alternative to a traditional mixture, a recycled one and its materials should perform as well as the first or even better. Up to the present time, the performance of recycled mixtures has been evaluated in different conditions: with different percentages of RAP incorporation, with or without different types of rejuvenator and with or without fractionation of the RAP.

In general, those evaluations verify that increasing RAP content on recycled bituminous mixtures results in increasing stiffness, thus lowering cracking resistance and workability but increasing resistance to permanent deformation. The addition of a rejuvenator or a softer binder can enhance the mixture's workability, as they reduce the bitumen's/mixture's viscosity and stiffness, and, in the case of the rejuvenated mixtures, still keeping a high permanent deformation resistance [5, 6].

In spite of the positive results in those evaluations, the RAP recycling practice has been limited mostly due to the lack of confidence in the recycled pavement's performance, the increased complexity of the operation, the variability of the material and the unknown degree of mobilization of the aged binder [2, 3, 7–9].

In essence, it is fundamental to deepen the knowledge on recycled bituminous mixtures in order to overcome the previously listed challenges and maximize the utility of RAP, bringing a contribution to the transition to a circular economy.

The main objective of this paper is to analyze the performance, in laboratory, of a recycled bituminous mixture incorporating high RAP content after going through an ageing process. This mixture contained 75% RAP content, treated with a commercial vegetable rejuvenator. The ageing process was simulated through Short Term Oven Ageing (STOA) and Long Term Oven Ageing (LTOA) procedures. The performance tests were carried out on two recycled mixtures incorporating RAP (before and after the ageing process), and one reference bituminous mixture produced only with virgin aggregates.

2 Methodology

2.1 Materials

The recycled bituminous mixtures that were produced in laboratory incorporated RAP milled from a high trafficked road, whose age was unknown as well as bitumen origin and nature. From the recovered RAP aggregates, it is possible observe that they have multiple origins as limestone, basalt and granitic. The RAP was treated with a commercial rejuvenator derived from crude tall oil, a by-product of the paper industry.

The virgin materials were basalt aggregates in the 10/16, 4/12 and 0/4 mm fractions, limestone aggregates in the 0/4 mm fraction and a virgin bitumen with a nominal penetration grade of 35/50.

2.2 Testing Methods

The tests performed throughout this study can be divided into three categories: (i) RAP characterization; (ii) mix design; and (iii) performance evaluation (Table 1).

An adequate characterization of the RAP and its components (bitumen and aggregates) was essential since the studied bituminous mixture incorporated a high percentage of RAP. The mix design tests included those used to determine the mixtures' volumetric and mechanical characteristics and its performance was evaluated through stiffness, fatigue resistance, and permanent deformation tests.

Table 1 Test methods

Standard	Test method	Scope	Cat
EN 933-1	Particle size distribution	Characterization of the aggregates	(i)
EN 12697-1	Soluble binder content	Determination of the RAP binder content	(i)
EN 12697-3	Rotary Evaporator method	Recovery of the RAP binder	(i)
EN 1426	Determination of needle penetration	Characterization of the binder	(i)
EN 1427	Determination of the softening point	Characterization of the binder	(i)
EN 12697-5	Maximum density	Determination of the volumetric properties	(ii)
EN 12697-6	Bulk density	Determination of the volumetric properties	(ii)
EN 12697-34	Marshall test	Determination of the Marshall properties	(ii)
EN 12697-26	Stiffness	Characterization of the stiffness	(iii)
EN 12697-24	Resistance to fatigue	Characterization of the fatigue behaviour	(iii)
EN 12697-22	Wheel tracking	Determination of the resistance to permanent deformation	(iii)

2.3 Bituminous Mixture Production and Ageing

The experimental study has concerned a dense graded hot bituminous mixture with a maximum aggregate dimension of 14 mm and a bitumen with a nominal penetration grade 35/50: AC 14 surf 35/50. Two mixtures of this type were analyzed in the laboratory and compared: a reference bituminous mixture containing only virgin materials and a mixture incorporating 75% of RAP and a rejuvenator, both produced and compacted in the laboratory.

The bituminous mixture for the specimens was prepared following the rejuvenator's manufacturer guidelines and compacted according to EN 12697-30 with a target temperature for compaction of 165 °C (specified by EN 12697-35 for a 35/50 penetration paving grade bitumen). The specimen production procedure was the following:

Step 1 All the materials were heated:

- The RAP was heated at 130 °C for 2h30.
- The virgin aggregates were heated at 205 °C for 4h00.
- The virgin bitumen was heated at 165 °C for 3h00.

Step 2 The RAP and rejuvenator were added to the mixer and mixed for 30 s.

Step 3 The virgin binder and virgin aggregates were added to the mixing bowl and mixed for 2 min 30 s.

Step 4 The mixture was poured into the mold and compacted.

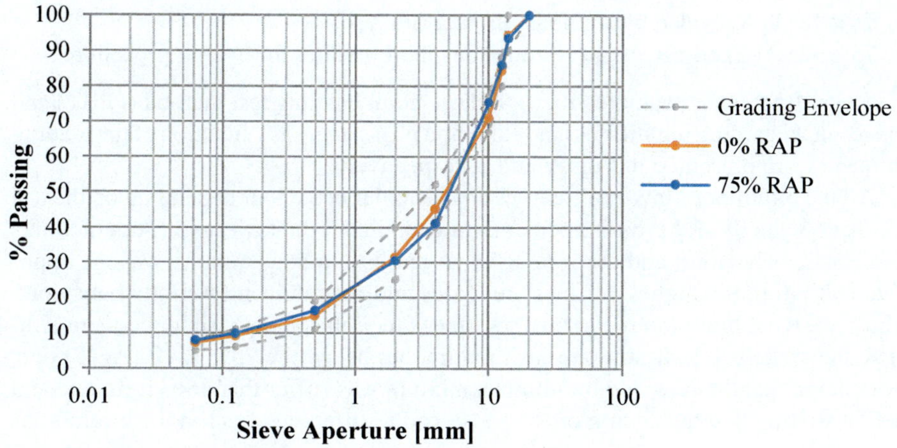

Fig. 1 Reference and recycled bituminous mixtures' gradation curve

The percentage of rejuvenator to mix with the aged bitumen to recover its properties was defined through the assessment of the penetrations and softening temperatures of bitumen samples mixed with different dosages of rejuvenator. The selected dosage was 4.5% of rejuvenator per weight of aged binder, whose sample properties were similar to those of a 35/50 grade bitumen, defined in EN 12591.

The 75% RAP recycled mixture was designed using the Marshall method. It was set that the mixture would be composed of 25% of virgin aggregates and 75% of RAP and the amount of each fraction was determined accordingly, through a trial and error process, in which the gradation curve was fit between the upper and lower limits for the aggregate gradation of an AC 14 surf, defined in the Portuguese Road authority's specifications [10]. The gradation curve, along with the upper and lower limits defined in the specifications, is presented in Fig. 1.

To analyze the performance of the recycled mixture after going through an ageing process, there were test specimens whose production integrated short-term oven ageing (STOA) and were further aged through long-term oven ageing (LTOA). This laboratory ageing procedures were based on AASHTO R30, which consists of the STOA, accounting for the ageing that occurs during the mixing, storage, transportation, and compaction of the mixture until it cools down; and the LTOA aiming to simulate the ageing that occurs during the bituminous mixture's service life.

3 Results and Discussion

This section presents the results of the performance tests identified in Table 1. The analyzed bituminous mixtures' designations are the following:

- 0% RAP: Virgin bituminous mixture.

- 75% RAP: Recycled bituminous mixture (unaged).
- 75% RAP (Aged): Recycled bituminous mixture after the ageing procedure.

The stiffness was evaluated through the 4-point bending test method (4PBT) and the results regarding the stiffness modulus and phase angle as a function of the loading frequency are presented in Figs. 2 and 3, respectively.

All the bituminous mixture's curves show similar behaviour depending on the test frequency: the 0% RAP bituminous mixture exhibited the lowest stiffness modulus and elastic behaviour; and the 75% RAP (Aged) bituminous mixture was the opposite, exhibiting the highest stiffness modulus and predominantly elastic behaviour. The 75% RAP bituminous mixture's stiffness modulus and phase angle were situated above the 0% RAP bituminous mixture, but below the 75% RAP (Aged) one, evidencing that the recycled bituminous mixture was stiffer than the virgin one and demonstrating that the ageing process also had a stiffening effect in the bituminous mixture.

The fatigue performance was analyzed by the strain values that induce specimens' failure after 10,000 (ε_4), 100,000 (ε_5) and 1,000,000 (ε_6) loading cycles, calculated

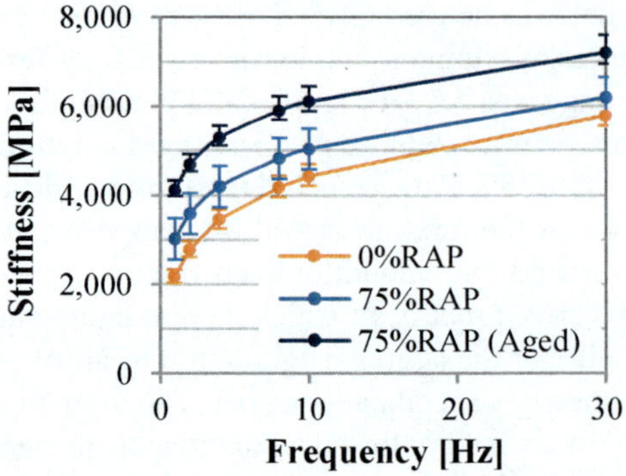

Fig. 2 Stiffness modulus versus frequency

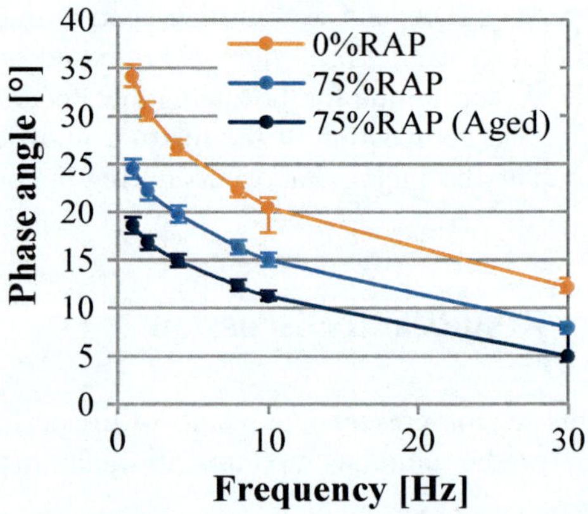

Fig. 3 Phase angle versus frequency

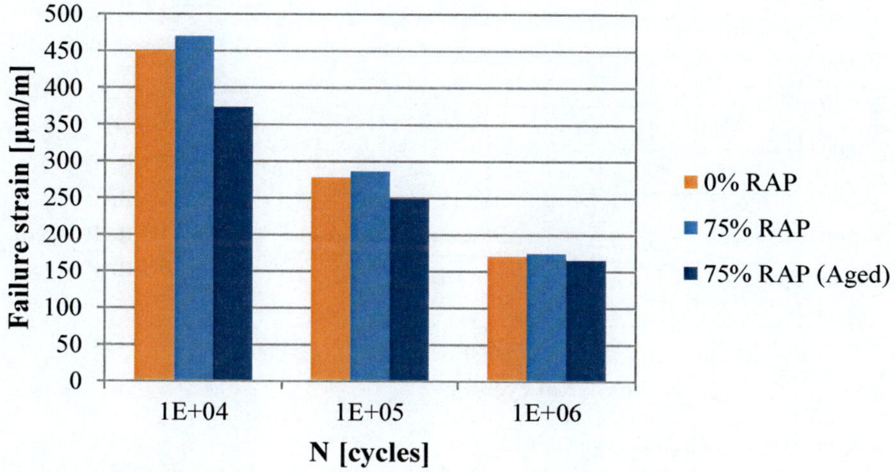

Fig. 4 Strain at fatigue failure

from each bituminous mixture's fatigue law. These parameters are represented in Fig. 4.

Overall, the 75% RAP bituminous mixture has the highest failure strain in all the parameters. Also, for the 10,000 and 100,000 loading cycles, the 75% RAP (Aged) bituminous mixture has a lower fatigue behaviour than the other bituminous mixtures. However, it is more relevant to evaluate the fatigue performance for a high number of loading cycles. As such, for the 1,000,000 cycles, the difference in failure strain between the bituminous mixtures is minimal: between the recycled bituminous mixtures, it is a difference of about 5%, while between the unaged bituminous mixtures it is of about 2%.

The rut depth progression of the bituminous mixtures is presented in Fig. 5. The recycled bituminous mixtures had similar rut depth progression, presenting a tendency to stabilize, while the virgin mixture had an increasingly higher value.

These results demonstrate that the recycled bituminous mixture is stiffer than the virgin one, making it less susceptible to rutting: an expected behaviour due to the presence of aged binder in the recycled material. The further lowering of the rut depth from the 75% RAP bituminous mixture to the 75% RAP (Aged) also demonstrates the stiffening effect of the ageing process.

4 Conclusions

The main objective of this study was to analyze the performance and ageing behaviour of a recycled bituminous mixture incorporating high RAP content (75%), whose age

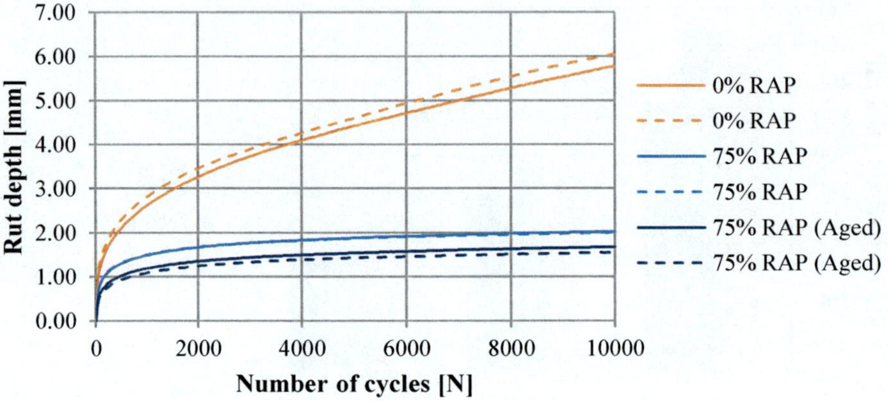

Fig. 5 Progression of the rut depth

was unknown. The performance was assessed in laboratory through stiffness, fatigue resistance, and permanent deformation tests.

The summary of the test results is:

- The stiffness modulus and phase angle had similar behaviour with the increase of loading frequency for all the bituminous mixtures. The recycled materials exhibited higher stiffness than the virgin one and had a predominantly elastic behaviour, being the 75% RAP (Aged) mixture the one with the highest stiffness.
- The 75% RAP bituminous mixture exhibited the highest failure strain in all of the parameters, yet, in the 1,000,000 cycles, the failure strain was similar in all the materials.
- The aged 75% RAP bituminous mixture exhibited the highest permanent deformation resistance, followed by the 75% RAP mixture, being the virgin bituminous mixture the one with the worst performance.

Through the analysis and comparison of the performance tests, it was possible to point out the following main conclusions:

- The recycled bituminous mixture presented higher stiffness than the virgin one, which was reflected in the stiffness and permanent deformation test results.
- The ageing process had a stiffening effect, reflected on the highest stiffness and permanent deformation resistance and lowest fatigue resistance.

The performance results obtained from this study contribute to the demonstration of the viability of RAP recycling and the introduction of this type of material in the paving industry. Yet, as the laboratory bituminous mixture production and ageing process do not simulate all the exact conditions of production in a plant, compaction on-site and ageing throughout its service life, a full-scale trial would be the only way to assess this type of bituminous mixture's performance in real circumstances.

Acknowledgements The authors are grateful to the Portuguese Foundation for Science and Technology, IP for the financial support provided through grant SFRH/BD/114715/2016 financed by the Portuguese Government budget. The authors also want to express their gratitude to the companies that provided the materials and respective technical information needed to develop this study.

References

1. European Commission: Closing the loop—An EU action plan for the Circular Economy. In: Communication from the Commission to the European Parliament, the Council, the European Economic and Social Committee of the Regions (2015)
2. EAPA: Asphalt the 100% recyclable construction product EAPA Position paper (2014)
3. Zaumanis, M., Mallick, R.B., Frank, R.: 100% hot mix asphalt recycling: challenges and benefits. Transp. Res. Proc. **14**, 3493–3502 (2016)
4. EAPA: Recommendations for the Use of Rejuvenators in Hot and Warm Asphalt Production. European Asphalt Pavement Association (2018)
5. Mogawer, W., Bennert, T., Daniel, J.S., Bonaquist, R., Austerman, A., Booshehrian, A.: Performance characteristics of plant produced high RAP mixtures. Road Mater. Pavement Des. **13**(SUPPL. 1), 183–208 (2012)
6. Zaumanis, M., Mallick, R.B., Frank, R.: Evaluation of different recycling agents for restoring aged asphalt binder and performance of 100% recycled asphalt. Mater. Struct. Constr. **48**(8), 2475–2488 (2015)
7. Karlsson, R., Isacsson, U.: Material-related aspects of asphalt recycling—State-of-the-art. J. Mater. Civ. Eng. **18**(1), 81–92 (2006)
8. Zaumanis, M., Mallick, R.B.: Review of very high-content reclaimed asphalt use in plant-produced pavements: state of the art. Int. J. Pavement Eng. **16**(1), 39–55 (2015)
9. Lo Presti, D., Jiménez Del Barco Carrión, A., Airey, G., Hajj, E.: Towards 100% recycling of reclaimed asphalt in road surface courses: Binder design methodology and case studies. J. Clean. Prod. **131**, 43–51 (2016)
10. EP: 14.03—Paving: Material's Characteristics (in Portuguese). In: Caderno de Encargos Tipo Obra, Infraestruturas de Portugal (2014)

Strength and Microstructure Development of Fly Ash Geopolymer Binders Using Waste Glass Powder

Md. Nabi Newaz Khan, Jhutan Chandra Kuri, and Prabir Kumar Sarker

Abstract This study investigated the effect of using waste glass powder (GP) as partial replacement of fly ash on the strength and microstructure development of geopolmer binders. GP was used at different rates varying from 1 to 5% mass of fly ash and 8 molar NaOH solution was used as the activator. Geopolymer paste specimens of 50 mm cube were heat cured at 80 °C for the first 24 h after casting and then left in ambient conditions. The 28-day compressive strength of the fly ash geopolymer without GP was 23 MPa. The inclusion of 2% GP improved the 28-day compressive strength of heat cured fly ash geopolymer by 18%. This is attributed to the increment of soluble silica from the GP. However, compressive strength declined with further increase of GP content beyond 2%. The decline of strength is attributed to the presence of excess unreacted GP particles. X-ray diffraction (XRD) analysis indicated that the presence of higher amount of crystalline Na-type chabazite and zeolite phases for using 2% GP. Scanning electron microscopy (SEM) images showed denser microstructure of geopolymers containing 1–2% GP. The elemental composition of gel formed in these binders consisted of higher percentages of Si, Al and Na, which indicates the formation of higher amount of sodium aluminosilicate gel (N–A–S–H).

Keywords Fly ash geopolymer · Waste glass powder · Compressive strength · Sodium aluminosilicate gel

1 Introduction

Research on alkali activated binders or geopolymer binders is increasing in the recent time since it has a huge potential to reduce the carbon footprint of construction sector by replacing Portland cement. Alkali activated binders or geopolymer binders are produced by the chemical reaction between aluminosilicate materials and an alkali activator. In this binder system, two most common industrial by-products such as

M. N. N. Khan (✉) · J. C. Kuri · P. K. Sarker
School of Civil and Mechanical Engineering, Curtin University, GPO Box U1987, Perth, WA 6845, Australia
e-mail: m.khan107@postgrad.curtin.edu.au

V. M. C. F. Cunha et al. (eds.), *Proceedings of the 3rd RILEM Spring Convention and Conference (RSCC 2020)*, RILEM Bookseries 35,
https://doi.org/10.1007/978-3-030-76543-9_5

fly ash and ground granulated blast furnace slag (GGBFS) can be used as the main aluminosilicate sources. Besides, sodium hydroxide, sodium carbonate and sodium silicate solutions are the most effective and common alkaline activators used for activation of the solid aluminosilicate source materials [1, 2]. Previous studies showed that the amount of available reactive silica in the source materials has a great impact on the geopolymerization process [3, 4]. Therefore, some studies added a certain amount of silica rich materials such as nanosilica and silica fume to improve the geopolymerization process [5, 6]. For instance, Deb et al. [5] found that the use of 2% nanosilica significantly improved the workability, compressive strength, sorptivity and acid resistance of fly ash based geopolymer mortars. Similarly, Ramezanianpour and Moeini [6] noticed that the inclusion of 5% silica fume enhanced the mechanical and durability properties of alkali activated slag mortars. Besides, some researchers studied the effects of rice husk ask, sugar cane bagasse ash and palm oil fuel ash in geopolymers due to the availability of reactive silica in these materials [7–9].

The current research trend indicates that waste glass powder (GP) has a potential to be used as a supplementary binder material in cementitious systems due to the abundance of reactive silica [10, 11]. Usually, GP is produced by crushing the waste glasses. However, studies related to the use of GP as a source of reactive silica in geopolymers is very limited. Only a few studies were conducted to determine the feasibility of GP in geopolymers as a source of reactive silica [12–14]. Therefore, the primary objective of this study is to determine the effect of glass powder on the strength and microstructural development of fly ash geopolymer binders. In this regard, compressive strength and microstructural development of fly ash geopolymer paste samples with 0, 1, 2, 3, 4 and 5% GP were studied.

2 Materials and Methods

2.1 Materials

In this study, fly ash was used as the main aluminosilicate source and waste glass powder (GP) was used as a partial replacement of fly ash. Fly ash was collected from the Gladstone power station in Queensland, Australia and glass powder was produced by crushing waste glass cullet using a laboratory ball mill. As seen in Table 1, the fly ash used in this study is classified as class F according to ASTM C618 [15] since the total of SiO_2, Al_2O_3 and Fe_2O_3 contents exceeded 70% and the amount of SO_3 was only 0.25%. GP was mainly consisted of 70.61% SiO_2, 10.90% CaO and 12.8% Na_2O. The particle diameters of fly ash and GP are presented in Table 2. It can be seen that the average particle sizes (d_{50}) of fly ash and GP were 13.62 μm and 10.74 μm, respectively which indicates that waste glass powder used in this study was finer than the fly ash. The photographs of fly ash and GP are shown in Fig. 1.

Table 1 Chemical compositions of fly ash and GP

Constituents	Fly ash	GP
SiO_2	60.03	70.61
Al_2O_3	22.75	1.43
CaO	3.80	10.90
Fe_2O_3	6.78	2.49
K_2O	1.28	0.34
MgO	1.29	0.69
Na_2O	0.54	12.8
P_2O_5	0.89	0.02
SO_3	0.25	0.11
TiO_2	1.06	0.06
MnO	0.07	0.03
SrO	0.05	0.01
Cr_2O_3	0.01	0.06
ZnO	–	0.01
Loss on ignition	1.15	0.11

Table 2 Particle diameters of fly ash and waste glass powder

Characteristic diameter	Fly ash (μm)	GP (μm)
d_{80}	45.50	28.93
d_{50}	13.62	10.74
d_{10}	1.42	1.86

(a) Fly ash (b) GP

Fig. 1 **a** Fly ash and **b** GP

2.2 *Preparation of the Testing Samples*

The binder compositions of the geopolymer paste mixes are provided in Table 3. The mix proportions were selected based on the previous work reported by Deb et al. [5]. Fly ash was used as the main binder with 0–5% GP as fly ash replacement and 8 M

Table 3 Mixture proportions of geopolymer binders

Mix ID	Constituent (% mass)		A/B[a]	Molar ratios (calculated)	
	Fly ash	GP		Si/Al	Si/Na
FGP0	100	0	0.4	2.24	3.58
FGP1	99	1	0.4	2.26	3.53
FGP2	98	2	0.4	2.29	3.49
FGP3	97	3	0.4	2.32	3.45
FGP4	96	4	0.4	2.34	3.41
FGP5	95	5	0.4	2.37	3.37

[a]Activator to binder mass ratio

NaOH solution was used as the alkaline solution. The number at the end of a mix ID in Table 3 represents the mass percentage of GP in the solid binder. The amount of alkaline solution was kept constant at 40% of the total binder. First, all the dry ingredients were placed in the Hobart mixer and then 8 M NaOH solution was added as the mixing continued to prepare the geopolymer paste mixture. Later, this freshly prepared geopolymer paste were poured into the 50 mm acrylic plastic cube moulds. The moulds were placed in an oven after casting and cured it for the next 24 h at 80 °C. After the desired heat curing period, the specimens were demolded and stored in room temperature of 20 ± 2 °C with a relative humidity of 70 ± 10% until testing age.

The compressive strength of paste specimens was determined at 28 days using a 300 kN Shimadzu machine with a constant loading rate of 0.9 kN/s as recommended in ASTM C109 [16]. Scanning electron micrographs (SEM) and energy dispersive X-ray (EDX) spectroscopy were conducted using a NEON 40EsB (ZEISS) microscopy instrument on small cut paste sample to examine the morphology and microstructure of the paste specimens. The samples were coated with carbon before SEM and EDX analysis. X-ray diffraction (XRD) was performed on powdered paste samples at 28 days using a D8 advance (Bruker AXS) instrument to investigate the reaction products of the paste samples.

3 Results and Discussion

3.1 Compressive Strength

The effect of GP content on the compressive strength of fly ash geopolymers is shown in Fig. 2. It has been noticed that compressive strength increased for 1–2% GP in fly ash geopolymers and the maximum compressive strength was achieved for 2% GP. For instance, FGP1, FGP2 and FGP3 geopolymer paste samples showed about 13, 18 and 5% higher compressive strength as compared to the FGP0 geopolymer paste

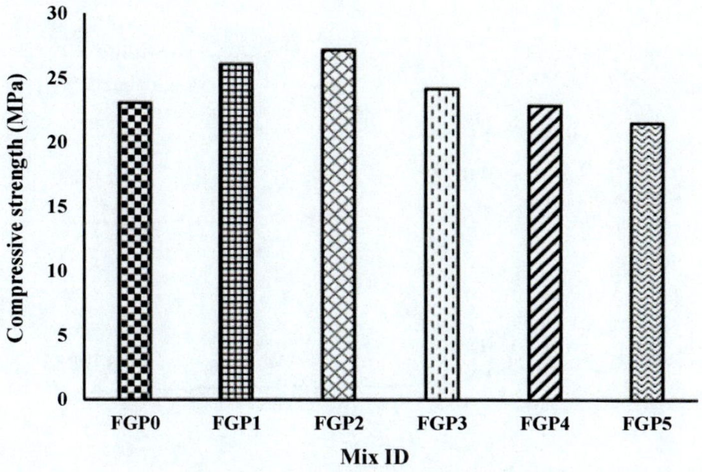

Fig. 2 28-days compressive strength of the fly geopolymer pastes with glass powder

sample. Besides, it is noted that compressive strength declined when the amount of GP content exceeded to 2%. This can be attributed to the decrease of Si/Na ratio and increase of Si/Al ratio by the inclusion of GP, as shown in Table 2. It can be seen that the amounts of SiO_2, and Na_2O were higher in GP than in fly ash, whereas the amount of Al_2O_3 was significantly higher in fly ash than in GP. As a result, when a small percentages of GP (about 1–3%) was used to replace fly ash, it improved the geopolymerization process by adding additional soluble Si, Al and Na. However, the use of GP beyond 3% could not produce additional geopolymeric product. This is attributed to the reduced amount of reactive Al. Therefore, compressive strength declined with the increase of GP content above 2%. These results show good agreement with the results reported by Torres-Carrasco and Puertas [12] and Deb et al. [5].

3.2 *Microstructural Investigation*

3.2.1 XRD Analysis

The XRD results of fly ash geopolymers containing 0, 2 and 5% glass powder are shown in Fig. 3. In all cases, quartz, mullite, chabazite and zeolite are the main detected crystalline phases. The peaks of crystalline quartz and mullite phases are considered to be due to the presence of unreacted fly ash particles in these mixes, whereas crystalline chabazite and zeolite phases represent the formation of sodium aluminosilicate hydrate (N–A–S–H) gel. It is seen that the presence of chabazite and zeolite phases in FGP2 paste are more prominent than the FGP0 and FGP5

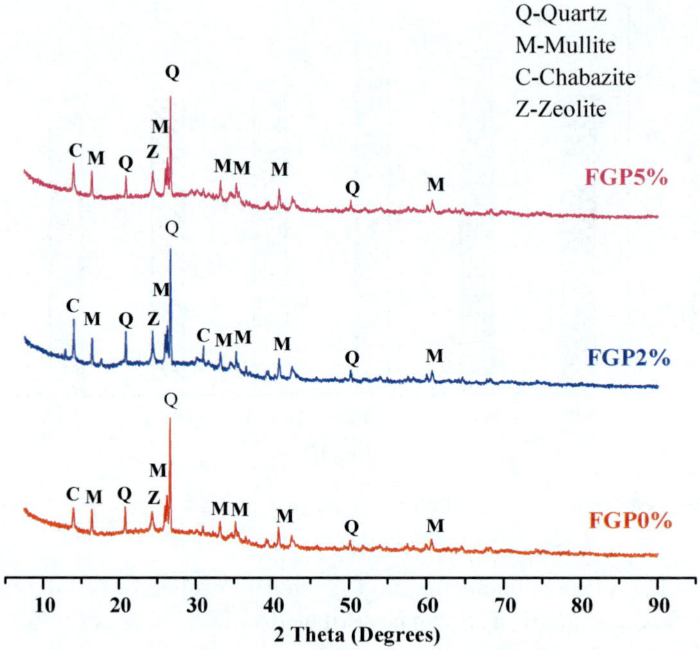

Fig. 3 XRD results of the fly geopolymer pastes with glass powder

paste samples which indicate the formation of a greater amount of N–A–S–H gel in the FGP2 paste mix. The formation of this greater amount of N–A–S–H gel in FGP2 paste mix is mainly attributed to the dissolution of additional Si and Na that was present in GP. Therefore, FGP2 showed the maximum compressive strength. Besides, the obtained peaks for quartz and mullite in FGP5 paste is higher than the FGP0 paste which indicates the presence of more unreacted particles in FGP5 paste. As a result, FGP5 showed lower compressive strength as compared to FGP0 paste specimen.

3.2.2 SEM and EDX

The SEM and EDX of FGP0, FGP2 and FGP5 paste specimens are presented in Figs. 4, 5 and 6, respectively. By comparing Figs. 4 and 5, it is seen that the elemental compositions of the reaction product observed in FGP0 and FGP2 pastes are quite similar. This reaction product is mainly sodium aluminosilicate gel which is consisted of primarily Si, Al and Na as confirmed by EDX. However, the presence of unreacted fly ash particles and voids are noticeably higher in FGP0 specimen than the FGP2 specimen. In addition, it is also noticed that the unreacted fly ash particles in FGP2 paste are well connected with the reaction products, whereas a loose connection is observed between the unreacted fly ash particles and reaction product in FGP0 paste.

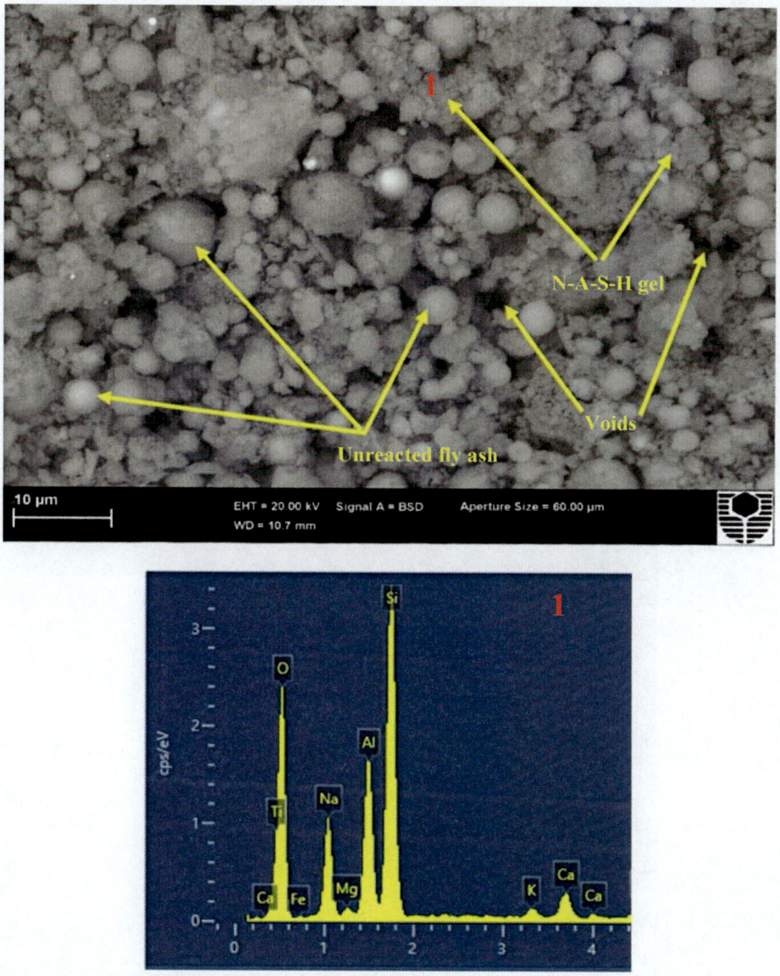

Fig. 4 SEM and EDX of FGP0 paste specimen

Furthermore, it is noticed that the microstructure of FGP5 contained greater amount of unreacted fly ash particles with visible cracks and internal voids. Besides, the peaks for Si, Al and Na in the reaction products are comparatively lower in FGP5 paste than in FGP0 and FGP2 pastes. This shows a good agreement with the XRD results reported in the earlier section as well as with some previous studies [5, 12].

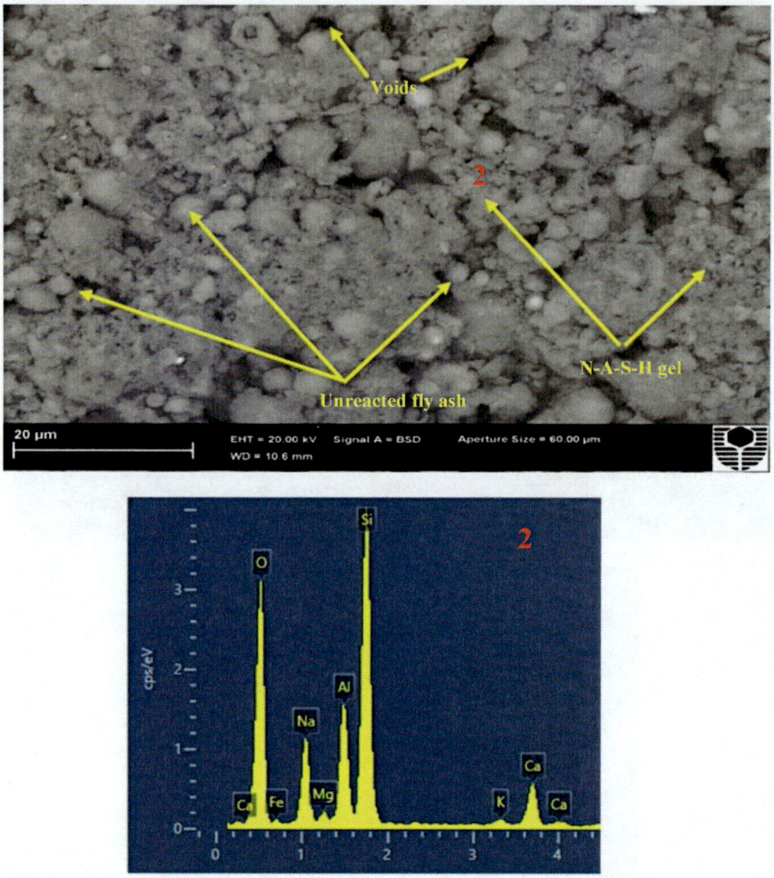

Fig. 5 SEM and EDX of FGP2 paste

4 Conclusions

Strength and microstructure development of fly ash geopolymers with 0–5% GP were studied. The following conclusions can be drawn based on the results of the experimental study:

1. Compressive strengths of the fly ash geopolymers increased with the use of GP up to 2% level and strength declined with the further increase of GP. The geopolymer paste containing 2% GP showed about 18% higher compressive strength than the geopolymer paste without GP.
2. XRD analysis showed that fly ash geopolymer with 2% GP had higher peaks for crystalline chabazite and zeolite phases, which indicates the formation of higher amount of sodium aluminosilicate hydrate (N–A–S–H) gel. This is attributed to the dissolution of additional Si and Na from 2% GP.

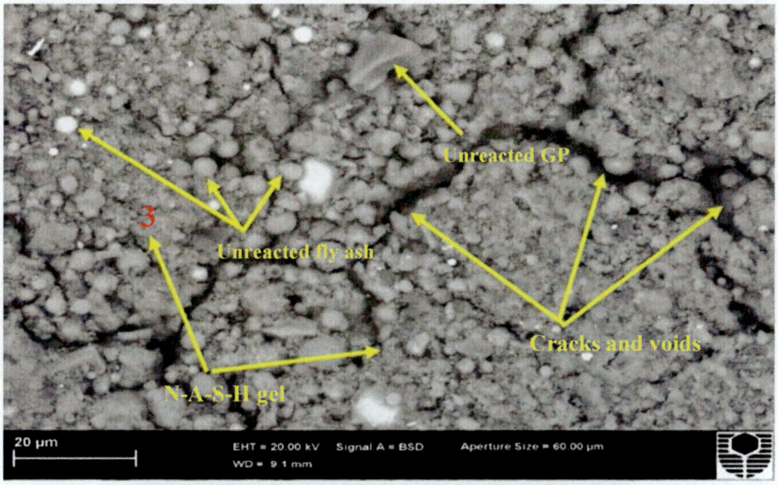

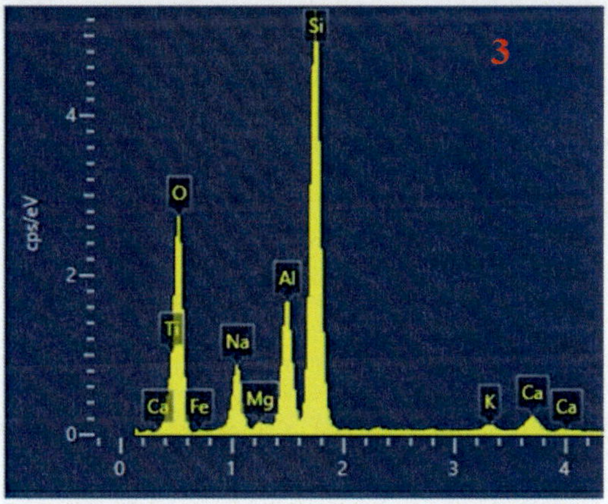

Fig. 6 SEM and EDX of FGP5 paste

3. The SEM images and associated EDX analysis indicated that the microstructure of fly ash geopolymer containing 2% GP had fewer unreacted fly ash particles and less voids as compared to the microstructure of fly ash geopolymer without GP.
4. Therefore, it is concluded that the use of 2% GP in fly ash geopolymers improved compressive strength and microstructure development due to the improvement of reaction kinetics by providing additional soluble Si and Na.

Acknowledgements The first author received Curtin International Postgraduate Research Scholarship (CIPRS) and Curtin Strategic International Research Scholarship (CSIRS) from Curtin University, Australia. The authors would like to thank SKM Recycling, Victoria, Australia for providing the waste glass cullet for this study. Besides, the assistance of John de Laeter Centre at Curtin University to the microstructural analysis is gratefully acknowledged.

References

1. Provis, J.L.: Geopolymers and other alkali activated materials: why, how, and what? Mater. Struct. **47**(1–2), 11–25 (2014)
2. Karim, M.R., Hossain, M.M., Khan, M.N.N., Zain, M.F.M., Jamil, M., Lai, F.C.: On the utilization of pozzolanic wastes as an alternative resource of cement. Materials **7**(12), 7809–7827 (2014)
3. Chen-Tan, N.C., Riessen, Ly, C.V., Southam, D.C.: Determining the reactivity of a fly ash for production of geopolymer. J. Am. Ceramic Soc. **92**(4), 881–887 (2009)
4. Criado, M., Fernandez, J.A., Torre, A.G., Aranda, M.A.G., Palomo, A.: An XRD study of the effect of the SiO_2/Na_2O ratio on the alkali activation of fly ash. Cem. Concr. Res. **37**(5), 671–679 (2007)
5. Deb, P.S., Sarker, P.K., Barbhuiya, S.: Effects of nano-silica on the strength development of geopolymer cured at room temperature. Constr. Build. Mater. **101**, 675–683 (2015)
6. Ramezanianpour, A.A., Moeini, M.A.: Mechanical and durability properties of alkali act vated slag coating mortars containing nanosilica and silica fume. Constr. Build. Mater. **163**, 611–621 (2018)
7. Meija, J.M., Mejia de Gutierrez, R., Puertas, F.: Rice husk ash as a source of silica in alkali-activated fly ash and granulated blast furnace slag systems. Mater. de Construccion **63**(311), 361–375 (2013)
8. Moraes, J.C.B., Tashima, M.M., Akasaki, J.L., Melges, J.L.P., Monzo, J., Borrachero, M.V., Soriano, L., Paya, J.: Effect of sugar cane straw ash (SCSA) as solid precursor and the alkaline activator composition on alkali-activated binders based on blast furnace slag (BFS). Constr. Build. Mater. **144**, 214–224 (2017)
9. Toniolo, N., Boccaccini, A.R.: Fly ash-based geopolymers containing added silicate waste. A review. Ceramics Int. **43**(17), 14545–14551 (2017)
10. Khan, M.N.N., Saha, A.K., Sarker, P.K.: Reuse of waste glass as a supplementary binder and aggregate for sustainable cement-based construction materials: a review. J. Build. Eng. **28**, 101052 (2020)
11. Khan, M.N.N., Sarker, P.K.: Alkali silica reaction of waste glass aggregate in alkali activated fly ash and GGBFS mortars. Mater. Struct. **52**(93), 1–17 (2019)
12. Torres-Carrasco, M., Puertas, F.: Waste glass in the geopolymer preparation. Mechanical and microstructural characterization. J. Clean. Prod. **90**, 397–408 (2015)
13. El-Naggar, M.R., El-Dessouky, M.I.: Re-use of waste glass in improving properties of metakaolin-based geopolymers: mechanical and microstructure examinations. Constr. Build. Mater. **132**, 543–555 (2017)
14. Liu, Y., Shi, C., Zhang, Z., Li, N.: An overview on the reuse of waste glasses in alkali activated materials. Resour. Conserv. Recycl. **144**, 297–309 (2019)
15. ASTM C618: Standard Specification for Coal Fly Ash and Raw or Calcined Natural Pozzolan for Use in Concrete. ASTM International, West Conshohocken, PA (2012)
16. ASTM C109/C109M: Standard Test Method for Compressive Strength of Hydraulic Cement Mortars (Using 2-in. or [50-mm] Cube Specimens), ASTM International, West Conshohocken, PA (2013)

The Influence of Polymers Impregnation on Bending Behaviour of *Phyllostachys pubescens* (Mosso) Bamboo

Lucas Muniz Valani, Fabrício de Campos Vitorino, Adriana Paiva de Souza Martins, and Romildo Dias Toledo Filho

Abstract As the consumption of natural resources in today's industrialized society is increasing and the supply of these resources tends to deplete, the search for alternative and environmentally friendly raw materials becomes mandatory. Bamboo is a very abundant resource and presents short growth, carbon sequestration ability, easy processing and good mechanical properties. However, bamboo is hygroscopic material and presents functional gradation of properties along the culm wall thickness. Cracks may occur in bamboo's wall as a consequence of a radial moisture gradient, that generate stress perpendicularly to natural fiber alignment, and this occurrence may compromise the durability and structural performance of the material. Aiming to minimize these shortcomings and to enhance physical properties, different polymers were used to impregnate samples of *Phyllostachys pubescens* bamboo through immersion: styrene butadiene rubber (SBR), carboxylated styrene butadiene rubber (XSBR) and polyvinyl alcohol (PVOH). Three temperatures of impregnation were investigated (23, 60 and 100 °C, and the most effective in relation to polymer penetration and retention was identified. Results of water absorption tests showed that PVOH resulted in better physical behaviour and smaller water absorption values, compared with non impregnated samples. Three-points bending test was carried out employing samples with 20 × 2 × 0.5 cm. The samples were extracted from the internodes of the middle portion of a 3 years old culms. The obtained parameters (bending strength, stiffness and toughness) were compared among the 4 conditions (plain, SBR, XSBR, PVOH) in order to understand how the impregnation process affected the mechanical performance under bending loads.

Keywords Bamboo · Polymers impregnation · Bending behavior · Physical behavior · Low impact construction materials

---

L. M. Valani · F. de Campos Vitorino (✉) · A. P. de Souza Martins · R. D. Toledo Filho
Center of Education and Research on Low Environmental Impact Materials and Technologies on Sustainable Construction, Federal University of Rio de Janeiro, Rio de Janeiro, Brazil
e-mail: fabriciovitorino@coc.ufrj.br

A. P. de Souza Martins
e-mail: adriana@coc.ufrj.br

V. M. C. F. Cunha et al. (eds.), *Proceedings of the 3rd RILEM Spring Convention and Conference (RSCC 2020)*, RILEM Bookseries 35,
https://doi.org/10.1007/978-3-030-76543-9_6

1 Introduction

The planet faces a severe climatic change due to the increasingly greenhouse gases emission. Up to 39% of these gases emission can be attributed to conventional construction and building industry [1], therefore, the use of environmentally friendly materials has become a necessity. In addition, the growing need for forest resources has provoked annual declines in the availability of logs and high-quality timber. In this context the search of alternative materials for the replacement of hard and soft wood has become increasingly critical [2].

Bamboo has been deemed as one of the most promising alternatives for replacing wood in some applications. It is a very abundant resource in tropical and subtropical countries, and presents short growth, carbon sequestration and storage ability and good thermo-hygro-mechanical properties [3–6]. In construction area, it has many applications, such as panels, furniture, fiber reinforcement, structural elements and more.

Despite the many advantages, bamboo is a unidirectional fibrous material which is very susceptible to longitudinal splitting [7] and presents hygroscopic behavior and dimensional instability [8, 9].

Aiming to minimize these shortcomings and to enhance physical properties, different treatments can be carried out. In Ref. [10] impregnated bamboo with polyvinyl alcohol and clay nanoparticles, and they obtained penetration of polymer in the lumens and cell walls, resulting in improved flexural behavior. The hydroxyl groups, responsible for the hydrophilic behavior of bamboo, were reduced. Studies developed by Anwar et al. [11] using bamboo samples impregnated with formaldehyde resin showed a reduction in equilibrium humidity from 20 to 5% and optimal drying time at 60 °C after the impregnation equals to 9 h. The percentage mass gain after drying was up to 11%. Deka et al. [8] performed the treatment of bamboo with thermo-rigid resins and obtained increasings in the flexural strength and dimensional stability, as well as and reductions in water absorption, compared to non-impregnated samples. Liese [12, 13] emphasizes that the anatomical structure of bamboo offers greater resistance to penetration of solutions when compared to wood, and only low molecular weight compounds can penetrate in the cell walls. Another hindering aspect is that the metaxylem vessels, the main impregnant penetration pathway, occur only in small proportions in the bamboo tissue. An impregnation method employing temperature of 40 °C, lower than those normally adopted, in the range of 100–200 °C, was proposed by [14], leading to higher moisture resistance, higher dimensional stability and higher surface hardness. Furuno et al. [15] have shown that low molecular weight resins are able to penetrate cell walls, while high molecular weight resins fill only the lumens of cells and do not contribute to the improvement of dimensional stability. He et al. [16] performed wood impregnation at room temperature using an environmentally friendly impregnant (tung oil). Reductions on the hygroscopicity and dimensional instability were observed, evidencing the environmental benefits of the method, with the use of non-toxic products and energy saving.

There are many studies in the literature that address the influence of the polymer impregnation on the physical and mechanical behavior of the wood, as well on the biodegradation resistance. However, bamboo, as a functionally graded material (FGM), has its anatomical structure very different compared with wood, justifying further studies to better understand the implications of this peculiarity. The main objective of this work was to verify the influence of polymer impregnation on the physical and flexural behavior of *Phyllostachys pubescens* bamboo. The selected polymers were styrene butadiene rubber (SBR), carboxylated styrene butadiene rubber (XSBR) and polyvinyl alcohol (PVOH). The selection was based on the easiness of obtainment in Rio de Janeiro region and on the low cost. These studies may contribute to overcome the natural intrinsic deficiencies of the material and encourage its wider use.

2 Materials and Methods

2.1 Bamboo Material

The bamboo samples were extracted from the region between the nodes, in the middle portion of a 3 years old culms, provided by the company "Takê Cortinas". After the extraction, the culms dried slowly until reach the equilibrium moisture, and then they were treated against insects and fungi by fumigation.

In the laboratory, the culms with 3 m length were subdivided into longitudinal and curve sections, then internodes segments were obtained. The outer and inner parts of the segments were removed, and strips measuring $20 \times 2 \times 0.5$ cm were produced.

2.2 Impregnants

Three impregnants were selected: polyvinyl alcohol (PVOH), styrene butadiene rubber (SBR) and carboxylated styrene butadiene rubber (XSBR). These compounds were selected based on low cost, easiness of acquiring and good properties (thermal stability, elasticity, water tightness, sunlight and ultraviolet rays resistance). Polyvinyl alcohol presents molecular weight of 15,000 g/mol and pH equals to 4.8 (21 ºC). Styrene butadiene rubber and carboxilated styrene butadiene have solids content and pH equals to 34.5% and 11; 49.4% and 9, respectively. Plastic viscosity values (at the temperature of 23 ºC) of SBR, XSBR and PVOH are 0.19; 0.842 and 3.793 Pa s, respectively.

2.3 Impregnation Process by Immersion

The bamboo strips were previously dried in an oven (103 ºC) before the impregnation, until the constancy of mass. After that, they were immersed in the polymer contained in a sealed recipient, and put in an oven adjusted to a predefined temperature until mass stabilization. Two impregnation temperatures were tested: 23 and 60 ºC. Then the samples were removed from the oven and left to cool until the equilibrium with room temperature. After that they returned to the oven (103 ºC) to dry the polymer until the mass constancy. The mass gain after this step corresponded to the polymer impregnation load. Water absorption tests after this step evidenced which temperature led to smaller water absorption.

2.4 Bending Tests

Three points bending tests were performed using a Shimadzu AGX electromechanical equipment with displacement control at a rate of 1 mm/min. A load cell with capacity of 100 kN was coupled to the sample. Eight samples for each of the four conditions (non impregnated and impregnated with PVOH, SBR and XSBR) were tested. As the region with the lowest concentration of fibers is more porous, consequently presenting a higher polymer load, we sought to position this region of the samples at the bottom, aiming to better evidence the influence of the impregnation under flexion loads. The distance between the extreme supports was 160 mm and the displacements were measured through a LVDT positioned in the center of the span. Figure 1 shows the setup of the test.

The flexural strength was obtained through Eq. (1), according ASTM D1037 [17]:

(a) (b)

Fig. 1 **a** Setup of the bending tests, **b** detail of the setup

$$\sigma_{max} = \frac{3PL}{2bh^2} \quad (1)$$

where

σ_{max}	maximum bending stress (MPa),
P	maximum load (N),
L	distance between the extreme supports,
b	width of the sample (mm),
h	thickness of the sample (mm).

3 Results and Discussion

3.1 Physical Tests

The curves of polymers impregnation at different temperatures as a function of time can be seen in Fig. 2. In the green area is possible to see the moment the samples were placed in the 103 ºC oven until mass constancy, where it was possible to see the effective polymer mass load. Reference curve refers to bamboo sample impregnated only with water (water absorption); the negative mass value found (– 3%) can be attributed to the loss of water absorbed plus water soluble extractives [18, 19]. In the case of polymers impregnation, it is noticeable that temperature is an important factor on its effectiveness, where the highest polymer mass gain where reached under 60 ºC. SBR for instance, reached the higher impregnation mass in bamboo (43%), followed by XSBR (28%) and PVOH (15%). This trend can be attributed to the lower

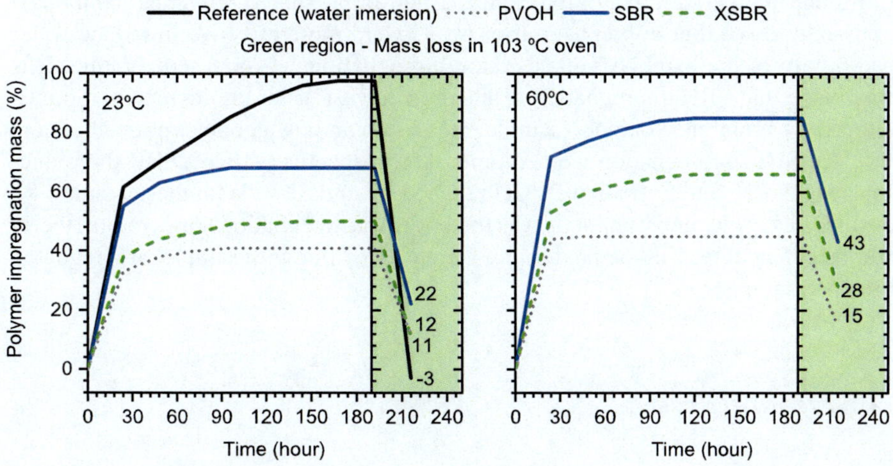

Fig. 2 Polymers impregnation in different temperature conditions

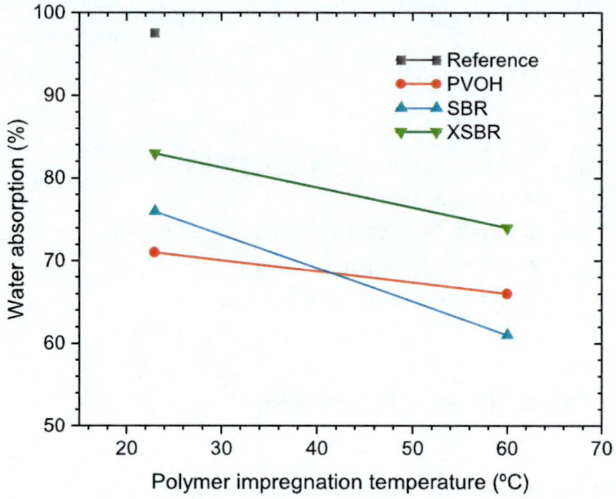

Fig. 3 Water absorption of bamboo at different polymers impregnation temperatures

viscosity of SBR (0.19 Pa s) followed by XSBR (0.842 Pa s) and PVOH (3.79 Pa s). In turns, the rise in temperature diminishes even more polymers plastic viscosity and as a sealed container was used there was also a rise in the internal container pressure which may force polymers throughout the bamboo microstructure.

It is also noticeable that, the rate of polymer (and water) absorption in bamboo is maximum up to 24 h, after that the absorption rate starts to decline until reach mass constancy. At room temperature, the water mass gain in the reference bamboo reached constancy at 170 h, followed by XSBR (144 h), PVOH (96 h) and SBR (96 h). At 60 °C, SBR reached mass constancy at 120 h, followed by XSBR (120 h) and PVOH (48 h).

The graph in Fig. 3 shows the water absorption of bamboo samples impregnated at different temperature conditions and the non-impregnated reference bamboo. It is possible to see that impregnations at 60 °C were more effective in reduce water absorptivity of the bamboo samples than impregnations at room temperature. It is remarkable that SBR impregnation reached the lowest water impregnation at 60 °C. Compared to reference bamboo sample, the reduction in water absorption at 23 and 60 °C for SBR impregnation were 22 and 38%, respectively. For XSBR the reductions reached 15 and 24%; and PVOH reached 27 and 32%. With these results, we consider that 60 °C impregnations was the optimal impregnation temperature, therefore, mechanical test, in the next topic, were carried out with samples impregnated at temperature.

3.2 *Flexural Tests*

Figure 4 shows the stress–deflection typical curves of the 60 °C impregnated samples

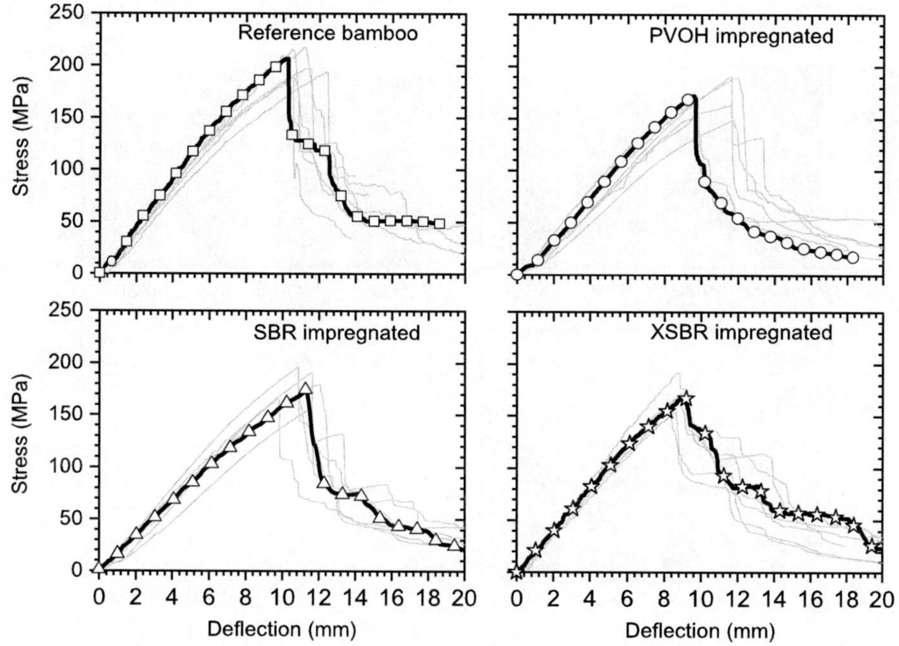

Fig. 4 Stress-deflection behavior of impregnated and reference bamboo. Branched area is the standard deviation

and the reference bamboo. At least seven samples were tested and the grey curves represent each tested sample. Black line is the typical curve.

It is possible to see that there is a linear increase of stress until a peak stress, after that bamboo fails and stress start to decrease abruptly fallowed by a point where stress decrease in a controlled mode. At this point, deflection level is already very advanced (around 12 mm) and therefore tests were stopped at 17 mm deflection. It is also noticeable that, XSBR impregnated bamboo had a controlled stress failure right after peak stress as can be seen by the less abrupt stress decrease at this point.

Figure 5 shows the peak stress value and respective standard deviation bar for impregnated and reference bamboos. It is seen that impregnations lead to decrease peak stress, however, this loss was considered acceptable. Max stress loss was for XSBR impregnation (− 17%) while the SBR impregnation reached 13% loss.

Regarding to peak deflection, it is possible to see that PVOH and SBR impregnations tend to have a slightly higher deflection capacity, however this difference can be considered statistically equals. In the other hand, XSBR impregnation reached reduction in peak deflection of about 17%.

Elastic modulus also tended to decrease, the maximum reduction was found for PVOH (− 22% compared to reference) while XSBR presented the highest value amongst the impregnated ones (− 7% compared to reference).

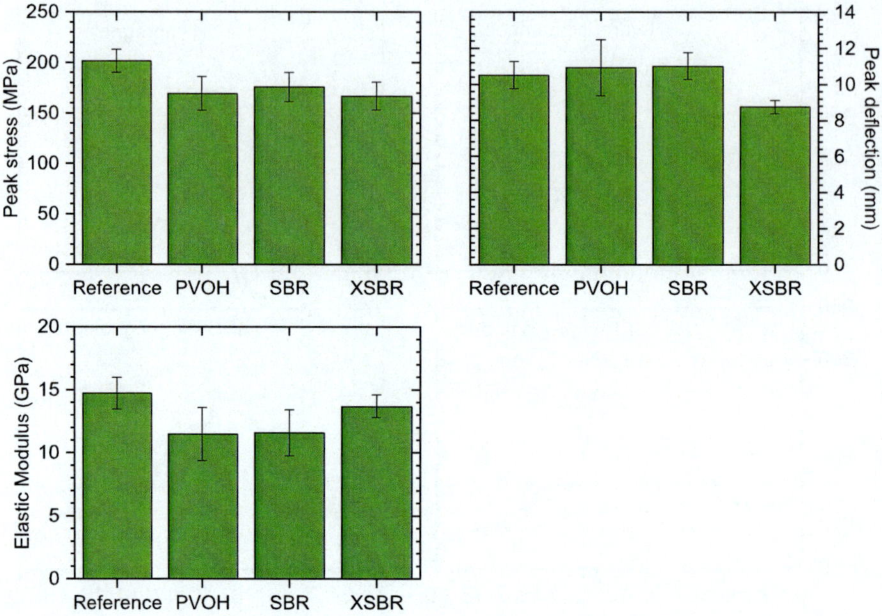

Fig. 5 Peak stress and deflection; and elastic modulus of the impregnated and reference bamboo

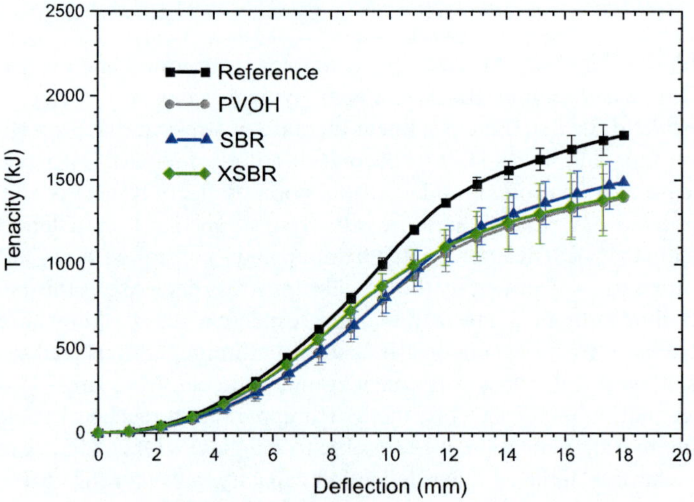

Fig. 6 Tenacity of impregnated and reference bamboo

Figure 6 presents the tenacity curves as function of deflection for impregnated and reference bamboos. Standard deviation bars of each can also be seen.

It is possible to see that reference bamboo reached higher tenacity energy levels than all the impregnated ones. It is also noticeable that XSBR impregnation reached values similar as that of the reference before bamboo failure. At 9 mm, XSBR reached a tenacity level 10% lower than reference bamboo. After failure, SBR impregnation reached higher tenacity levels than other impregnations. At 18 mm, SBR impregnation found a tenacity level 16% lower than reference bamboo.

The reductions either in peak stress, deflection capacity and tenacity can be attributed to a thermal stress generated at the step where the samples were place in to the 103 °C oven to bring samples to mass constancy. That may have generated a thermal gradient that in turns generated microcracks in the samples. Therefore, efforts to mitigate such problems and enhance the impregnation process will be carried out.

4 Conclusions

With this study the following remarks can be drawn:

- The impregnation of bamboo by means of sealed container and temperature is feasible up to 60 °C. Above this temperature polymer sets.
- The optimal impregnation reached reductions in water absorption of about 38%, 24% and 32% for PVOH, SBR and XSBR respectively.
- Impregnated samples found reductions in peak stress flexural tests. Compared to non-impregnated samples, the PVOH, SBR and XSBR found reductions of about 16%, 13% and 17% respectively.
- The peak deflection was not impacted with impregnations except XSBR which found reduction of 17%. However, it has a soften stress reduction after failure in opposition to reference and other impregnations.
- The elastic modulus reduced for polymer impregnated bamboo. The reductions found reached 22%, 21% and 7%, for PVOH, SBR and XSBR respectively.
- The tenacity of the samples tended to be similar up to 7 mm deflection, after that impregnated samples found tenacity values lower than reference bamboo.
- The reductions in peak stress, elastic modulus and tenacity found by polymer impregnated samples can be attributed to a pre-cracking caused by thermal gradient. Therefore, more studies have to be done in the impregnation process.

References

1. UN—United Nations Environment Programme: UN Environment Homepage. https://www.unenvironment.org/resources/report/global-status-report-2018. Last accessed 2020/01/02
2. Yang, T., Lee, C., Lee, C., Cheng, Y: Effects of different thermal modification media on physical and mechanical properties of Moso bamboo. Constr. Build. Mater. **119**, 251–259 (2016)

3. Mahdavi, M., Clouston, P.L., Arwade, S.R.: Development of laminated bamboo lumber: review of processing, performance, and economical considerations. J. Mater. Civ. Eng. **23**(7), 1036–1042 (2011)
4. Huang, Z., Sun, Y., Musso, F.: Assessment of bamboo application in building envelope by comparison with reference timber. Constr. Build. Mater. **156**, 844–860 (2017)
5. Azadeh, A., Ghavami, K.: The influence of heat on shrinkage and water absorption of *Dendrocalamus giganteus* bamboo as a functionally graded material. Constr. Build. Mater. **186**, 145–154 (2018)
6. Fei Lv, H., Ma, X., Zhang, B., Chen, X., Liu, X., Fang, C., Fei, B.: Microwave-vacuum drying of round bamboo: a study of the physical properties. Constr. Build. Mater. **211**, 44–51 (2019)
7. Mitch, D., Harries, K.A., Sharma, B.: Characterization of splitting behavior of bamboo culms. J. Mater. Civ. Eng. **22**(11), 1195–1199 (2010)
8. Deka, M., Das, P., Saikia, N.: Studies on dimensional stability, termal degradation and térmite resistant properties of bamboo (*Bambusa tulda* Roxb.) treated with thermosetting resins. J. Bamboo Rattan **2**, 29–41 (2003)
9. Anwar, U.M.K., Paridah, M.T., Hamdan, H., Sapuan, S.M., Bakar, E.S.: Effect of curing time on physical and mechanical properties of phenolic-treated bamboo strips. Ind. Crops Prod. **29**, 214–219 (2009)
10. Adamu, M., Rahman, M.R., Hamdan, S.: Formulation optimization and characterization of bamboo/polyvinyl alcohol/clay nanocomposite by response surface methodology. Compos. B **176**, 1–9 (2019)
11. Anwar, U.M.K., Paridah, M.T., Hamdan, H., Zaidon, A., Bakar, E.S.: Impregnation of bamboo (*Gigantochloa scortechinii*) strips with low-molecular-weight phenol formaldehyde resin. J. Bamboo Rattan **5**, 159–165 (2006)
12. Liese, W.: INBAR Technical Report 18: The Anatomy of Bamboo Culms. Beijing, INBAR (1998)
13. Liese, W.: Conference presented at International Symposium on Guadua Bamboo, Pereira, Colombia. Available in http://apuama.org/wp-content/uploads/2016/10/Preserva%C3%A7%C3%A3o-do-colmo-do-bambu.pdf. Last accessed 2020/01/02 (1994)
14. Croituru, C., Patachia, S., Lunguleasa, A.: A mild method of wood impregnation with biopolymers and resins using 1-ethyl-3-methylimidazolium chloride as carrier. Chem. Eng. Res. Des. (2014) (in Press). https://doi.org/10.1016/j.cherd.2014.04.031
15. Furuno, T., Imamura, Y., Kajita, H.: The modification of wood by treatment with low molecular weight phenol-formaldehyde resin: a properties enhancement with neutralized phenolic-resin and resin penetration into wood cell walls. Wood Sci. Technol. **37**, 349–361 (2004)
16. He, Z., Qian, J., Qu, L., Yan, N., Yi, S.: Effects of tung oil treatment on wood hygroscopicity, dimensional stability and thermostability. Ind. Crops Prod. **140**, 1–6 (2019)
17. ASTM—American Society for Testing and Materials: Standard Test Methods for Evaluating Properties of Wood-Base Fiber and Particle Panel Materials: D1037. Philadelphia, USA (2012)
18. Pinto, M.C.: Influência da Temperatura e de Tratamentos Alcalinos na Remoção dos Extrativos da Biomassa de Madeira e seus Efeitos na Hidratação de Pastas de Cimento e de Bioconcreto. Masters dissertation/Universidade Federal do Rio de Janeiro, Brazil (2019)
19. Mendes, S., Hugen, L.N., dos Santos, R.D., Filho, R.D.T., Rocha, S.F.: Further analyses on hornification promotion by washing cycles on natural vegetable fibers. In: Proceedings of 4th Brazilian Conference on Composites Materials. Pontifícia Universidade Católica do Rio de Janeiro, 2018, pp. 298–305

Recovering of Clinker Minerals from Hydrated Portland Cement Paste

Semion Zhutovsky and Andrei Shishkin

Abstract Recycling of concrete construction and demolition waste is necessary for the increase of sustainability and reducing the environmental impact of concrete construction because of the increasing rate of such waste production and its accumulation. Coarse aggregates can be partially recovered from concrete waste. However, it is not clear whether hydrated cement paste can be converted back to clinker. However, concrete waste fines, which are a mix of fine aggregates, coarse aggregate debris, and the hydrated cement paste, are currently not a part of the recycling process. The ability of hydrated cement paste to be recovered back to clinker minerals that have binder properties has not been studied systematically. In the current research, the phase transitions in hydrated cement paste heated to a temperature in the range from 600 and 1450 °C were investigated by means of X-ray diffractometry and thermal analysis. The experimental results demonstrate that hydrated Portland cement paste can be recovered back to clinker minerals. The recovered cement paste contains all the main clinker minerals similarly to the initial cement. The results provide evidence for the possibility of recycling hydrated cement and concrete into the cement clinker. The recycled clinker will potentially have a lower carbon footprint in comparison to original Portland cement.

Keywords Recycling · Clinker · Portland cement · Concrete waste fines

1 Introduction

The amount of concrete construction and demolition wastes is increasing constantly. Recycling of construction and demolition waste is necessary for saving natural resources, reducing the waste dump and landfill. The industry of concrete recycling has been steadily developed. European countries have goals to increase the reusing of waste construction materials up to 70% [1].

S. Zhutovsky (✉) · A. Shishkin
National Building Research Institute, Technion—Israel Institute of Technology, Haifa, Israel
e-mail: semionzh@technion.ac.il

V. M. C. F. Cunha et al. (eds.), *Proceedings of the 3rd RILEM Spring Convention and Conference (RSCC 2020)*, RILEM Bookseries 35,
https://doi.org/10.1007/978-3-030-76543-9_7

The waste concrete contains hydrated cement paste and aggregates, but only the coarse aggregates are generally extracted now. Concrete waste fines which are a mix of fine aggregate, the residue of coarse aggregates and hydrated cement paste are out of recycling in industrial plants [2, 3]. The quality of recycled concrete aggregates is lower than that of natural ones, because of residual hydrated cement [4]. Using recycled aggregates has a negative effect on the properties of concrete [5, 6] that limit their maximum content in concrete.

On the other hand, recycled concrete fines could be used as a raw material in Portland cement production. Portland cement is used as a binder in construction for over a hundred years. Its production has a high environmental impact due to the exhaustion of natural resources and emissions of carbon dioxide. The use of recycled concrete fines for clinker production can potentially reduce its carbon footprint and save natural resources. The recycled concrete fines can replace up to 15% of raw materials, according to chemical and mineralogical limitations [7–10].

The fine aggregates often contain minerals used for cement production such as calcite and quartz. However, it is not clear whether hydrated cement paste can be recovered back to clinker minerals. It possible that the hydrated cement paste will prevent the formation of good clinker in the production process. Although the chemical composition of hydrated cement paste is equal to the chemical composition of Portland cement with the addition of water, the addition of pure hydrated cement paste to raw material at rates over 30% caused problems [4, 11].

Portland cement clinker is manufactured by heating of the raw materials in the rotary kiln to 1450 °C. In this high-temperature treatment process, major clinker phases are obtained: alite (C_3S) 50–70%, belite (C_2S) 15–30%, tricalcium aluminate (C_3A) 5–10%, and tetracalcium aluminoferrite (C_4AF) 5–15%. The main clinkering chemical reactions and their temperature ranges are well studied [12]. Alite forms at a temperature above 1300 °C in the liquid phase, tricalcium aluminate forms above 1000 °C through the intermediate phases, belite, and tetracalcium aluminoferrite form already near 700 °C. However, there is no clear answer to what phase transformations in hydrated Portland cement paste take place during high-temperature processing in the rotary kiln.

The hardened Portland cement paste is a product of hydration reaction, i.e. the reaction of cement with water. It mainly consists of calcium silicate hydrate (C–S–H) which is often referred to as cementitious gel and Portlandite ($Ca(OH)_2$, which is in cement chemistry notation CH). The cementitious gel has variable stoichiometry ($mCaO \cdot SiO_2 \cdot nH2O$, where m is about 1.7, and $1.4 < n < 4$) with impurities and inclusions of aluminum, iron and sulfur.

There are many studies on phase transformations in heat-treated concrete and hydrated cement paste at temperatures under 1000 °C [13, 14]. At these temperatures, the main decompositions stages take place with no further detectable phase transformations. The main decomposition stages include [12–15]: release of free evaporable water and partial decomposition of physically bound water between 30 and 120 °C; in the range between 110 and 170 °C decomposition of ettringite, gypsum, carboaluminate hydrates; between 180 and 600 °C dehydration of C–S–H; between 450 and 550 °C portlandite dehydration; and in the range between 700 and 900 °C calcium carbonate decomposition. Nevertheless, there is no sufficient information about phase transformations in hydrated cement at the temperature between 1000 and 1450 °C, i.e. in the range of cement clinker mineral phases formation. One exception is maybe anhydrite decomposition with the SO_2 gas evaporation, which is known to take place at a temperature over 1200 °C and can be clearly seen in a thermogravimetric analysis [16].

There is some controversy in the literature about the high-temperature phase transitions in hydrated cementitious materials. The high-temperature X-ray diffraction (XRD) showed that C–S–H starts to decompose near 620 °C which is related to significant loss of strength. The characteristic amorphous C–S–H hump in XRD range of 25–40° 2θ completely disappears at 690 °C and a new strong crystalline peak of orthorhombic α′-Belite and Quartz appear [17].

Concrete structures exposed to fire have high mechanical damage and also chemical transformations [18]. It was reported that from 500 to 800 °C, due to the volume change of calcium silicate hydrate, cracks take place in the cement paste matrix and at about 1000 °C the hydrates including C–S–H and CH are transferred into crystalline phases completely [19]. The pore system increases rapidly after 400 °C with a graduate mass loss. The cement binder in concrete subjected to high temperatures save a partial regeneration ability [20]. At temperatures above 1200 °C, the binder in the concrete completely disintegrated [21].

Another study reports that the hump related to C-S–H around 29.35° 2θ disappeared above 500 °C accompanied by increased intensities of β-Belite (β-C_2S) peaks in paste samples after exposure to various temperatures for 6 h [22].

Some researchers claim that at the temperatures over 200 °C calcium silicate hydrate begins to form a "new nesosilicate", with CaO/SiO_2 ratio near 2, similar to a structure of Belite, but less crystalline [23]. At 750 °C C–S–H gel completely replaced mainly by this new phase. The process is reversible and a new C–S–H gel forms back due to hydration. The heating of 5-year old cement paste to various temperatures under 1000 °C, indicated β-belite and some alite are formed above 500 °C [24].

The high-temperature transformations of synthetic calcium silicate hydrate indicated that C–S–H transforms into β-Wollastonite ($CaSiO_3$) at 800–900 °C and into α-Wollastonite at 1220–1280 °C [25]. If CaO/SiO_2 ratio > 1 the competitive formation of Belite (C_2S) and Rankinite (C_3S_2) is increasing together with Wollastonite (CS).

Cement paste transformations in concrete during heating were also studied. C_2S polymorphs were found after thermal processing up to 900 °C in laboratory-prepared pastes [26]. Materials obtained at temperatures near 740 °C achieved higher early-age strengths due to the higher relative concentration of α′-belite, compared to the less reactive β-belite. The opposite results were obtained for 2-year-old laboratory-prepared cement paste pieces burned for 8 h under different temperatures [27]. High-temperature treatment (1100 °C) transformed the C–S–H into Wollastonite with a low content of Belite.

It can be seen that the information about high-temperature phase transitions in hydrated Portland cement paste is not sufficient and controversial. This is especially true for the temperatures above 1000 °C were the main clinker phases are forming.

The objective of the current research was to study the phase transitions in hydrated cement paste in the temperature range of 600–1450 °C. The study was carried out by means of X-ray diffractometry (XRD) and thermal analysis. The experimental results clearly demonstrate that hydrated Portland cement paste can be recovered back to minerals of clinker indicating high potential for recycling of hydrated cementitious materials.

2 Materials and Test Methods

2.1 Materials

In the present study, cement paste with water to cement ratio of 0.68 was investigated. It was produced using standard Portland cement of CEM I 52.5 N type. The cement paste was mixed in a pan mixer and cast into 1-inch cube molds, which were demolded at the age of 1 day. The curing was carried out in a lime solution until 28 days.

2.2 Methods

The hydrated cement paste was crushed into pieces of approximately 15 mm and burned in a high-temperature bottom-loading laboratory furnace. Burning was performed to the temperatures between 600 and 1450 °C. The samples were air quenched after removing them from the furnace.

Thermal analysis included thermogravimetry (TG), differential thermogravimetry (DTG), and differential scanning calorimetry (DSC). Thermal analysis was carried out in the Netzsch STA 449 F5 Jupiter instrument. The oxidizing atmosphere of dry air at the flow rate of 80 ml/min was used. For thermal analysis samples were ground below 45 μm size. The ground samples were put in a platinum crucible and tested in the temperature range from 25 to 1500 °C at the heating rate of 10 °C/min.

The mineral composition was determined by means of XRD. For this purpose Malvern PANAlytical EMPYREAN X-ray diffractometer was used with the following configuration: Goniometer radius of 240 mm, an X-ray source was CuKα1,2 ($\lambda = 1.5408$ Å) with the X-Ray generator operated at a voltage of 45 kV and a current of 40 mA. The optical configuration of XRD was as follows: the incident beam optics included a 10 mm mask, 0.04 rad Soller slit along with ¼° divergence and 1° anti-scatter fixed slits; the diffracted beam optics consisted of 8 mm anti-scatter fixed slit and 0.04 rad Soller slit. The detector was PIXcel 3D detector used in a 1D continuous scan mode. The scan was performed using Brag-Brentano θ–θ geometry, between 10 and 70° 2θ. Timestep of 80.32 s with a step size of 0.013° 2θ was used resulting in a total measurement time of 25.22 s. The quantitative analysis was performed by means of Rietveld refinement using HighScore Plus software.

3 Results

The results of thermal analysis of hydrated cement paste, which include TGA and DSC are shown in Figs. 1 and 2, respectively. Four mass loss steps can be identified in the TG curve in Fig. 1, with the corresponding DTG peaks. The first step in the range of 90–300 °C is associated with the decomposition of AFt and AFm phases partially overlapping with the dehydration of calcium-silicate-hydrate (C–S–H) gel. The second step is associated with portlandite dehydration in the range of 400–500 °C. The third step in the temperature range of 600–750 °C is caused by the decomposition of calcite and accompanied by the emission of CO_2 gas. The last fourth step between 1250 and 1400 °C is associated with the final decomposition of sulfates with the emission of gaseous SO_2. The DSC peaks corresponding to these steps can be seen in Fig. 2. In addition to these endothermic peaks, the solid–liquid transition can be identified in the DSC heating curve, as well as the liquid–solid transition at 1250 °C that is clearly seen in the cooling curve in Fig. 2. Some additional small DSC peaks that are associated with clinkerization reaction can be seen in the vicinity of 1300 °C.

According to TGA the Portlandite content of 14.4% and calcite of less than 3% was determined. The calculated amount of decomposed anhydride according to TGA

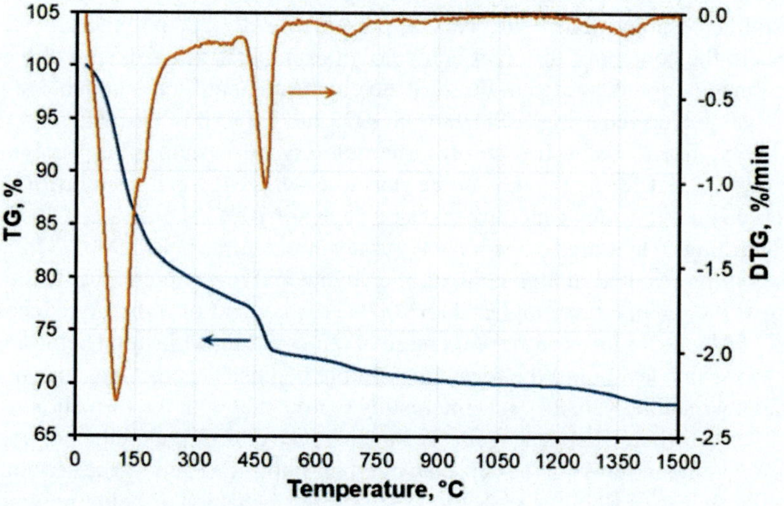

Fig. 1 Thermogravimetric analysis of hydrated cement paste

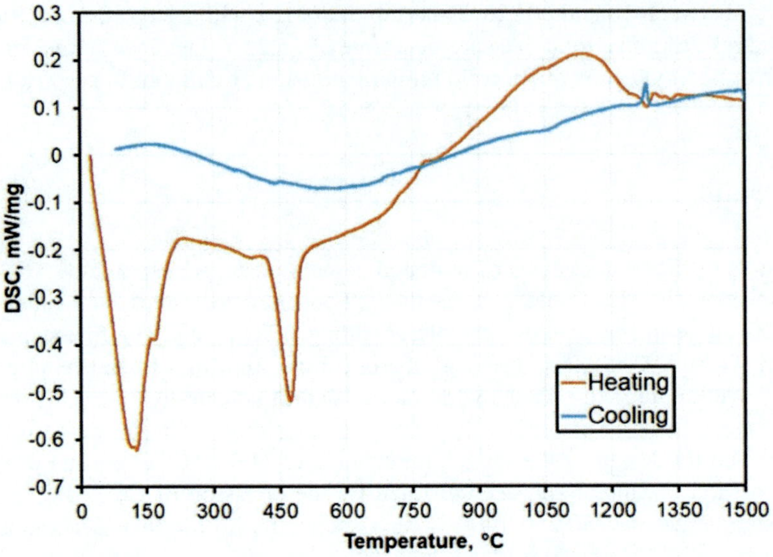

Fig. 2 Differential scanning calorimetry of hydrated cement paste

was about 2.0 that indicates that some sulfates from the original cement still remained in the burned material in form of anhydrite.

The change of the content of alite, belite and free lime with the burning temperature as determined by XRD is shown in Fig. 3. Belite and fee lime were formed in the process of C–S–H decomposition. It can be seen in Fig. 3 that alite starts to form after 1200 °C, at which point the amount of belite and free lime decreases correspondingly.

It should be noted that belite can be present in different polymorphic forms. Changes in the content of different belite polymorph types as a function of burning temperature are demonstrated in Fig. 4. Rietveld refinement for α′ and β belite polymorphs was performed using ICSD 98-008-1097 and ICDD 04-007-9746 references, respectively. The α′-belite has an orthorhombic crystal system. The β-polymorph is monoclinic and forms from α′-belite during cooling. Reversible transformation between this polymorphs generally takes place near 670 °C [12].

The burning of hydrated cement paste at the temperatures below 900 °C followed by fast cooling resulted in high content of α′-belite and low content of β-belite. The burning at the temperatures higher than 900 °C the content of α′-belite reduces and content of β-belite reduces with the increase of temperature, while total belite content stays at the same level until the formation of alite begins. This fact was also noticed in some other studies [26, 28]. It is interesting to note that with the formation of alite after 1200 °C α′-belite appears again. The appearance of different belite polymorphs at different temperatures can be caused by certain impurities and by the certain rates of cooling that may stabilize α′-belite [28]. Similar behavior of belite polymorphs heated to 1250 °C with the subsequent fast cooling was reported in the literature [29].

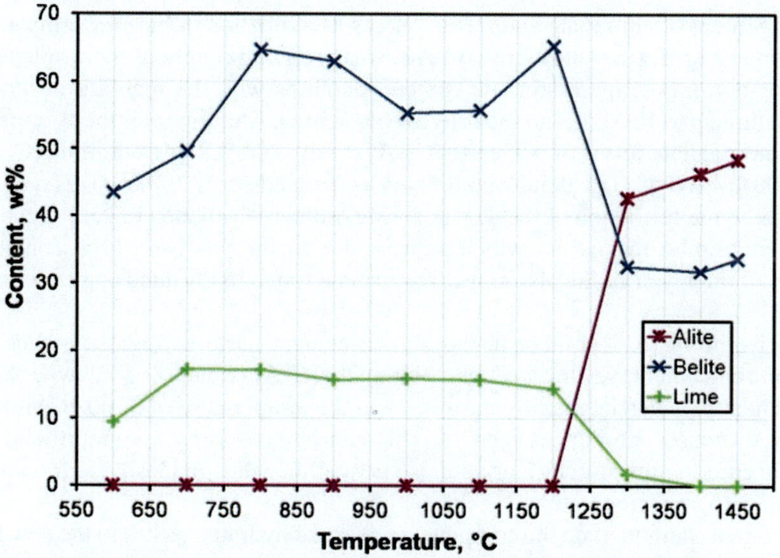

Fig. 3 Content of alite, belite and free lime as a function of burning temperature

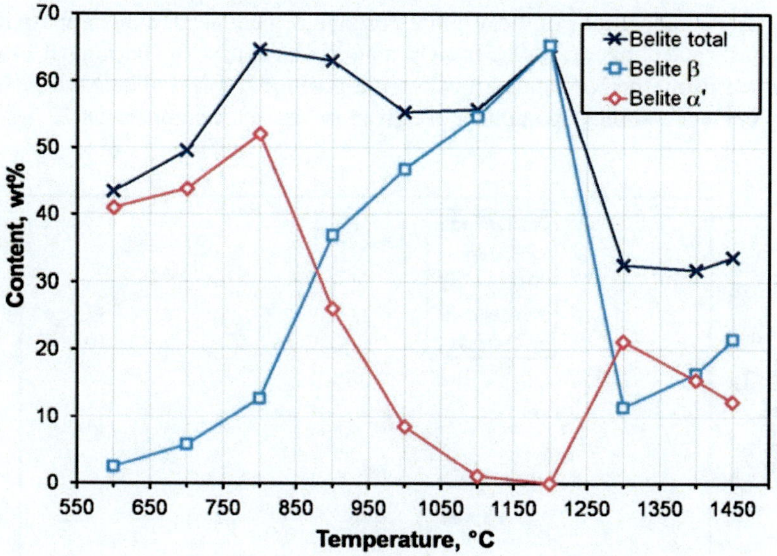

Fig. 4 Changes of belite type as a function of burning temperature

The content of aluminate and ferrite phases as a function of burning temperature is shown in Fig. 5. Brownmillerite (C_4AF) was present throughout the whole tested range of burning temperatures. Its content increases with the increase of burning temperature up to 1300 °C and then slightly decreases. Aluminates initially appeared in the intermediate forms of mayenite ($C_{12}A_7$) until 1200 °C and ye'elimite ($C_4A_3\$$) from 900 to 1200 °C. Tricalcium aluminate C_3A formed after 1200 °C. This sequence is similar to the formation of the phases during ordinary Portland clinker production, except ye'elimite formation, which appears due to the relatively high content of sulfates, though ye'elimite also sometimes can be found in Portland cement clinkers [13].

The ferrite (C_4AF) content in the burned cement paste is slightly higher than in the initial cement, while tricalcium aluminate (C_3A) content is somewhat lower. C_3A content reaches its maximum of 3.1% at the temperature of 1300 °C and then slightly decreases similarly to ferrite. This can be caused by aluminum and iron partially entering the structure of alite. No crystalline silica oxides were found after heating up to any temperature.

The results demonstrate the complete recovery of hydrated pastes to the new Portland cement clinker by thermal treatment. Numerous phase transformations occur throughout the studied temperature range during the burning. But the major clinker phases have been formed in this process indicating high potential for recycling. The mineral compositions of the original Portland cement used in the preparation of cement paste and clinker recycled from hydrated cement paste by burning at 1450 °C as obtained using the XRD quantification are given in Table 1. The XRD scans of original cement and recycled clinker are given in Fig. 6 for comparison.

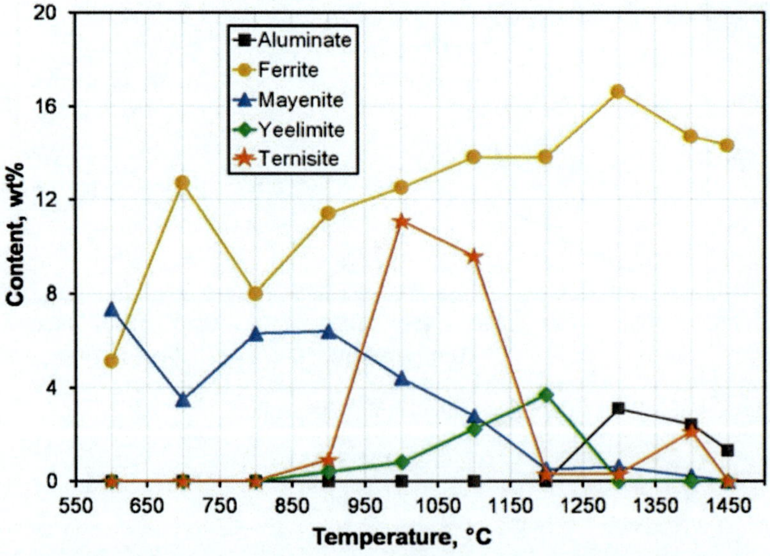

Fig. 5 Content of aluminate and ferrite phases as a function of burning temperature

Table 1 Mineral composition of original cement and recycled clinker, % wt

	Alite C_3S	Belite C_2S	Ferrite $C_4(AF)$	Alumnate C_3A	Anhydrite C$	Bassanite C$$H_{0.5}$
Original cement	55.2	20.3	13.9	4.4	0.0	6.6
Recycled clinker	48.4	33.7	14.3	2.4	2.3	0.0

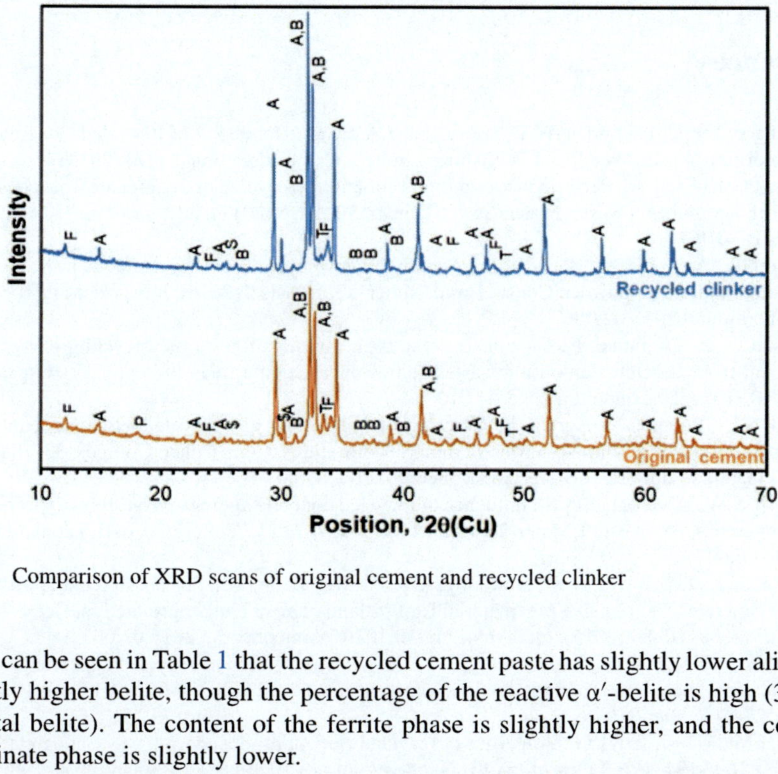

Fig. 6 Comparison of XRD scans of original cement and recycled clinker

In can be seen in Table 1 that the recycled cement paste has slightly lower alite and slightly higher belite, though the percentage of the reactive α′-belite is high (36.2% of total belite). The content of the ferrite phase is slightly higher, and the content aluminate phase is slightly lower.

It can be clearly seen that the clinker recovered from the hydrated cement paste has high quality. A certain amount of anhydrite is left in the recycled clinker from the decomposition of basanite that was present in the original cement, but this may be corrected by the adjustment of the added gypsum at the stage of cement production.

4 Conclusions

The paper presents the study of phase transformations in hydrated cement paste during burning at the temperatures of 600–1450 °C. The results of the research demonstrate that clinker minerals can be recovered from hydrated cement paste by heat treatment. This indicated a high potential for recycling of concrete and mortar waste. Since hydrated cementitious materials have lower embedded carbonated content than a limestone used in clinker production, such recycling can reduce CO_2 emissions in the process of Portland cement production.

Acknowledgements This work is partly supported by the Israeli Ministry of Housing and Construction. The research grant is greatly appreciated.

References

1. Pacheco-Torgal, F., Tam, V.W.Y., Labrincha, J.A., et al.: Handbook of Recycled Concrete and Demolition Waste. Woodhead Publishing Limited, Cambridge, Francis Do (2013)
2. Katz, A., Kulisch, D.: Performance of mortars containing recycled fine aggregate from construction and demolition waste. Mater. Struct. Constr. **50**, 1–16 (2017). https://doi.org/10.1617/s11527-017-1067-x
3. Kwon, E., Ahn, J., Cho, B., Park, D.: A study on development of recycled cement made from waste cementitious powder. Constr. Build. Mater. **83**, 174–180 (2015). https://doi.org/10.1016/j.conbuildmat.2015.02.086
4. Gastaldi, D., Canonico, F., Capelli, L., et al.: An investigation on the recycling of hydrated cement from concrete demolition waste. Cem. Concr. Compos. **61**, 29–35 (2015). https://doi.org/10.1016/j.cemconcomp.2015.04.010
5. Kou, S.C., Poon, C.S., Agrela, F.: Comparisons of natural and recycled aggregate concretes prepared with the addition of different mineral admixtures. Cem. Concr. Compos. **33**, 788–795 (2011). https://doi.org/10.1016/j.cemconcomp.2011.05.009
6. Tabsh, S.W., Abdelfatah, A.S.: Influence of recycled concrete aggregates on strength properties of concrete. Constr. Build. Mater. **23**, 1163–1167 (2009). https://doi.org/10.1016/j.conbuildmat.2008.06.007
7. Schoon, J., De. Buysser, K., Van Driessche, I., De. Belie, N.: Fines extracted from recycled concrete as alternative raw material for Portland cement clinker production. Cem. Concr. Compos. **58**, 70–80 (2015). https://doi.org/10.1016/j.cemconcomp.2015.01.003
8. Diliberto, C., Lecomte, A., Mechling, J.-M., et al.: Valorisation of recycled concrete sands in cement raw meal for cement production. Mater. Struct. **50**, 127 (2017). https://doi.org/10.1617/s11527-017-0996-8
9. Ai, H.M., Wei, J., Lu, P.G.: Properties of recycled cement produced from waste concrete. Adv. Mater. Res. **194–196**, 1170–1175 (2011). https://doi.org/10.4028/www.scientific.net/amr.194-196.1170
10. De. Schepper, M., De. Buysser, K., Van Driessche, I., De. Belie, N.: The regeneration of cement out of completely recyclable concrete: clinker production evaluation. Constr. Build. Mater. **38**, 1001–1009 (2013). https://doi.org/10.1016/j.conbuildmat.2012.09.061
11. Bordy, A., Younsi, A., Aggoun, S., Fiorio, B.: Cement substitution by a recycled cement paste fine: Role of the residual anhydrous clinker. Constr. Build. Mater. **132**, 1–8 (2017). https://doi.org/10.1016/j.conbuildmat.2016.11.080
12. Taylor, H.F.W.: Cement Chemistry, 2nd edn. Thomas Telford, London (1997)
13. Scrivener, K., Snellings, R., Lothenbach, B.: A Practical Guide to Microstructural Analysis of Cementitious Materials. CRC Press, Boca Raton, FL (2016)
14. Collier, N.C.: Transition and decomposition temperatures of cement phases—a collection of thermal analysis data. Ceram. Silikaty **60**, 338–343 (2016). https://doi.org/10.13168/cs.2016.0050

15. Alarcon-Ruiz, L., Platret, G., Massieu, E., Ehrlacher, A.: The use of thermal analysis in assessing the effect of temperature on a cement paste. Cem. Concr. Res. **35**, 609–613 (2005). https://doi.org/10.1016/j.cemconres.2004.06.015
16. Fukami, T., Tahara, S., Nakasone, K., Yasuda, C.: Synthesis, crystal structure, and thermal properties of $CaSO_4 \cdot 2H_2O$ single crystals. Int. J. Chem. **7** (2015)https://doi.org/10.5539/ijc.v7n212-20
17. Song, H., Jeong, Y., Bae, S., et al.: A study of thermal decomposition of phases in cementitious systems using HT-XRD and TG. Constr. Build. Mater. **169**, 648–661 (2018). https://doi.org/10.1016/j.conbuildmat.2018.03.001
18. Ma, Q., Guo, R., Zhao, Z., et al.: Mechanical properties of concrete at high temperature-a review. Constr. Build. Mater. **93**, 371–383 (2015)
19. Zhang, Q., Ye, G.: Microstructure analysis of heated Portland cement paste. Proc. Eng. **14**, 830–836 (2011). https://doi.org/10.1016/j.proeng.2011.07.105
20. Chromá, M., Vo, D., Bayer, P.:Concrete rehydration after heating to temperatures of up to 1200 °C. Int. Conf. Durab. Build. Mater. Components. 1633–1639 (2011)
21. Arioz, O.: Effects of elevated temperatures on properties of concrete. Fire Saf. J. **42**, 516–522 (2007). https://doi.org/10.1016/j.firesaf.2007.01.003
22. Lim, S., Mondal, P.: Micro- and nano-scale characterization to study the thermal degradation of cement-based materials. Mater. Charact. **92**, 15–25 (2014). https://doi.org/10.1016/j.matchar.2014.02.010
23. Alonso, C., Fernandez, L.: Dehydration and rehydration processes of cement paste exposed to high temperature environments. J. Mater. Sci. **39**, 3015–3024 (2004). https://doi.org/10.1023/B:JMSC.0000025827.65956.18
24. Sabeur, H., Saillio, M., Vincent, J.: Thermal stability and microstructural changes in 5 years aged cement paste subjected to high temperature plateaus up to 1000 °C as studied by thermal analysis and X-ray diffraction. Heat Mass Transf. und Stoffuebertragung (2019). https://doi.org/10.1007/s00231-019-02599-w
25. Tajuelo Rodriguez, E., Garbev, K., Merz, D., et al.: Thermal stability of C–S–H phases and applicability of Richardson and Groves' and Richardson C–(A)–S–H(I) models to synthetic C–S–H. Cem. Concr. Res. **93**, 45–56 (2017). https://doi.org/10.1016/j.cemconres.2016.12.005
26. Serpell, R., Zunino, F.: Recycling of hydrated cement pastes by synthesis of α′H-C2S. Cem. Concr. Res. **100**, 398–412 (2017). https://doi.org/10.1016/j.cemconres.2017.08.001
27. Wang, J., Mu, M., Liu, Y.: Recycled cement. Constr. Build. Mater. **190**, 1124–1132 (2018). https://doi.org/10.1016/j.conbuildmat.2018.09.181
28. Serpell, R., Lopez, M.: Properties of mortars produced with reactivated cementitious materials. Cem. Concr. Compos. **64**, 16–26 (2015). https://doi.org/10.1016/j.cemconcomp.2015.08.003
29. Dashkevich, R.Y., Aleksandrov, A.V.: Resonance character of polymorphic transformations for the phase transition α′ L → β-Ca2SiO4. Russ. J. Non-Ferrous Met. **48**, 404–406 (2008). https://doi.org/10.3103/s1067821207060053

Recycling of Slightly Contaminated Demolition Waste—Part 2: PAH

Lia Weiler and Anya Vollpracht

Abstract For the purposes of environmental protection, resource efficiency and to foster a sustainable development it is advisable to recycle also slightly contaminated CDW and keep it from ending up in landfills. A high-quality option is the use as aggregate for concretes (RCA). In doing so, it must be guaranteed, that no harms for health and environment are caused by potential emissions. The use of crushed demolition waste as a concrete aggregate is technically state-of-the-art. Investigations on the leaching behaviour of recycled concrete showed a good integration of inorganic contaminants. However, PAH were found to be leached in noticeable amounts even for moderately contaminated CDW, complying with the German threshold of 25 mg/kg for RCA. For this contribution different test specimens, containing moderately PAH-contaminated aggregates, were tested in leaching tests following DIN CEN/TS 16637-2. The results were evaluated according to the current German evaluation procedure and compared to limit values for PAH in groundwater. PAH leaching from concrete proved to be a relevant issue, the permitted release of 0.22 mg/m^2 and therefore the threshold for groundwater was exceeded in many cases. The PAH content of the recycled aggregate was no suitable indicator for an environmental compatible concrete, the most determining factor, besides the total aggregate amount added, is the contaminated fines share.

Keywords PAH contamination · Leaching · Demolition waste · Concrete aggregate · Recycling

1 Introduction

The use of crushed CDW as a concrete aggregate is technically state-of-the-art and has recently been investigated in a large joined research project in Germany [1]. The project aimed at expanding the actually quite restrictive regulations in terms of the maximum allowable amount, the particle size added as aggregate, and the

L. Weiler (✉) · A. Vollpracht
RWTH Aachen University, Aachen, Germany
e-mail: weiler@ibac.rwth-aachen.de

V. M. C. F. Cunha et al. (eds.), *Proceedings of the 3rd RILEM Spring Convention and Conference (RSCC 2020)*, RILEM Bookseries 35,
https://doi.org/10.1007/978-3-030-76543-9_8

permitted applications. In this context also the limits for possible contaminations were reconsidered and leaching tests following DIN CEN/TS 16637-2 were conducted on concrete with recycled aggregate. The investigations showed a good integration of inorganic contaminants (see also Part 1: "Recycling of Slightly Contaminated Demolition Waste—Inorganic Constituents" in this proceeding). However, organic contaminations, especially PAH, were found to be leached in noticeable amounts even for moderate contaminated CDW, complying with the German threshold of 25 mg/kg for recycled aggregates [2]. They were subsequently examined in more detail.

1.1 PAH in CDW

PAH are natural components of coal and crude oil as well as a product of incomplete combustion processes. Traces of PAH are to be found in all mineral building materials, the content is equivalent to the overall environmental background concentration levels [3]. By sealing materials and adhesives, containing tar or bitumen, higher amounts of PAH are introduced to building materials, in this case CDW. Other sources of contamination are domestic furnaces, but also tire abrasion or industrial use [4–7].

PAH are a group of more than 10.000 organic compounds, consisting of at least two aromatic rings, and bearing PBT (persistent, bio accumulative, toxic) characteristics; which makes them of concern for health and environment.

For environmental assessments the so called 16 EPA-PAH (PAH_{16}) are often analysed as representatives for all compounds. The PAH_{16} are easily detectable, toxicological considerable and cover the range of two to six rings, resulting in the different characteristics pictured in Fig. 1 [8].

Due to their adhesive behavior and lower volatility higher molecular PAH are likely to be present in higher amounts in CDW; investigations accompanying this work and Butera [3] confirm a predominance of PAH compounds with higher molecular weight in CDW.

Fig. 1 Overview PAH characteristics

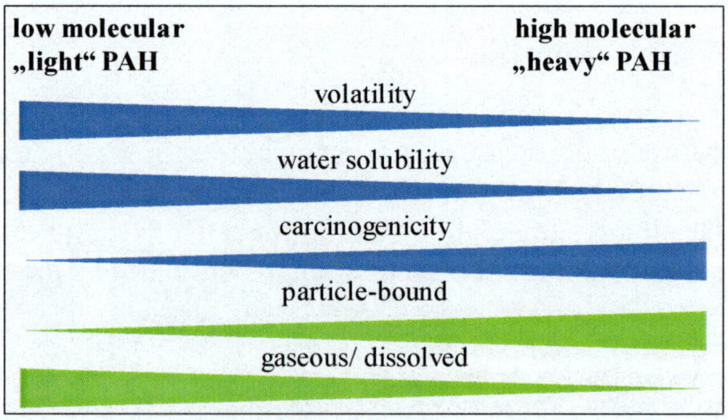

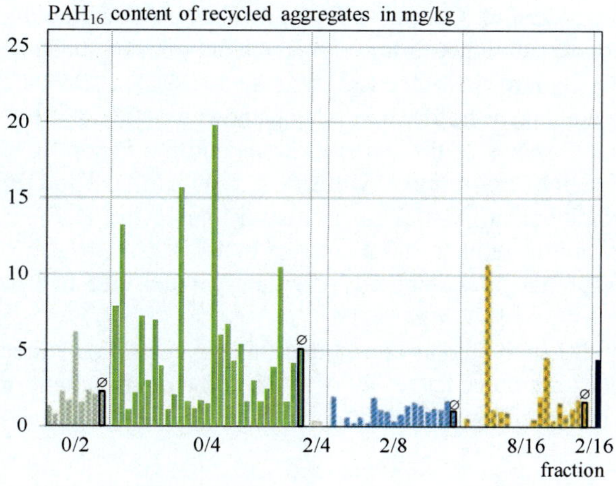

Fig. 2 PAH_{16} content of 75 recycled aggregate samples taken from 2015 to 2017 at a German recycling facility; data from [9]

PAHs preferentially bind to dust, soil, and soot particles and are therefore mainly found in the fines fraction [8]. Figure 2 illustrates this fact, showing the PAH_{16} content determined for 75 recycled aggregate samples taken from 2015 to 2017 at a German recycling facility classified by grain size fractions.

1.2 Use of CDW Containing PAH as a Concrete Aggregate

In Germany a threshold of 25 mg/kg PAH_{16} is defined for the use of CDW as RCA (see DIN 4226-101, [10]). According to a study of the Federal Environment Ministry of Baden-Wuerttemberg on PAH-contents of 62 samples from recycling facilities in Germany, 4.8% of delivered CDW do not comply with this value [11]. The Danish study described in [3] investigated 33 samples and confirms this range, finding 3% above 25 mg/kg.

Yet the threshold is not based on toxicological data or actual release tests and therefore does not necessarily guarantee an environmentally harmless product when used as RCA. On the one hand PAH are known to not be integrated in cementitious matrices [6, 12], but on the other hand their solubility in water is very low; therefore it is unclear whether critical leaching occurs in contact with groundwater for example.

An official limit value for PAH leaching from concrete does currently not exist in Germany or in the Netherlands or—to our knowledge—in any other country. This missing threshold complicates the assessment of PAH leaching; therefore it was necessary to derive a limit, based on the general concept currently used in Germany.

This assessment concept for the leaching behaviour of construction products and materials is used in the context of technical approvals for new building materials and the assessment of potentially relevant standardized building materials like artificial aggregates or fly ashes. The concept is described in [13] and was published by the "German Centre of Competence for Construction" (DIBt). According to this concept the maximum permissible amount of substance released within 64 days in the Dynamic Surface Leaching Test (DSLT, CEN/TS 16637-2) is depending on the specific threshold value for groundwater (GFS) of the German Working Group on water issues of the Federal States and the Federal Government represented by the Federal Environment Ministry (LAWA) [14]. The GFS for PAH_{15} (PAH_{16} without Naphthalene) is 0.2 μg/L. According to the DIBt concept this corresponds to a permitted release of 0.22 mg/m^2 in the DSLT. Because of its comparably low toxicity Naphthalene is assessed individually with a GFS of 1 μg/L and therefore a permitted release of 1.22 mg/m^2.

It has to be considered that this concept is very much on the safe side because it does not consider decay of PAH in the environment or adsorption on soil particles after leaching.

2 Materials and Methods

PAH contaminated concrete specimens with different compositions were tested in two different leaching tests. The results were evaluated according to the current German evaluation procedure described in Sect. 1.2.

2.1 *Leaching Tests*

Batch Test. Shake bottle tests were conducted according to EN 12457-4. For the leaching no grain reduction was made, so that a maximum size of 16 mm instead of 10 mm was used. 180 g of the sample were leached in double determination in 2 L PE-bottles with 1800 ml of deionised water, resulting in a liquid to solid ratio (L/S) of 10. The bottles were rotated with 5 rpm for 24 h. Before sampling the particles were allowed to settle for 15 min and the supernatant was centrifuged.

The pH value, electrical conductivity and temperature were immediately determined, the PAH were analysed by GC–MS.

DSLT. The tank leaching tests were carried out in double or triple determination based on the European harmonised standard DIN CEN/TS 16637-2 [15]. The test specimens were positioned on spacers in glass chromatography vessels as pictured in Fig. 3. Afterwards 2.3 L of deionised water were added, resulting in a ratio of eluate volume divided by the surface of the test specimens (V/A) of 79.9 L/m^2. The sanded glass lid was closed and additionally sealed with adhesive tape.

The leaching water was renewed in intervals of 0.25, 1, 2.25, 4, 9, 16, 36 and 64 days, as specified by DIN CEN/TS 16637-2. From each eluate fraction two samples were taken, one 100 ml PE bottle for the determination of pH and electrical conductivity and one 1 L glass bottles for the PAH analyses. The 1 L samples were stored protected from light at 4 °C until their analyses for PAH_{16} by GC–MS.

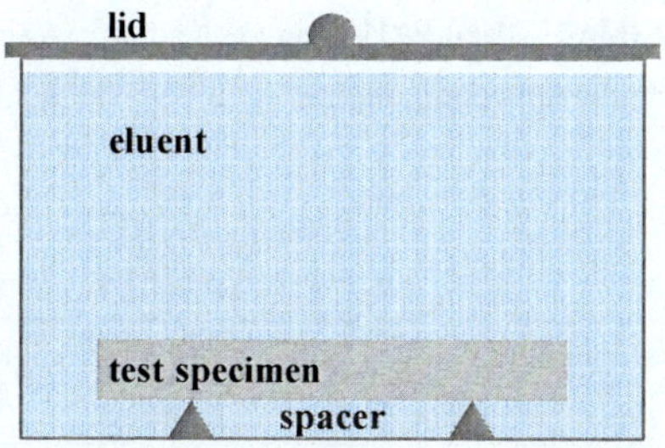

Fig. 3 Schematic drawing of the DSLT setup

Based on the determined PAH concentrations in the eluate, the emissions and the cumulated release was calculated using Eq. (1) of Part 1: "Recycling of Slightly Contaminated Demolition Waste—Inorganic Constituents" in this proceedings.

2.2 Recycled Aggregates

Five different aggregates were used and characterized in terms of grain size distribution in accordance to DIN EN 12620, grain bulk density and water uptake following DIN EN 1097-6. The PAH_{16} content measured by GC–MS after hexane extraction from the solid phase (DIN ISO 18287) is given in Table 1.

In addition the RCA were tested according to the batch test (see Sect. 2.1). The results are shown in Table 2.

2.3 Investigated Concrete Mixtures

Different concrete mixtures were used to produce test specimens with the PAH contaminated RCA described in Sect. 2.1. All test specimens were produced with 280 kg/m^3 of the same CEM I 32.5 R and a water to cement ratio of 0.6. Table 3 gives an overview of the aggregates used for the different concretes.

Additional water was added corresponding to the water absorption of the respective recycled aggregates. Natural aggregate was added to complement the recycled aggregate to an A16/B16 grading curve.

The recycled aggregate was stored in the mixing water at room temperature (20 ± 2 °C) for 24 h prior to mixing. Mixing was carried out in small, 2–8 L batches. The mixture was then given into untreated silicone formwork (160 · 40 · 40 mm^3) and the samples were demolded after 24 h. Until the beginning of the leaching tests at a sample age of 56 days, the samples were stored at 20 °C and 65% RH, wrapped airtight into PE foil.

Table 1 Total PAH content (hexane extraction, DIN ISO 18287) of the RCA

Parameter/component	a Mixed CDW	b Crushed masonry	c Crushed concrete	d Crushed concrete	e[a] Crushed concrete
	0/16 mm (mg/kg)	0/16 mm (mg/kg)	2/16 mm (mg/kg)	2/16 mm (mg/kg)	0/2 mm (mg/kg)
Naphtalene	0.0356	0.00202	0.0346	1.04	0.0346
Acenaphthylene	0.00892	0.00171	0.00936	0.338	0.00936
Acenaphthene	0.0378	0.00232	0.0273	2.01	0.0273
Fluorene	0.0252	0.00341	0.0119	0.77	0.0119
Phenanthrene	0.353	0.0349	1.75	6.62	1.75
Anthracene	0.088	0.00951	0.113	2.11	0.113
Fluoranthene	2.2	0.0825	2.76	10.4	2.76
Pyrene	1.63	0.0755	2.09	8.16	2.09
Benz(a)anthracene	0.118	0.0409	1.17	5.59	1.17
Chrysene	0.647	0.0448	1.16	4.12	1.16
Benz(b)fluoranthene	0.47	0.0381	1.00	4.78	1.00
Benz(k)fluoranthene	0.309	0.0212	0.563	2.01	0.563
Benz(a)pyrene	0.469	0.0366	0.773	4.17	0.773
Indeno (1,2,3,c,d)pyrene	0.268	0.0283	0.188	0.107	0.188
Dibenz(a,h)anthracene	0.111	0.0129	0.636	1.69	0.636
Benz(g,h,i)perylene	0.28	0.0263	0.537	1.43	0.537
Sum PAH_{15}	7.01	0.459	12.79	54.3	12.79
Sum PAH_{16}	7.05	0.461	12.82	55.3	12.82

[a]crushed version of aggregate c

3 Results and Discussion

3.1 Cumulative Release

The following figures show the cumulative release of the ***R***-concretes. To take the analytical inaccuracy due to detection limits (DTL) into account, a minimum and a maximum release are shown. Substances below the DTL where either counted as 0 for the minimum or as the DTL for the maximum value respectively.

The cumulated PAH_{15} release ranged from 0.08–0.15 mg/m^2 to 0.72–076 mg/m^2 (average 0.32–0.36 mg/m^2). Naphthalene is not shown in the figures, because the overall release was quite low considering the threshold and the lower toxicity of this compound (0.0138–0.29 mg/m^2).

In comparison to their initial content, low molecular PAH as for example phenanthrene or acenaphthene are leached in higher amounts, but despite their very low

Table 2 Batch test results of the RCA (EN 12457-4)

Parameter/component	a	b	c	d	e
	Mixed CDW	Crushed masonry	Crushed concrete	Crushed concrete	Crushed concrete
	0/16 mm (µg/L)	0/16 mm (µg/L)	2/16 mm (µg/L)	2/16 mm (µg/L)	0/2 mm (µg/L)
Naphtalene	0.73	0.048	0.23	1.28	0.895
Acenaphthylene	0.0591	0.01	0.00953	0.015	0.00954
Acenaphthene	2.17	0.12	0.414	1.17	0.861
Fluorene	0.207	0.0658	0.135	0.435	0.304
Phenanthrene	0.113	0.223	0.89	2.32	1.44
Anthracene	0.0408	0.0635	0.263	0.650	0.367
Fluoranthene	3.41	0.447	1.64	3.33	1.17
Pyrene	2.44	0.367	1.27	2.46	0.827
Benz(a)anthracene	0.303	0.0618	0.934	1.69	0.475
Chrysene	0.285	0.0584	0.84	1.53	0.42
Benz(b)fluoranthene	0.109	0.0311	1.17	2.065	0.5345
Benz(k)fluoranthene	0.0768	0.01	0.467	0.804	0.208
Benz(a)pyrene	0.11	0.0331	1.05	1.81	0.449
Indeno (1,2,3,c,d)pyrene	0.0647	0.01	0.538	1.05	0.242
Dibenz(a,h)anthracene	0.0208	0.01	0.105	0.165	0.0435
Benz(g,h,i)perylene	0.0596	0.0311	0.569	0.971	0.2485
Sum PAH_{15}	9.47	1.54	10.3	20.5	7.59
Sum PAH_{16}	10.2	1.59	10.5	21.7	8.49

Table 3 Information on the investigated concrete mixes

Label	Unit	A1	A2	A3	A4	A5	A6	A7
Recycled aggregate used (see Table 1)	–	a	b	c	c	d	e	d
Replacement of the natural aggregate	vol%	95	82	35	50	35	35	35
Share of fines (<2 mm) in recycled aggregate	wt%	27.7	38.2	0	0	40.0	100	0
PAH_{16}-content of the recycled aggregate	mg/kg	7.05	0.461	12.8	12.8	55.3	12.8	55.3

solubility in water, higher molecular PAH (pyrene, flouranthene) are released from the concrete as well.

Figure 4 exemplarily shows the release curves for PAH_{15} of A1 and A3 in comparison. The release course is the same for all sample compositions. It becomes apparent that the PAH release is higher at the beginning of the leaching test. The assessment

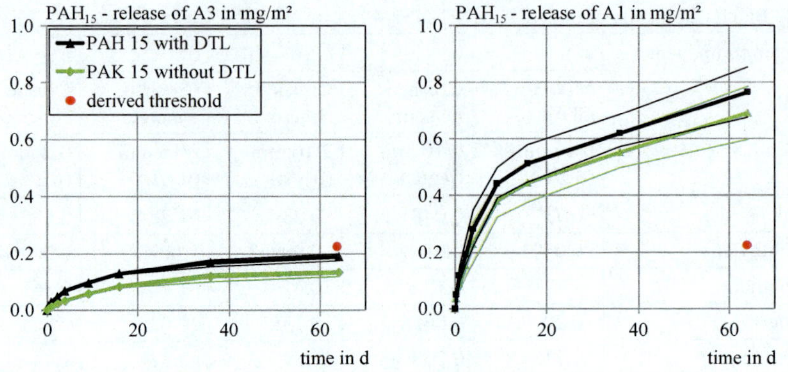

Fig. 4 Exemplary release curves of the samples A1 and A3

according to CEN/TS 16637-2 showed that the leaching process is mainly diffusion controlled whereby samples A3 and A7 are depleting (exemplarily shown in Fig. 5). For the samples A1, A2 and A4 no explicit leaching mechanism could be determined.

The cumulated minimum and maximum releases of all samples are compiled in Fig. 6. In all tests a relatively high proportion of the total PAH content of the RCA was released from the samples (average of 0.65%, maximum 2.4%). This seems to confirm the findings of [6, 12] that PAH are not incorporated in the cementitious matrix unlike most inorganic constituents.

Two to three out of seven concretes meet the threshold for groundwater and can be considered as environmental compatible. Concrete A2 with a very low PAH content (0.46 mg/kg) and concrete A3 (35 vol% RCA, 13 mg/kg PAH_{16}) are well below the threshold. Concrete A4 with 50 vol% RCA cannot be finally assessed due to the analytical inaccuracy.

3.2 *Influencing Factors*

Looking at all samples, a proportional relation from the total RCA content, or also PAH_{16} content of the RCA to the leached PAH_{15} amount is not recognisable. This is illustrated by Fig. 7 that shows the dependency of the cumulated PAH_{15} release on the legally regulated factor total PAH_{16} content and the factor PAH_{16} content in the fines fraction.

Other factors, e. g. the varied concrete composition, the RCA-fines fraction or the material composition of the CDW mix must have a major effect on the leaching process. It seems that concretes made with a high RCA-fines share leach stronger compared to such made with the same RCA share and PAH content but no contaminated fines. It is assumed, that a part of the PAH is solved in the mixing water, remains dissolved in the pore solution, and is therefore directly available for leaching processes through the concretes pores. High contents of fines results in a bigger

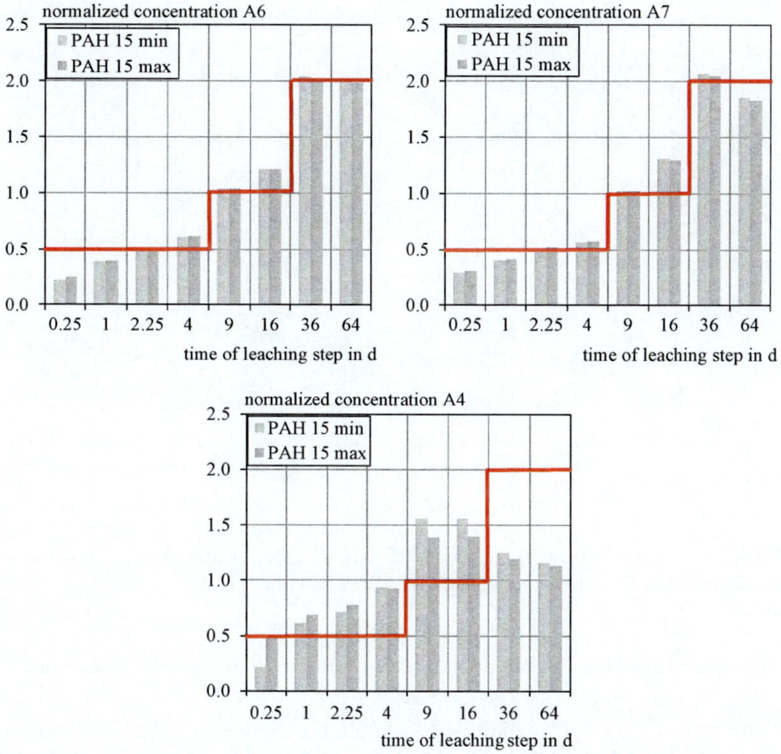

Fig. 5 Exemplary diagrams for diffusion controlled leaching in the DSLT; A6: diffusion controlled, A7: diffusion controlled with depletion, A4: no dominant mechanism discernable

Fig. 6 Minimum and maximum cumulated PAH_{15} release of all samples; comparison to the GFS-threshold value

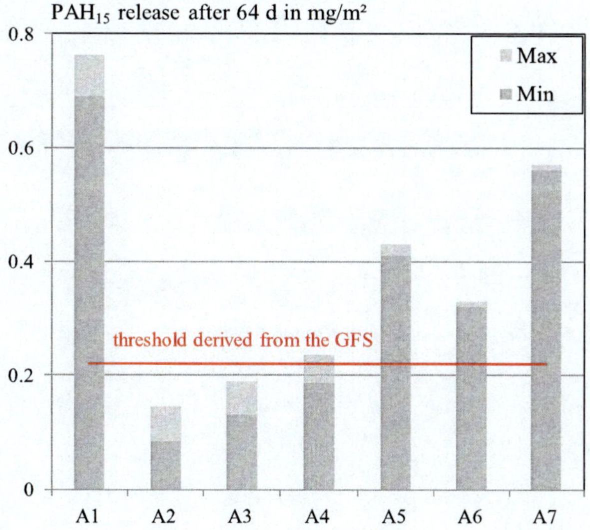

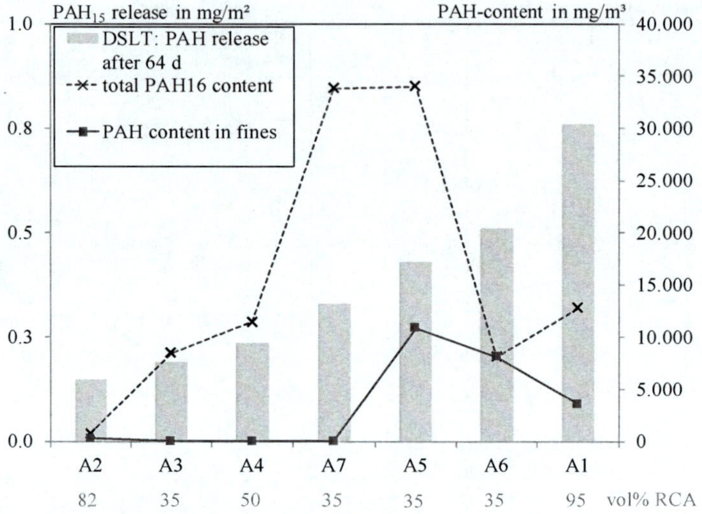

Fig. 7 Comparison of the PAH_{15}-release after 64 days in the DSLT; varieties of total PAH_{16} content in the concrete and PAH_{16} content contained in the fines fraction

surface area and higher solubility, especially of easier soluble, low molecular PAHs in this case.

This theory is supported by the results of batch tests shown in Fig. 8. Low molecular PAH were leached stronger from the crushed version (e) of the same material (c). At the same time heavy PAH are retained, probably due to bonding to the particles.

The total PAH content of the RCA showed an effect on the leached concentrations as well. Further testing is needed to broaden the database and define interdependencies.

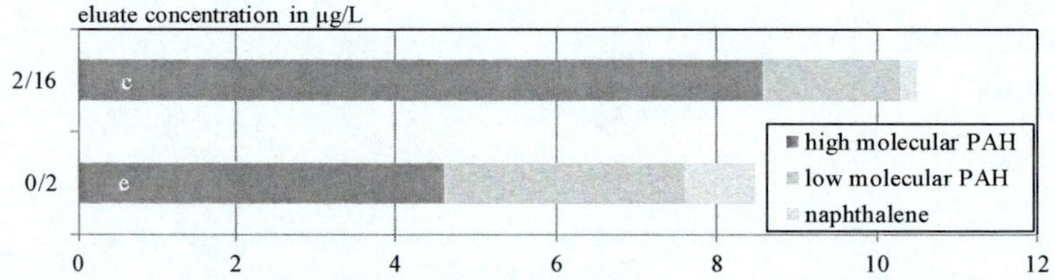

Fig. 8 Eluate concentrations in the batch test

4 Summary and Outlook

In this work different ***R***-concrete mixtures with RCAs that fulfill or exceed the German standard DIN 4226-101 have been leached in the dynamic leaching test DSLT. The results were then evaluated according to the current German evaluation procedure and compared to the derived limit values for PAH.

PAH leaching from concrete proved to be a relevant issue, especially when aiming at expanding the German regulations with respect to the permissible amount of RCA and the use of RC sand in concrete. For countries with less strict legal provisions, like the Netherlands with 50 mg/kg for example, it could also be questionable if environmental protection is ensured by their values. The derived maximum release of 0.22 mg/m^2 was exceeded in several cases. The concrete mixture A3 which complies with the German standard meets the threshold, but it has to be considered, that the used RCA contained only half of the allowed content of 25 mg/kg PAH_{16}.

On the other hand, for an environmental compatibility assessment, it must be taken into account, that the threshold approach is rigorous. Interactions with the soil are not considered and possible overestimations caused by the leaching with deionized water instead of less aggressive groundwater are neglected.

It can be concluded, that the PAH_{16} content of the RCA is no suitable indicator for an environmental compatible concrete. To predict the PAH_{15} release at least the amount of RCA added and the contaminated fines shares have to be taken into account. To fully understand the influencing processes and effectively assess the environmental compatibility of concretes containing PAH contaminated RCA, further investigation is needed.

Defining the interdependencies between the three investigated factors, will be a next step to prognosticate the leaching behaviour of the concrete and to confirm a reasonable PAH threshold for RCA that relates to the leaching of concrete made with RCA.

References

1. R-Beton Homepage (*Project Page Recycling Concrete*). https://www.r-beton.de/. Last accessed 2020/01/07
2. Vollpracht, A.: Umweltrelevante Merkmale – in der Regel kein Problem (*Environmental characteristics—no problem in general*). Beton **67**(9), 328–331 (2017)
3. Butera, S., Christensen, T.H., Astrup, T.F.: Composition and leaching of construction and demolition waste: inorganic elements and organic compounds. J. Hazard. Mater. **276**, 302–311 (2014)
4. Köhler, M., Mehnert, J., Weis, N.: Erfahrungen mit PAK-Belastungen durch Homogenasphaltplatten (*Experiences with PAH-Contaminations Through Homogenous Asphalt Sheets*), pp. 193–202. Umwelt, Gebäude & Gesundheit (2010)
5. Aatmeeyata, M.S.: Polycyclic aromatic hydrocarbons, elemental and organic carbon emissions from tire wear. Sci. Total Environ. **408**(20), 4563–4568 (2010)
6. Mulder, E., Brouwer, J.P., Blaakmeer, J., Frénay, J.W.: Immobilisation of PAH in waste materials. Waste Manage. **21**, 247–253 (2001)

7. Lindgren, A.: Road Construction Materials as a Source of Pollutants, Doctoral Thesis, Lund University, Department of Energy Sciences (1998)
8. Brandt, M., Einhenkel-Arle, D.: Polycyclic Aromatic Hydrocarbons Harmful to the Environment! Toxic! Inevitable? German Environment Agency, Dessau-Roßlau (2016)
9. Knappe, F., Reinhardt, J., Theis, S., Müller, C., Reiners, J., Böing, R.: Schlussberichte zum BMBF-Verbundforschungsvorhaben „*R*-Beton - Ressourcenschonender Beton – Werkstoff der nächsten Generation - Teil 3: Ökobilanz, Praxistest und Transfer (Final reports to the BMBF joint research project "*R*-concrete—resource-saving concrete—building materials of the next generation"—Part 3: life cycle assessment, field test and transfer). Berlin, Beuth GmbH Berlin publishing company, bulletin series of the DAfStb—to be published in 2020
10. German Institute for Standardization (ed.): DIN 4226-101:2017-08 Recycled Aggregates for Concrete in Accordance with DIN EN 12620—Part 101: Types and Regulated Dangerous Substances. Beuth Verlag GmbH (2017)
11. Schäfer, C., Gamm, A., Weiß, W.: Analytische Untersuchung von Bauschuttrecyclingmaterial auf Sulfat und PAK (*Analytical Investigations of Construction Recycling Material for Sulphate and PAH*). Az: 35-8980.11/Bauschutt, Federal Environment Ministry of Baden-Wuerttemberg, Stuttgart (2003)
12. Paria, S., Yuet, P.K.: Solidification–stabilization of organic and inorganic contaminants using Portland cement: a literature review. Environ. Rev. **14**(4), 217–255 (2006)
13. DIBt (ed.): Grundsätze zur Bewertung der Auswirkungen von Bauprodukten auf Boden und Grundwasser (*Principles of the Impact Assessment of Building Products on Soil and Groundwater*), Part II: Bewertungskonzepte für spezielle Bauprodukte (*Assessment Concepts for Specific Building Materials*), Chapter 1: Betonausgangsstoffe und Beton (*Concrete Constituents and Concrete*). Berlin (2011)
14. LAWA (ed.): *German Working Group on Water Issues of the Federal States and the Federal Government Represented by the Federal Environment Ministry*): Ableitung von Geringfügigkeitsschwellen für das Grundwasser (*Derivation of de minimis Thresholds for Groundwater Quality*). Stuttgart (2017)
15. Din, E.V. (ed.): DIN CEN/TS 16637-2:2014 DIN SPEC 18046-2, Construction Products—Assessment of Release of Dangerous Substances—Part 2: Horizontal Dynamic Surface Leaching Test; German Version, Beuth Verlag, Berlin (2014)

Recycling of Slightly Contaminated Demolition Waste—Part 1: Inorganic Constituents

Anya Vollpracht and Lia Weiler

Abstract Due to the high amount of construction and demolition waste (CDW) generated, the recycling and reuse of these materials is an important factor for the sustainability of the whole building sector. The use of CDW as recycled aggregate is a high-quality recovery; however, contaminations that may emerge during the service life of a building, have to be considered. Part 1 of this publication deals with inorganic contaminants. Within a large research project numerous CDW samples from a commercial recycling plant were investigated in granular form. Selected samples were used as aggregates and the leaching of the resulting concretes was tested. In addition a doped concrete was tested to simulate a higher degree of contamination. The results show, that inorganic compounds are well bound in the concrete. The German limit values for the leaching of unbound CDW guaranty a save use, however there is no direct correlation between the leaching of unbound CDW and concrete with CDW aggregate, because the leaching of concrete is dominated by the binder.

Keywords Reuse of demolition waste · Environmental compatibility · Leaching · Heavy metals and trace elements

1 Introduction

Construction and demolition waste (CDW) is the main waste stream generated in der EU. It adds up to 800 million tonnes per year and therefore accounts for about one third of all waste generated [1]. Producing 200 million tonnes of CDW per year, Germany is one of the main responsible countries.

The EU states CDW as a priority waste stream with a high potential for recycling and reuse, it sets the goal to a 70% recovery rate in 2020 by Article 11.2b of the Waste Framework Directive. With a rate of 88% in 2017 [1] this goal is easily reached in

A. Vollpracht · L. Weiler (✉)
Institute of Building Materials Research (ibac), RWTH Aachen University, Aachen, Germany
e-mail: weiler@ibac.rwth-aachen.de

A. Vollpracht
e-mail: vollpracht@ibac.rwth-aachen.de

V. M. C. F. Cunha et al. (eds.), *Proceedings of the 3rd RILEM Spring Convention and Conference (RSCC 2020)*, RILEM Bookseries 35,
https://doi.org/10.1007/978-3-030-76543-9_9

Germany, but most of the material is not reused in concrete or other construction materials but in road construction and earth works. This situation should be changed to reach real circular economy.

Some demolition wastes pose particular problems to the recycling process: During service life contaminations can be accumulated originating from paints and coatings, old tar sealing or residuals of industrial use. These contaminations can pose environmental risks and impede recyclability. Selective demolition is an effective method to preserve the main stream from contaminations, but firstly this is not always considered on site and secondly there will always remain a contaminated share. To improve resource efficiency and to foster a sustainable development, as much as possible of these materials should be kept from ending up in landfills. A high-quality option is the use as an aggregate for concretes.

Because of the potential contamination, the German standard DIN 4226-101 demands that recycled aggregates are tested prior to utilisation in concrete. The requirements include a short term batch test described in Sect. 2.2 and some limitations concerning the total content of specific organic constituents. Part 1 of this contribution deals with inorganic constituents (heavy metals and trace elements); organic contaminants are discussed in part 2 using the example of PAH.

Depending on the type of CDW and on the exposure conditions of the final concrete, only 35 or 45% of the aggregate can be replaced according to the current regulations in Germany [2]. Fine CDW particles (<2 mm) must not be used at all. In order to expand the currently quite restrictive regulations, a large joined research project has been carried out, called "***R***-concrete" (for resource-saving concrete) [3–5]. In this context also the limits for inorganic contaminations have been reconsidered and leaching tests following DIN CEN/TS 16637-2 have been conducted on concrete with recycled aggregate and doped concrete [4]. This contribution summarises the findings.

2 Leaching Tests

2.1 Preliminary Note

Leaching tests can be carried out on the original CDW sample prior to use or on the final product, namely the concrete with partial or complete replacement of natural aggregate by CDW. For quality control it is easier and faster to use the unbound CDW. However, it has to be considered that the results do not give any information on the environmental compatibility of the final product. Therefore two different tests were carried out in this research: a batch test (Sect. 2.2) and the dynamic surface leaching test (DSLT) according to DIN CEN/TS 16637-2 (Sect. 2.3).

2.2 Batch Test: Shake Bottle Test

The batch tests were carried out according to EN 12457-4. In deviation to the standard the original material with a maximum grain size of 16 mm was used without crushing of the fraction > 10 mm. 90 g of the sample were leached in a PE-bottle with 900 ml of deionised water (liquid/solid ratio, $L/S = 10$). The bottle was rotated with 5 revolutions/min for 24 h. Afterwards the particles were allowed to settle for 15 min. Then the eluate was decanted and filtrated (0.45 μm).

The following parameters were determined: pH value, electrical conductivity and the concentrations of chloride, sulfate, antimony, arsenic, barium, lead, boron, cadmium, chromium, cobalt, copper, molybdenum, nickel, mercury, selenium, thallium, vanadium and zinc. The anions were determined with ion chromatography and for the heavy metals and trace elements ICP-MS was used.

2.3 Dynamic Surface Leaching Test

The DSLT according to DIN CEN/TS 16637-2 was designed to test monolithic samples. In this project concretes with different CDW samples and different replacement levels were produces (see Sect. 3.2). The *R*-concrete samples (cylinders with $\varnothing = 150$ mm and $h = 100$ mm) were stored in airtight packaging until an age of 56 days. Each sample was put in a PE vessel with screw cap (see Fig. 1). To guarantee a sufficient distance between the sample and vessel, plastic bar spacers of 2 cm thickness were used. The water volume divided by the sample surface was set to 80 l/m^2; therefore 5 L of deionised water were used. At fixed intervals of 0.25, 1, 2.25, 4, 9, 16, 36, and 64 days, the eluate was sampled and changed.

The eluates were analysed for the same parameters as for the batch test (see Sect. 2.2). The test was conducted as two fold determination.

Based on the results, the cumulative release was calculated following Eq. (1).

$$R_n = \sum_{i=1}^{n} R_i = \sum_{i=1}^{n} c_i \frac{V}{A} \tag{1}$$

R_n cumulative release at the end of interval n in mg/m^2.

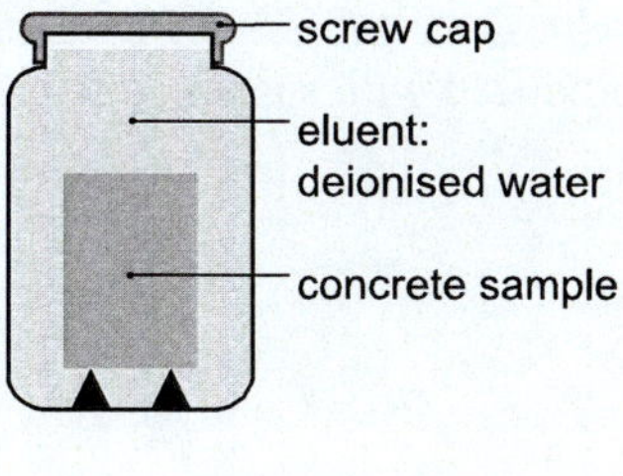

Fig. 1 Schematic illustration of the DSLT

i leaching interval (1–8).
R_i release during interval i in mg/m^2.
V volume of water in L.
A surface area of the sample in m^2; $V/A = 80$ l/m^2.

For the total release in the DSLT limit values were set by the "German Centre of Competence for Construction" (DIBt) for the following parameters: antimony, arsenic, barium, lead, cadmium, chromium, cobalt, copper, molybdenum, nickel, mercury, thallium, vanadium (currently suspended), zinc, chloride, fluoride and sulfate [6].

3 Materials

3.1 CDW Samples

The recycled aggregates were supplied by an industrial partner of the joint research project, Scherer and Kohl GmbH & Co KG. They regularly delivered samples out of their usual production. The project partner HeidelbergCement AG tested these 74 samples with respect to the total content and the concentrations in the batch test ($L/S = 10$) [5]. An overview of the results and the limit values for the batch test from DIN 4226-101 is given in Table 1. Most of the concentrations are well below the limits, only the sulfate concentration exceeds the limit in one case (printed in bold). High sulfate concentrations result from gypsum particles. These impurities cannot always be prevented during deconstruction. The high pH-values and electrical conductivities result from crushed concrete and are typical for uncarbonated demolition waste.

A representative share of five CDW samples was investigated in more detail including leaching tests on ***R***-concretes [4]. Two of these samples were sorted wastes: crushed concrete and crushed bricks. The other three are mixed CDW. The total content and the leaching in the batch test are given in Table 2.

The concentrations of the batch tests are also well below the limit values. Relatively high concentrations were determined for vanadium for the crushed bricks. However, there is no limit value for this element.

Although the joint project ran over a period of more than three years, no CDW sample was found that exceeded the limit values for heavy metals and trace elements. Since all the eluate concentrations of the batch tests are well below, the limit values already represent a slightly contaminated waste. To our knowledge these limit values were never verified in terms of the environmental compatibility of a concrete produced with such a contaminated CDW.

Table 1 Overview of the batch test results (based on EN 12457-4) for inorganic constituents, [5] and comparison to the limit values of DIN 4226-101

Sort of CDW	Parameter/component	Unit	Limit value	0/2 mm	0/4 mm	2/8 mm	8/16 mm
Demolished concrete	pH	–	12.5[1]	12.2–12.7	11.6–12.7	12.1–12.7	12.0–12.7
	Electr. conductivity	µS/cm	3000[1]	2800–6265	1383–6165	3100–6410	2370–6090
	Chloride	mg/L	150	5.9–14	3.1–24	2.7–10.2	3.0–9.6
	Sulfate		600	1.2–7.9	1.1–390	1.6–26	1.7–35
	Arsenic	µg/L	50	<9	<7	<9	<9
	Lead		100	<10	<6	<6	<6
	Cadmium		5	<0.5	<0.5	<0.4–1,7	<0.4–0.66
	Chromium		100	<0.4–10	1.5–35	<0.4–11.8	<0.4–13.4
	Copper		200	<0.6–7.5	4.0–25	<0.6–3.0	<0.6 –4.15
	Nickel		100	<3	<1–4.1	<3	<3
	Mercury		2	<0.2	<0.2	<0.2	<0.2
	Zinc		400	<0.8	<3	<3	0.52–1.1
Mixed CDW	pH	–	12.5[1]	12.2–12.4	11.2–12.5	12.2–12.5	12.3–12.6
	Electr. conductivity	µS/cm	3000[1]	2055–6200	932 –4475	2470–4665	2850–5410
	Chloride	mg/L	150	4.6–13	9.0–19	4.8–7.4	4.0–10
	Sulfate		600	0.78–25	1.9–**621**	1.3–29	1.3–19
	Arsenic	µg/L	50	<8	<6	<12	<11
	Lead		100	<6	<5	0.52–10	<6
	Cadmium		5	< 0.5	< 0.4	< 0.5	< 0.5
	Chromium		100	<0.4–10	4.4–46	<0.4–11.2	<0.4–10
	Copper		200	<0.2–17.5	3.5–38	<0.8–3.3	<0.2–2.6

(continued)

Table 1 (continued)

Sort of CDW	Parameter/component	Unit	Limit value	0/2 mm	0/4 mm	2/8 mm	8/16 mm
	Nickel		100	<3	<2–3.7	<3	<3
	Mercury		2	<0.2–0.33	<0.2–0.24	<0.2	<0.2
	Zinc		400	<0.8	<2–50	<0.8–5.5	<0.8–8.3

[1] Orientation value, exceeding values do not lead to exclusion

Table 2 Characterisation of the CDW samples tested with the DSLT

Investigation	Parameter/component	Unit	Demolished concrete (BBt)	Crushed bricks (ZBt)	Mixed CDW (RG2)	Mixed CDW (RG3)	Mixed CDW (RG4)
			0/16 mm	0/16 mm	2/16 mm	4/32 mm	0/16 mm
Total content (aqua regia digestion, EN 16174)	Antimony	mg/kg	0.975	0.796	0.690	0.704	0.481
	Arsenic		26.3	16.8	16.8	12.0	10.3
	Barium		107	231	83.3	117	106
	Lead		31	43.3	8.98	14.2	8.87
	Boron		22.9	38.2	12.0	12.2	16.4
	Cadmium		0.355	0.386	0.152	0.205	0.159
	Chromium		31.6	62.6	20.6	53.1	21.9
	Cobalt		17.3	20.1	32.4	37.3	39.6
	Copper		18.8	17.5	7.78	8.42	7.63
	Molybdenum		1.37	0.594	0.598	0.633	0.492
	Nickel		14.4	29.8	14.6	18.6	14.6
	Mercury		0.082	0.0896	0.270	0.309	0.346
	Selenium		2.1	1.54	0.805	0.995	0.885
	Thallium		0.186	0.0905	0.141	0.257	0.112
	Vanadium		42.7	53.9	19.6	19.8	19.7
	Zinc		76.3	291	60.3	46.8	49.4
Batch test (EN 12457-4)	pH	–	12.4	8.84	11.9	11.6	11.0
	Electr. conductivity	µS/cm	4635	1181	675	655	314
	Chloride	mg/L	2.40	20.8	2.50	3.15	2.70
	Sulfate		1.65	536	32.9	28.3	27.8

(continued)

Table 2 (continued)

Investigation	Parameter/component	Unit	Demolished concrete (BBt)	Crushed bricks (ZBt)	Mixed CDW (RG2)	Mixed CDW (RG3)	Mixed CDW (RG4)
			0/16 mm	0/16 mm	2/16 mm	4/32 mm	0/16 mm
	Antimony	µg/L	0.116	0.527	2.07	1.65	1.29
	Arsenic		<1	7.70	3.14	1.26	2.28
	Barium		199	77.6	15.4	34.5	6.58
	Lead		1.55	4.14	0.248	0.428	0.223
	Boron		8.26	456	52.7	28.8	56.5
	Cadmium		<0.1	<0.1	<0.1	<0.1	<0.1
	Chromium		14.3	54.1	4.14	4.75	4.33
	Cobalt		0.379	1.78	0.235	0.304	0.111
	Copper		2.1	3.37	2.09	2.60	1.46
	Molybdenum		2.62	6.99	0.661	4.24	0.774
	Nickel		0.551	4.48	<0.1	<0.1	<0.1
	Mercury		0.0670	0.0191	0.0142	0.0137	<0.01
	Selenium		<1	2.44	<1	<1	<1
	Thallium		<0.1	<0.1	<0.1	<0.1	<0.1
	Vanadium		0.832	151	6.92	7.18	17.2
	Zinc		3.50	24.7	2.8	2.2	2.2

3.2 R-concrete

The five CDW described in Table 2 were used to produce concrete with recycled aggregate using CEM I 32.5 (Z1). The contents of heavy metals and trace elements of the cement are given in Table 3. The cement content was 280 kg/m^3, the water/cement ratio was 0.6. The grading curve of the aggregates should stay in the range A16/B16 according to DIN 1045-2. With this limitation, as much recycled aggregate as possible was used:

BBt: 85 vol.%
ZBt: 85 vol.%
RG2: 95 vol.%
RG3: 43 vol.%
RG4: 82 vol.%

As a reference, concrete with natural aggregate was produced. This reference was tested twice.

Table 3 Content of trace elements and heavy metals of the cements (CEM I 32.5 R); determined after aqua regia digestion according to EN 16174

Parameter/component	Z1 mg/kg	Z2[a] mg/kg
Antimony	5.21	7.76
Arsenic	3.85	157
Barium	233	290
Lead	4.63	204
Boron	64.2	80.7
Cadmium	0.137	6.86
Chromium	50.2	60.8
Cobalt	6.14	11.7
Copper	88.2	218
Molybdenum	0.953	68.9
Nickel	23.1	51
Mercury	0.00408	0.0657
Selenium	1.24	0.596
Thallium	0.0121	4.67
Vanadium	62.5	103
Zinc	267	1010

[a]Used for the doped concrete

3.3 Doped Concrete

In addition to the ***R***-concretes it was decided to test a doped concrete to find out, if the limit values given in DIN 4226–101 are appropriate and guaranty a concrete with sufficiently low leaching characteristics. For these investigations another cement was used (Z2). The trace element analysis is included in Table 3.

For the calculation of the doping, a concrete with 100% recycled aggregate and an aggregate content of 1900 kg/m^3 was assumed to test the worst case. Table 4 shows the compounds used for doping and elucidates the calculation of the amount, which has to be added to the concrete. Natural aggregate was used for the doped concrete.

4 Results of the DSLT

4.1 Leaching of the R-concrete

Figure 2 shows exemplarily the release of three heavy metals of the ***R***-concretes in comparison to the reference concrete with natural sand and gravel. In most cases the release is in the same range as for the reference concrete. The ***R***-concrete with crushed brick (ZBt) sometimes has a lower, sometimes a bit higher release, but is in all cases way below the limit values. Even vanadium, which was quite mobile in the granular ZBt (see Table 2), is well bound in the cement matrix.

Consequently no correlation between the results of the batch test and the release on the resulting ***R***-concrete can be found (see Fig. 3). Therefore the results of the batch test give no indication on the leaching behavior of the final concrete. The leaching of ***R***-concrete is dominated by the binder.

4.2 Leaching of the Doped Concrete

Figure 4 shows the leaching of chloride and sulfate from the doped concrete compared to the reference natural sand and gravel. Due to the very high addition of these anions, the leaching increases, but since the toxicities of chloride and sulfate are a lot lower than for the heavy metals, the release is still less than 1/10 of the permissible release.

As examples for the heavy metals that were added to the concrete, Fig. 5 shows the releases of chromium and zinc. No relevant increase can be observed.

Some remarkable changes were observed for elements that were not added to the concrete. Figure 6 shows the release of molybdenum and thallium. It is obvious that both elements are leached to higher extend, although the total content is the same. This means in the reference concrete, these elements are bound to a higher extend than in the doped concrete. Probably molybdate is replaced by the added sulfate or chromate ions, which are slightly smaller, as can be seen in Fig. 7. This leads to a

Table 4 Calculation of the doping

Parameter/component	Limit value of DIN 4226-101 in mg/L	Equivalent content of the aggregate in mg/kg	Dosage to be added to the concrete in g/m^3	Compound used for doping	Proportion of the contaminant in the compound in wt%	Amount of doping compound added to concrete in g/m^3
Chloride	150	1500	2850	$\mathbf{NaCl}$	60.66	4690[2]
Sulfate	600	6000	11,400	$Na_2\mathbf{SO_4}$	67.63	16,840[2]
Arsenic	0.050	0.5	0.95	$\mathbf{As}_2O_3$	75.74	1.25
Lead	0.100	1	1.9	$N_2O_6\mathbf{Pb}$	62.56	3.04
Cadmium	0.005	0.05	0.095	$C_4H_6\mathbf{Cd}O_4 \cdot 2H_2O$	42.17	0.23
Chromium	0.100	1	1.9	$K_2\mathbf{Cr}O_4$	26.78	7.10
Copper	0.200	2	3.8	$\mathbf{Cu}Cl_2$	47.26	8.04
Nickel[a]	0.100	1	–	–	–	–
Mercury[b]	0.002	0.02	–	–	–	–
Zinc	0.400	4	7.6	$\mathbf{Zn}SO_4 \cdot 7H_2O$	22.73	33.43

[a]Not considered in the doping experiment

[b]Considering the chloride and sulfate contents of the other doping compounds

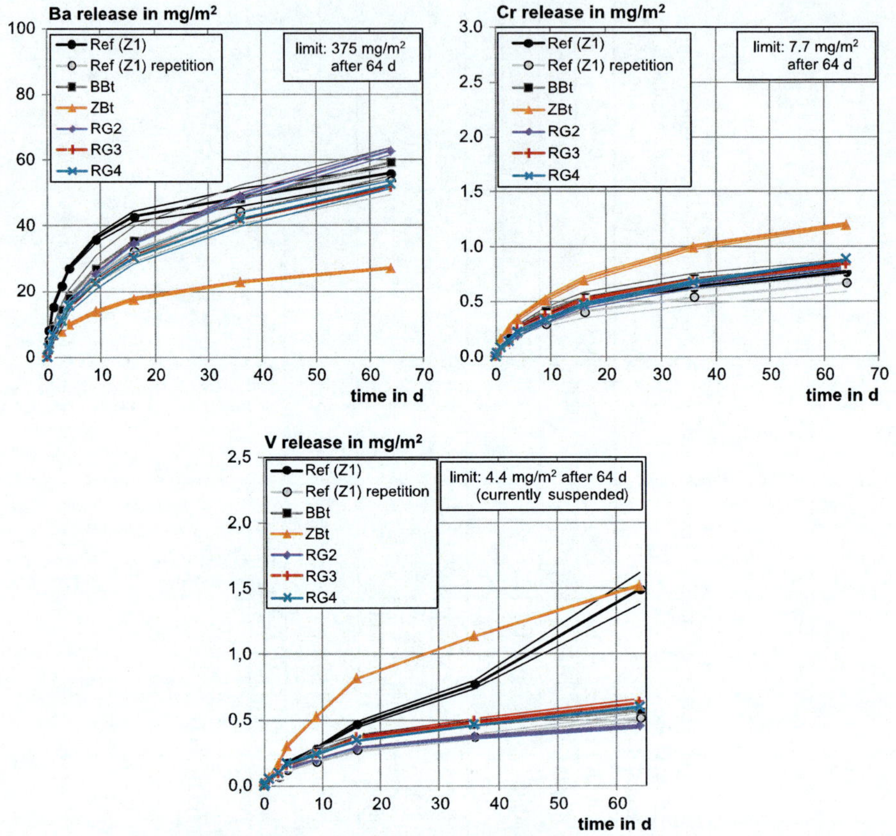

Fig. 2 Release of barium, chromium and vanadium from the ***R***-concretes and the reference concrete in the DSLT; limit values from [6]

Fig. 3 Correlation of batch test results and the leaching of the resulting ***R***-concrete

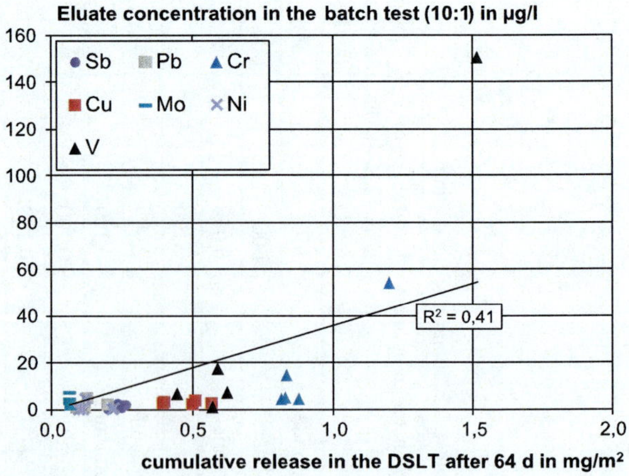

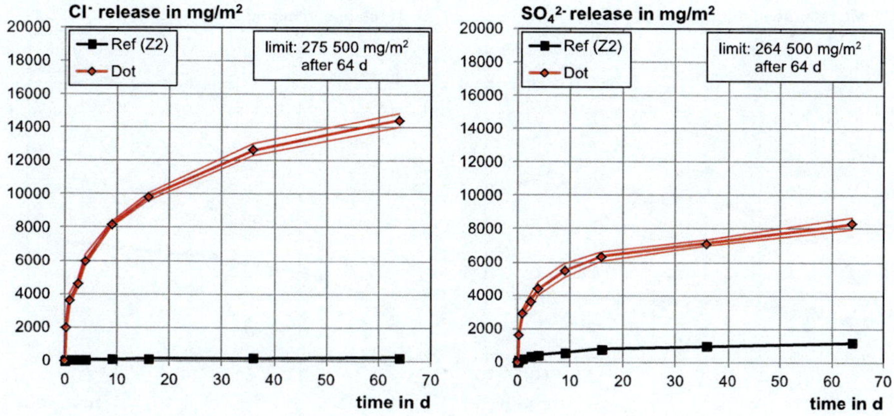

Fig. 4 Release of chloride and sulfate from the doped concrete and the reference in the DSLT; limit values from [6]

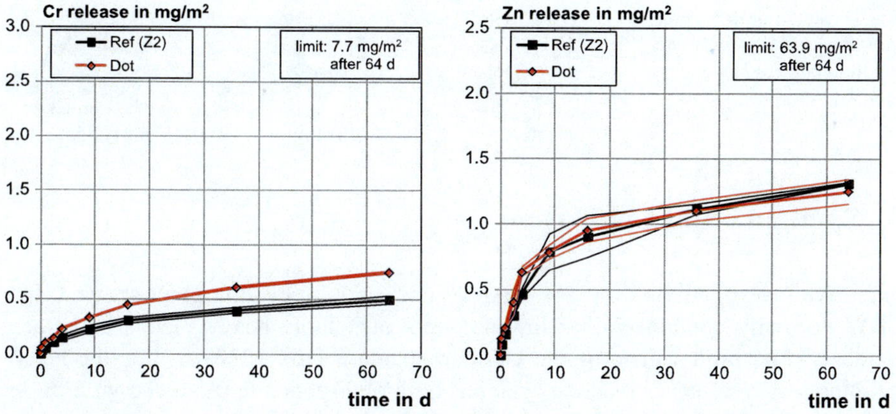

Fig. 5 Release of chromium and zinc from the doped concrete and the reference in the DSLT; limit values from [6]

higher leachability of the molybdate. The average molybdenum content of German cements is 2.7 mg/kg (personal communication by Dr. Spanka, VDZ). The cement used here has an exceptionally high Mo content (refer to Table 2). However, the release is still lower than the limit value.

Thallium could also be replaced by a doped element, e.g. arsenic. For thallium the leaching of the doped concrete exceeds the limit value from [6]. Z2 contains 4.67 mg/kg thallium (see Table 4). The average value of German cements is 0.4 mg/kg [8]. Thallium is very volatile in the cement kiln and therefore concentrates in the kiln dust [9]. Higher concentrations are only possible, when kiln dust is intermixed with the cement clinker—this seems to be the case here.

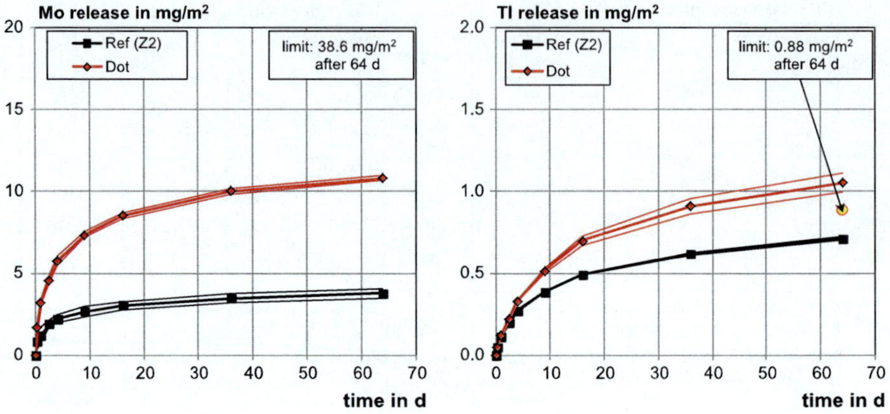

Fig. 6 Release of molybdenum and thallium from the doped concrete and the reference in the DSLT; limit values from [6]

Fig. 7 Sketch of the oxianions sulfate, chromate and molybdate, [7]

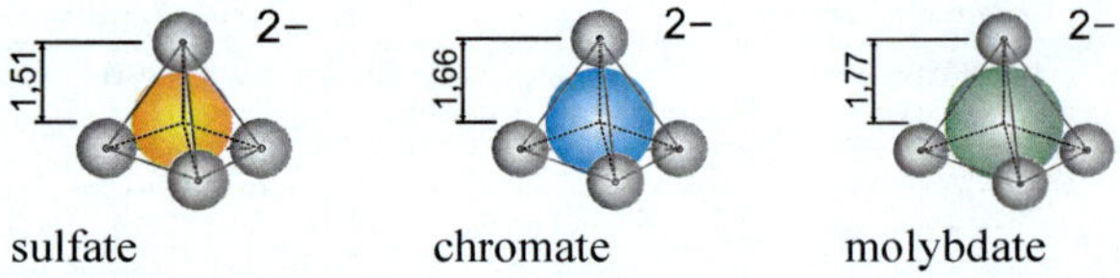

5 Summary and Conclusion

Extensive investigations on CDW from a commercial recycling plant showed that CDW normally contains rather low amounts of soluble heavy metals and trace elements. The limit values of the German standard DIN 4226-101 for the batch leaching test were never reached. The leaching of ***R***-concrete produced with these CDW was well below the German limit values for concrete used in groundwater [6]—which are the strictest of the world.

In order to check if the limit values given in DIN 4226-101 are appropriate and guaranty a concrete with sufficiently low leaching, a doped concrete was investigated. The amount of the doped constituents was calculated based on the limit values and 100% replacement of the aggregate. None of the added heavy metals or trace elements was leached to a significantly higher extend compared to the reference concrete with the same cement. Only chloride and sulfate leaching increased, but the release is still very low compared to the limit values.

The input of contaminants can lead to an increase of other elements, originally contained in the binder. For standardised cements this is usually no problem, but binders with high heavy metal contents should be checked in this respect.

References

1. German Federal Statistical Office (Destatis): Waste Report (2019)
2. Deutscher Ausschuss für Stahlbeton (German committee for reinforced concrete, DAfStb): DAfStb-Richtlinie Beton nach DIN EN 206-1 und DIN 1045-2 mit rezyklierten Gesteinskörnungen nach DIN EN 12620 (DAfStb Guideline for concrete according to DIN EN 206-1 and DIN 1045-2 with recycled aggregated according to DIN EN 12620), Beuth GmbH Berlin publishing company, Sept 2010
3. Knappe, F., Reinhardt, J., Theis, S., Kresser, S., Scheidt, J., Breit, W., Sachsenhauser, B., Müller, C., Severins, K., Vollpracht, A., Haufe, J.: Schlussberichte zum BMBF-Verbundforschungsvorhaben „***R***-Beton - Ressourcenschonender Beton – Werkstoff der nächsten Generation - Teil 1: Konzeptionierung der neuen Werkstoffe (Final reports to the BMBF joint research project „***R***-concrete – ressource-saving concrete – builing materials of the next generation" – Part 1: Conception of the new building materials). Berlin, Beuth GmbH Berlin publishing company, bulletin series of the DAfStb – to be published in 2020
4. Pancic, A., Schnell, J., Müller, C., Borchers, I., Seidel, M., Vollpracht, A., Weiler, L.: Schlussberichte zum BMBF-Verbundforschungsvorhaben „***R***-Beton - Ressourcenschonender Beton – Werkstoff der nächsten Generation - Teil 2: Praxisanforderungen an die neuen Werkstoffe (Final reports to the BMBF joint research project „***R***-concrete – ressource-saving concrete – builing materials of the next generation" – Part 2: Practical requirements on the new building materials). Berlin, Beuth GmbH Berlin publishing company, bulletin series of the DAfStb—to be published in 2020
5. Knappe, F., Reinhardt, J., Theis, S., Müller, C., Reiners, J., Böing, R.: Schlussberichte zum BMBF-Verbundforschungsvorhaben „***R***-Beton - Ressourcenschonender Beton – Werkstoff der nächsten Generation - Teil 3: Ökobilanz, Praxistest und Transfer (Final reports to the BMBF joint research project „***R***-concrete – ressource-saving concrete – builing materials of the next generation" – Part 3: life cycle assessment, field test and transfer). Berlin, Beuth GmbH Berlin publishing company, bulletin series of the DAfStb—to be published in 2020
6. Deutsches Institut für Bautechnik, DIBt (German Centre of Competence for Construction): Muster-Verwaltungsvorschrift Technische Baubestimmungen – Anhang 10: Anforderungen an bauliche Anlagen bezüglich der Auswirkungen auf Boden und Gewässer (ABuG) (model administrative regulation on technical construction regulations – annex 10: requirements on structural facilities regarding the impacts on soil and groundwater (ABuG)). Published by the DIBt (www.dibt.de), Jan 2017
7. Vollpracht, A., Brameshuber, W.: Binding and leaching of trace elements in Portland cement pastes. Cem. Concr. Res. **79**(1), 76–92 (2016). https://doi.org/10.1016/j.cemconres.2015.08.002
8. Verein Deutscher Zementwerke: VDZ (association of the German cement industry): Spurenelemente in deutschen Normcementen (Heavy metals in Geman standardized cements). Bau + Technik GmbH publishing company, VDZ, Düsseldorf (2001)
9. Bhatty, J.I.: Role of Minor Elements in Cement Manufacture and Use. Portland Cement Association, Research and Development Bulletin RD109T, Skokie, USA (1995)

Processed Municipal Solid Waste Incineration Ashes as Sustainable Binder for Concrete Products

Aneeta Mary Joseph, Natalia Alderete, Stijn Matthys, and Nele De Belie

Abstract After proper pre-treatment procedures, Municipal Solid Waste Incineration (MSWI) bottom ashes show all needed characteristics to substitute part of the Portland clinker in concrete. The use of these ashes greatly contributes to the establishment of a circular economy. As a by-product of waste-to-energy plants, they are abundantly and locally available. Ashes currently available after proprietary pre-treatment procedures contain elemental aluminium that deters their usability as supplementary cementitious material. An efficient pre-treatment option that is low-cost and avoids secondary pollution was recently proposed to remove elemental aluminium from ashes. This method consists of slow grinding and sieving, resulting in aluminium and other metals remaining in the coarse fraction facilitating their removal. In this research, the use of pre-treated Processed Incineration Ashes (PIA) in concrete was investigated. A concrete mix with replacement of 20% of the Portland cement CEM I 52.5 R by PIA was optimised to show similar compressive strength at 28 days to the CEM I 52.5 N reference, and even at 90 days compressive strength was 20% higher than for the CEM I reference concrete. Furthermore, equivalent performance with the benchmark concrete was found for the tested durability properties (open porosity, carbonation). Also leaching tests were performed to assess the potential use of PIA in the construction industry. Results provide evidence of the good overall performance of the PIA concrete.

Keywords Incineration ashes · Sustainability · Circular economy · Durability

A. M. Joseph · N. Alderete · S. Matthys · N. De Belie (✉)
Ghent University, Technologiepark Zwijnaarde 60, Campus Ardoyen, B-9052 Gent, Belgium
e-mail: nele.debelie@ugent.be

N. De Belie
Strategic Initiative Materials (SIM vzw), Project ASHCEM Within the Program "MARES", Technologiepark Zwijnaarde 48, B-9052 Ghent, Belgium

V. M. C. F. Cunha et al. (eds.), *Proceedings of the 3rd RILEM Spring Convention and Conference (RSCC 2020)*, RILEM Bookseries 35,
https://doi.org/10.1007/978-3-030-76543-9_10

1 Introduction

In our current society there is a daily worldwide waste production of around 1.2 kg per person and a total generation of around 1.3 billion tonnes per year [1]. Even though an integrated waste management approach including reuse, recycling, incineration, etc., is adopted in developed countries, a significant amount of reusable materials still gets landfilled there. The incineration of non-hazardous non-recyclable waste (namely municipal solid waste) can be a viable solution as alternative to landfilling. In industrialized countries, the percentage incinerated can reach up to 62% [2, 3]. In fact, the European waste legislation focusses on circular economy as one of the main policies and action plans aim to reduce landfilling to a minimum [4]. The residues from incineration of municipal solid waste, such as bottom ashes, have potential as binder for concrete in terms of quantity and demand if a few hurdles can be solved. The use of the generated bottom ash in concrete as partial cement replacement has the double advantage of reducing landfilling and at the same time reducing the Portland clinker content in concrete (largely responsible for the concrete carbon footprint).

The ASH-CEM project targeted the development of processed incineration ash for the production of novel clinkers, blended cements and carbonate binders, as well as their application in construction elements. This represents a great advancement in the use of ashes in the construction industry, as there is currently limited knowledge on the ultimate concrete products due to limited production volumes on lab scale. For this research, several treatments were evaluated to extract non-ferrous metals and to reduce the Al/Zn content below 1%. Details about the laboratory treatment can be found in [5]. Based on those results, 1.2 ton of municipal solid waste incineration (MSWI) bottom ashes (2/6 fraction) were treated to remove aluminium, ground and sieved (dv50 = 27 μm). Furthermore, previous research [6, 7] on reactivity screening by the conventional Chapelle test, as well as by the R^3 method, indicate that the processed incineration ashes (PIA) have a pozzolanic reactivity similar to the one of coal combustion fly ash and natural pozzolans. These promising results encourage the use of PIA as supplementary cementitious material (SCM) in concrete.

This paper presents the results of experiments regarding the use of PIA in concrete as partial replacement of cement. A concrete mix with replacement of 20% of the Portland cement CEM I 52.5 R by PIA was optimised to show similar compressive strength at 28 days to the CEM I 52.5 N reference. Furthermore, durability properties such as open porosity (OP), carbonation and air permeability and leaching tests were assessed to evaluate the potential use of PIA in the construction industry.

2 Materials and Methods

Two types of 'reference' concrete were studied with CEM I 52.5 N and CEM II B-V 32.5 R (denoted as Mix 1 and Mix 2, respectively). Mix 2 was included because blended cement in the current study was manufactured by blending PIA with CEM

Table 1 Composition of mixes 1, 2, and 3

Material	Mix 1	Mix 2	Mix 3	Mix 4
Sand 0/1 (kg/m^3)	271	271	271	271
Sand 0/4 (kg/m^3)	541	541	541	542
Gravel 4/16 (kg/m^3)	997	997	997	999
CEM I 52.5 N (kg/m^3)	360	–	270	–
CEM II B-V 32.5 R (kg/m^3)	–	360	–	–
CEM I 52.5 R (kg/m^3)				296
PIA 2/6 (kg/m^3)	–	–	90	74
Water (kg/m^3)	173	174	172	166.5
Plasticiser (% kg of binder)	1.8	1.8	2.7	3.9

I 52.5 N; and CEM II B-V 32.5 R is a commercially manufactured blended cement with comparable amount of SCMs ($\approx$32% fly ash) and used in regular constructions. Naturally, the highest amount of cement replacement possible without reducing the performance of the concrete was desired. For this purpose, 25% (in weight) of the CEM I 52.5 N was replaced by PIA, and this mix was denoted as Mix 3. Another objective was to manufacture an 'optimized' mix which has the same compressive strength as that of Mix 1 at 28 days. This new mix was designed by changing the cement type to CEM I 52.5 R to provide high enough early strength even though SCMs are included, varying the water to binder ratio (w/b) and the amount of cement replacement by PIA. After testing 8 different mix designs, the final composition of Mix 4 was chosen.

The composition of Mixes 1–4 is shown in Table 1. For all concrete mixes the same gravel, sand, water and a plasticizer based on modified natural substances were used. The mixing procedure was as follows: first cement (+PIA in case of Mix 3 and Mix 4), fine and coarse aggregates were dry-mixed for a minute; then the water was added, and the mixing continued for 2 more minutes, finally the plasticiser was added and mixed for 1 extra minute. Samples were cured in a conditioned room at 20 ± 2 °C and 95 ± 5% relative humidity (RH).

A higher amount of plasticiser was used for Mix 3 (with PIA) than in Mix 1 or Mix 2 (Table 1). This was needed because a similar workability was desired, and therefore the same slump was targeted. Due to presence of PIA, workability was slightly reduced, and it was necessary to increase the plasticiser content.

2.1 *Mechanical Properties*

Compressive strength. Compressive strength tests were carried out on cubes of 150 mm according to the standard NBN B 15-220 (1990). Three samples per mix were tested at 2, 7, 28, and 91 days of age.

Flexural and tensile splitting strength. The flexural and splitting tensile strength tests were carried out according to the standards NBN EN 12390-5 (2009) and NBN EN 12390-6 (2009), respectively. First the prisms of 150 mm side and 600 mm length were subjected to three point bending to determine the flexural strength. Then, with the remaining halves of the samples the splitting tensile test was performed. Three samples per mix were tested at 28 and 91 days of age.

2.2 *Durability Properties*

Open porosity. Cylinders with 100 mm in diameter and 200 mm in height were demoulded after 1 day of casting and cured in a climate room at a temperature of 20 ± 2 °C and a relative humidity (RH) higher than 95% until 28 and 90 days of age.

Five samples from the lowest 54 mm were cut and used for testing. Those slices were subjected to vacuum for 2 h and then water was drawn into the vacuum chamber until the samples were fully immersed. After 24 h samples were removed and weighed, this weight was denoted as saturated mass in air (m_{sa}). The samples were also weighed in water, and this weight was denoted as saturated mass in water (m_{sw}). Then, samples were subjected to drying in an oven at 50 °C until the change in mass was lower than 0.1% in a 24 h period and denoted as dry mass (m_d). The open porosity (OP) was calculated as = (m_{sa} − m_d)/(m_{sa} − m_{sw}).

Carbonation. Samples used to assess the carbonation depth were cylindrical slices drilled and sawn from cubes of 100 mm side. Two layers of epoxy paint were applied on the whole sample except the freshly cut surface, to ensure one-dimensional penetration of CO_2. After painting, samples were subjected to the same preconditioning procedure as the specimens for CIR. Three samples per mix were exposed to accelerated carbonation (1 vol.% CO_2 at 20 °C and 60% RH) and another three samples were exposed to natural carbonation (≈0.04 vol.% CO_2 at 20 °C and 60% RH). After different exposure times, samples were split in half and the carbonation depth was determined by spraying phenolphthalein solution on the freshly cut surface. The carbonation depth was measured at eight different places (10 mm distance between the measurements) to the nearest millimetre.

Leaching. The environmental safety of use of the mixes for leaching of heavy metals was also tested. Around 2 kg of each of the mixes 1, 2, 3 and 4 was ground to a size < 4 mm. Leaching of the samples was conducted by two step shake test (CMA 2/II/A 9.4) on around 200 g of sample. The heavy metals in the eluates were determined using inductively coupled plasma atomic emission spectroscopy (CMA/2/I/B.1).

3 Results and Discussions

Compressive strength is shown in Fig. 1. Early age strength is reduced when replacing the cement by PIA. This is observed by comparing Mix 1 and 3: compressive strength for Mix 3 is reduced with 43–27% in comparison to Mix 1 as the age increases from 2 to 28 days. Similarly, Li et al. [8] reported that compressive strength of mortar samples tested at 3 days and 28 days decreases gradually with increasing percentage of MSWI bottom ash. Furthermore, results of Mix 3 are comparable to the ones obtained from Mix 2. It can be seen that the compressive strength increase in Mix 3 at 91 days is relatively higher than the increase in Mix 1. This shows the delayed pozzolanic reaction of PIA. That is the reason why CEM I 52.5 R was chosen as binder together with PIA in the optimized Mix4, to obtain higher strength results at early ages. At all ages, Mix 4 has a comparable or higher compressive strength than Mix 1. This chosen 'optimized' mix was designed by replacing 20% of cement by PIA and with a *w*/*b* of 0.45. The small reduction in the PIA replacement level and in the *w*/*b* was successful to achieve comparable or even better results than the reference concrete Mix 1, with CEM I 52.5 N as only binder.

Results from flexural and tensile strength tests at 28 and 91 days are shown in Fig. 2a, b, respectively. In accordance with the compressive strength results, flexural strength values obtained with Mix 4 show at both ages a better performance than Mix 1. Regarding the tensile strength experiments, the optimized Mix 4 has better results than Mix 1, except for the outlier result of Mix 4 at 91 days.

Results of OP at 28 and 90 days are displayed in Fig. 3a, b respectively. At 28 days, results show that Mix 2 and Mix 3 have similar OP, whereas Mix 4 has similar OP as Mix 1. As Mix 2 and 3 have certain amount of clinker replacement by SCMs, they show a relatively higher value of OP. At 90 days, only a slight reduction in OP was found.

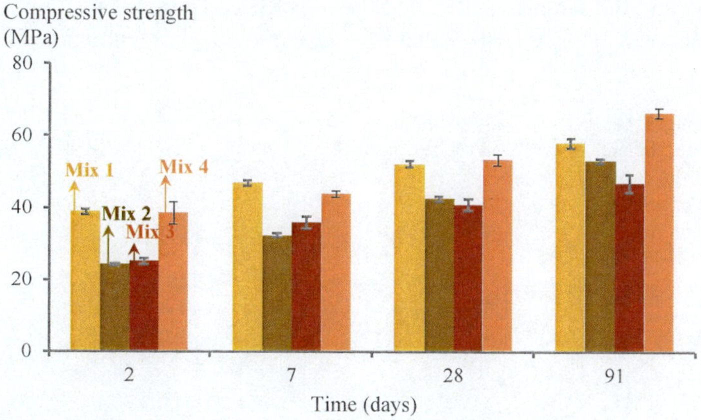

Fig. 1 Compressive strength results of mixes 1–4 at 2, 7, 28 and 91 days

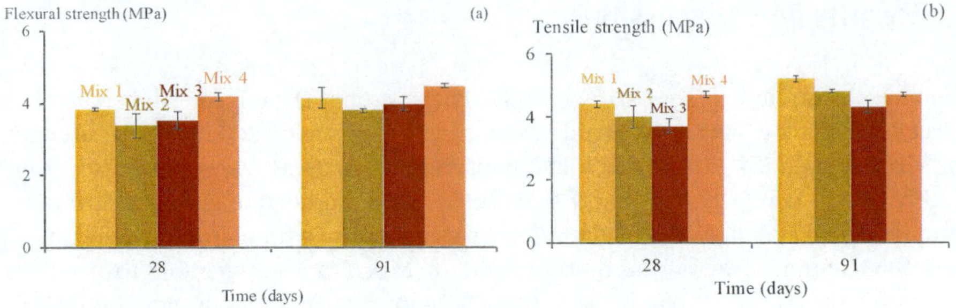

Fig. 2 Flexural (**a**) and tensile (**b**) strength of mixes 1–4 at 28 and 91 days

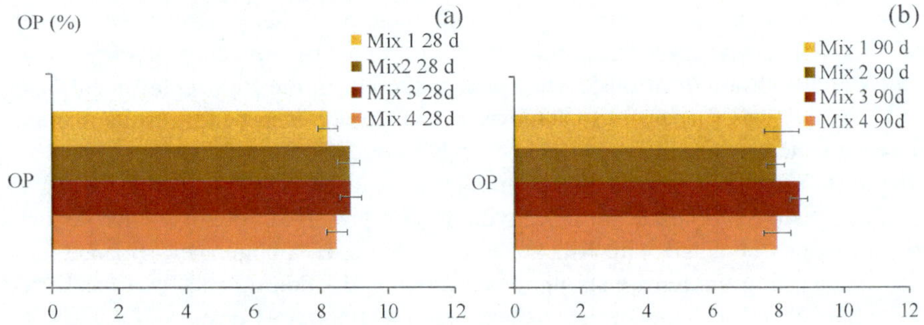

Fig. 3 Open porosity of mixes 1–4 at 28 (**a**) and 90 (**b**) days

The carbonation coefficient *A* (mm/$\sqrt{}$day) was calculated considering the penetration depth × (mm) as a function of the square root of the exposure time *t* ($\sqrt{}$days). Figure 4 shows the values obtained for both accelerated and natural conditions of carbonation and for samples at 28 and 90 days of age. These results indicate again a similar behavior between Mix 2 and Mix 3, as previously mentioned. The lowest

Fig. 4 Change of the carbonation coefficient, *A*, as a function of the curing period when exposed to accelerated and natural conditions

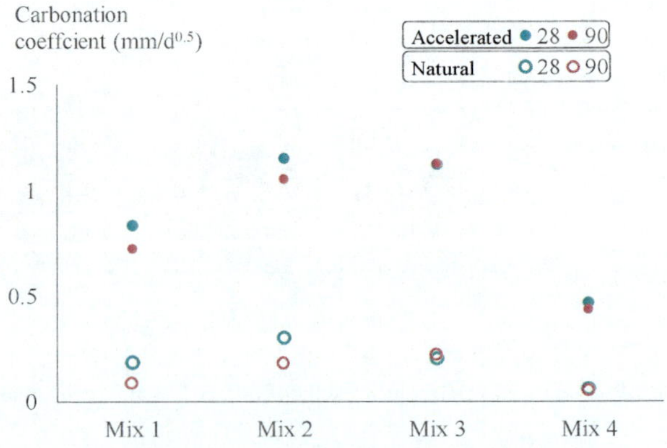

Table 2 Heavy metal content in eluates from two step shake test

Parameters	Mix 1	Mix 2	Mix 3	Mix 4	VLAREMA limits
	mg/kg dry mass				
Arsenic	0.00	0.00	0.00	0.00	0.80
Cadmium	0.00	0.00	0.00	0.00	0.03
Chromium	0.00	0.00	0.03	0.03	0.50
Copper	0.00	0.00	0.04	0.04	0.50
Mercury	0.00	0.00	0.00	0.00	0.02
Lead	0.00	0.00	0.22	0.03	1.30
Nickel	0.00	0.00	0.00	0.00	0.75
Zinc	0.00	0.03	0.00	0.00	2.80

carbonation coefficient was obtained with Mix 4 in natural and accelerated conditions, which confirms again the better performance of this mix with PIA. Overall, curing age larger than 28 days does not seem to considerably affect the resistance to carbonation in mixes with PIA (Mix 3 and 4).

The results from the leaching test are shown in Table 2. The leaching of heavy metals is very low in all the mixes and within the VLAREMA limits. This indicates that the tested PIA is safe for use.

4 Conclusions

This research describes the properties of different concrete mixes designed with processed incinerations ashes (PIA). Two 'reference' mixes (Mixes 1 and 2) and two PIA mixes (Mixes 3 and 4) were evaluated. Results obtained show that it is possible to replace 20% of the cement content by PIA and obtain comparable results with a reduction in w/b of 0.05. Moreover, results from the optimised Mix 4 show a higher value of compressive strength especially at later ages. Naturally, the use of CEM I R instead of CEM I N increases the strength at early ages. However, this occurred when 20% of the binder content was replaced by PIA, showing that excellent results can be obtained with a blended cement with PIA. Even more, results at later ages were similar or higher to the ones of Mix 1.

Moreover, results of the durability tests show no significant differences between Mix 2 and Mix 3. The 25% replacement of PIA (Mix 3) leads to a comparable durable behaviour as when using CEM II 32.5 B/V R cement (Mix 2). The optimised Mix 4 (with 20% replacement of cement by PIA and $w/b = 0.45$) showed an enhanced performance and comparable or better results in relation to Mix 1. This was particularly seen for the case of the advance of the carbonation front. Results from Mix 4 indicate a lower porosity and a better performance in comparison to the other mixes. This provides evidence of the good overall performance of the PIA concrete.

In addition to mechanical and durability properties, leaching results conformed to the VLAREMA limits making it environmentally safe for commercial applications.

The experiments carried out during this research exhibit the potential use of PIA in concrete as supplementary cementitious material and act as a basis for further research to optimize and standardise its use for commercial applications.

Acknowledgements The authors are grateful to VV/M cements for providing cements required for research, and Mr. Andres Van Brecht from Indaver for helping in conducting the leaching test. This research is a part of the ASHCEM project, which in itself is a part of bigger program MaRes aimed at creating and demonstrating an operational, flexible toolbox to recover metals and valorize residual matrix into building materials, financed by the Strategic Initiative Materials (SIM) in Flanders and VLAIO (Flanders Innovation and Entrepreneurship). The financial support from the foundations for this study is gratefully appreciated.

References

1. Hoornweg, D., Bhada-Tata, P.: What a waste: a global review of solid waste management. Urban Dev. Ser. Knowl. Pap. 15:116 (2012)
2. Cheng, H., Hu, Y.: Municipal solid waste (MSW) as a renewable source of energy: Current and future practices in China. Bioresour. Technol. **101**, 3816–3824 (2010). https://doi.org/10.1016/J.BIORTECH.2010.01.040
3. H. Belevi, H. Moench, Factors determining the element behavior in municipal solid waste incinerators. 1. Field studies. Environ. Sci. Technol. **34**, 2501–2506 (2000). https://doi.org/10.1021/es991078m
4. EU, European Commission. EUWaste Legislation. Available Online: http://Ec.Europa.Eu/Environment/Waste/Legislation/Index.Htm (2016). Accessed on 1 Dec 2018
5. Mary Joseph, A., Snellings, R., Nielsen, P., Matthys, S., De Belie, N.: Mitigation of Al expansion of MSWI bottom ash enabling use in cement and concrete. Unpubl. Prep (2018)
6. Mary Joseph, A., Snellings, R., Nielsen, P., Matthys, S., De Belie, N.: Pre-treatment and utilization of MSWI bottom ashes towards a circular economy article. Constr. Build. Mater. (2020, submitted). https://doi.org/10.1017/CBO9781107415324.004
7. Mary Joseph, A., De Belie, N., Matthys, S.: Reactivity of municipal solid waste incineration ashes as a supplementary cementitious material. In: 15th International Congress on the Chemistry of Cement (ICCC 2019), Prague, p. 149 in book of abstracts, 10 p on USB (2019)
8. Li, X.-G., Lv, Y., Ma, B.-G., Chen, Q.-B., Yin, X.-B., Jian, S.-W.: Utilization of municipal solid waste incineration bottom ash in blended cement. J. Clean. Prod. **32**, 96–100 (2012). https://doi.org/10.1016/j.jclepro.2012.03.038
9. Bertolini, L., Carsana, M., Cassago, D., Curzio, A.Q., Collepardi, M.: MSWI ashes as mineral additions in concrete. Cem. Concr. Res. **34**, 1899–1906 (2004). https://doi.org/10.1016/j.cemconres.2004.02.001

Autogenous Shrinkage in Structural Concrete Made with Recycled Concrete Aggregates

Mayara Amario, Caroline S. Rangel, Marco Pepe, Enzo Martinelli, and Romildo D. Toledo Filho

Abstract Sustainable alternatives should be employed in the construction industry to minimize environmental impacts, including high consumption of natural resources and high waste generation. The recycling of waste to be used as aggregates for concrete is an alternative with great potential. Although many studies have been carried out to investigate the influence of the use of Recycled Concrete Aggregates (RCAs) on the mechanical properties of concrete, the relationship between the intrinsic properties of aggregates and the main types of long-term deformation of Recycled Aggregate Concrete (RAC) still needs to be better understood. As a matter of the principle, autogenous shrinkage can cause significant damage to concrete structures, mainly due to the induction of cracking that can impair the durability of the structure. In this context, the present paper presents the preliminary results of an experimental investigation aimed at evaluating the influence of RCAs on the autogenous shrinkage of normal and high strength RACs (35 and 60 MPa). The RCA were derived from demolition concrete waste and used in a 9.5–19 mm particle size fraction. In addition, the concepts of scientific mix-design for the optimization of dry granular mixture were utilized through the Compressible Packing Model (CPM).

Keywords Recycled concrete aggregate · High strength concrete · Autogenous shrinkage

1 Introduction

It has been known for some time that the impact of the concrete industry on the environment can be reduced by introducing a sustainable material into concrete production: the Recycled Concrete Aggregates (RCAs). Therefore, maximizing the amount

M. Amario · C. S. Rangel · R. D. T. Filho
COPPE/Federal, University of Rio de Janeiro, Rio de Janeiro, Brazil

M. Pepe (✉) · E. Martinelli
Department of Civil Engineering, University of Salerno, Fisciano, Italy
e-mail: mapepe@unisa.it

TESIS Srl, Fisciano, SA, Italy

V. M. C. F. Cunha et al. (eds.), *Proceedings of the 3rd RILEM Spring Convention and Conference (RSCC 2020)*, RILEM Bookseries 35,
https://doi.org/10.1007/978-3-030-76543-9_11

of recycled materials in concrete composition is a very effective and promising approach to stimulate sustainable construction [1].

However, it is also known that aggregates are the largest components in concrete (sometimes representing up to 80% of the volume of concrete) and their characteristics have a significant effect on the properties of the final composite. Thus, in recent decades, several studies have sought to discover and explain the performance of concrete containing RCAs. A review of the existing literature on Recycled Aggregate Concrete (RAC) makes it evident that most authors investigate short-term behavior and only a limited number of studies have been reported so far on the long-term behavior of RACs [2].

Experimental data available on recycled aggregates from different sources show that the quality of the RCA depends mainly on the quality of the original demolished concrete used for recycling. Despite the high heterogeneity of RCAs, it has been shown that the main properties of recycled particles are directly related to it: particles characterized by larger amount of Attached Mortar (AM) also have higher water absorption capacity and lower density [3].

Autogenous shrinkage of concrete refers to the reduction of apparent volume or length of cement-based materials in seal and isothermal conditions. This phenomenon is caused by further cement hydration after the formation of the initial cement matrix structure [4]. Usually, concrete deformations due to early-stage autogenous shrinkage occur within the first 24 h after the contact of cement with water. The importance of this property is explained by the fact that autogenous shrinkage can cause micro fissures in the cementitious matrix, which leads to reduced structure durability.

A well-known theory is that autogenous shrinkage occurs due to the appearance of an internal pressure on the surface of the pores, caused by the change in saturation of these pores. Free water is gradually reduced due to the cement hydration reactions and, consequently, the condition of internal humidity is also reduced. Thus, in order to restore the equilibrium pressure on the pore surface, a capillary tension arises, which is responsible for generating the concrete deformation [4]. For Liu et al. [5], internal curing can effectively reduce the start and end of autogenous shrinkage of high-strength concretes after the start of the setting, regardless of the type of cement used. In addition, for the authors, the type, composition and size of the porous aggregate used can affect the effectiveness of internal healing.

Therefore, the pore structure plays an important role in the autogenous shrinkage of concrete. According to Jensen & Lura [6], for internal curing in cementitious materials, it is necessary: (a) possibility of transporting water from the reservoir (in this case, the porous aggregate) to all regions where cement hydration reactions will occur; (b) the relative humidity is close to 100%. In this context, studies have been carried out with the objective of using the greater porosity of certain aggregates as reservoirs for the internal cure of concretes.

Regarding the effect of RCA on this property, it can be highlighted that Manzi et al. [1] reported that concrete shrinkage deformation is negatively influenced by the use of recycled concrete aggregates. Gholampour and Ozbakkaloglu [2] obtained similar conclusion in their study, in which RACs containing RCAs from concrete residue of 40 MPa develop significantly higher shrinkage strain than conventional natural concrete.

On the other hand, Medjigbodo et al. [7] obtained lower values for autogenous shrinkage of recycled concrete, compared to natural concrete. Similar results were obtained by Maruyama and Sato [8], in which the autogenous shrinkage of high-strength recycled concrete was 40% less than that of the corresponding reference concrete. This phenomenon was attributed to the internal curing capacity generated by the recycled aggregates. In this way, water would be gradually released from the pores of the aggregates, slowing down the decrease in internal humidity and easing the magnitude of capillary stresses and the autogenous shrinkage of these concretes.

In this context, as the results obtained in the literature are often contradictory, this paper investigates the possibility of internal curing by employing recycled coarse concrete aggregates generated from demolition concrete waste on the autogenous shrinkage of new structural concrete for normal and high strength.

2 Materials and Methods

2.1 Materials

The natural aggregates used were: a fine fraction, composed of quartz sand, with nominal diameter smaller than 4.75 mm; a coarse aggregate (named "NAT_C0"), composed of coarse granite rocks, with nominal diameter between 4.75 and 9.5 mm; and a coarse aggregate (named "NAT_C1"), also composed of coarse granite rocks, with a nominal diameter between 9.5 and 19 mm.

The Recycled Concrete Aggregate (RCA) was derived from a demolition waste recycling plant. The concrete waste was composed of a mixture of different demolition concrete waste; therefore, the properties are unknown and the age is indeterminate. Four processing steps produced the Recycled Concrete Aggregates (RCAs): crushing, drying, sieving and homogenization. In the first step, a jaw crusher was used to reduce particle size by crushing. After the crushing step, the particles were subjected to air drying and separation in an industrial mechanical sieve. The recycled material of nominal diameter between 9.5 and 19 mm was classified as coarse aggregate 1, named "RCA_D_C1". In the last step, the recycled aggregate was homogenized by the longitudinal blending bed technique, which consists of alternately and in opposite directions spreading the same amount of material along a pile. Figure 1 show the coarse aggregates used in this study.

The aggregates characterization was performed through several tests. Specific gravity and water absorption were performed in the coarse aggregates according to NBR NM 53 [9] and in the fine aggregate according to NBR NM 52 [10] and NBR NM 30 [11], respectively. Table 1 reports the results for the mentioned properties.

The cement used in this study was a "high initial strength Portland cement" according to NBR 5733 [12] characterized by a compressive strength of 40 MPa

(a) NAT_C0 (b) NAT_C1 (c) RCA_D_C1

Fig. 1 Coarse aggregates

Table 1 Properties of the natural and recycled aggregates

Properties	Aggregates			
	Sand	NAT_C0	NAT_C1	RCA_D_C1
Maximum grain size (mm)	4.8	9.5	19.0	19.0
Specific gravity (kg/m^3)	2447	2662	2636	2255
Water absorption rate (%)	0.5	1.5	1.3	6.1
Attached mortar content (%)	–	–	–	35.1

at 28 days and a specific gravity of 3181 kg/m^3. A superplasticizer "MC Powerflow 1180" with a solid concentration content of 35% and specific mass of 1.070 g/cm^3 was used for workability control.

2.2 *Mixture Proportioning*

The scientific mix-design of the concrete mixtures was carried out according to the Compressive Packing Model (CPM), proposed by de Larrard [13]. This method is based on the properties of the constituents to evaluate the resulting concrete properties, assuming that the overall compactness achieved by the dry granular skeleton determines the resulting concrete performance [14].

Specifically, for each strength class (i.e., C35 and C60) two mixtures were designed: a reference concrete containing only natural aggregates (i.e., Cxx-NAT) and a RAC including D-waste by replacing 100% of coarse fraction C1 (i.e., Cxx-D-C1). Table 2 shows the composition of the four concrete mixtures.

The recycled materials were added in dry condition to the mixture and the absorption was considered in the calculation of the composition: the value used was 50% of total water absorption (obtained experimentally). This value of 50% is based on the studies developed by Amario et al. [14] and Pepe et al. [15] in which the authors

Table 2 Mixture proportioning

Mixture	Materials (kg/m^3)						
	NAT_C1	RCA_D_C1	NAT_C0	Sand	Cement	Effective water	Total water
C35-NAT	452	0	457	868	325	196	212
C35-D-C1	0	384	453	862	341	194	216
C60-NAT	448	0	452	860	448	145	150
C60-D-C1	0	382	451	857	463	137	147

verified that during the concrete mixing process the RCAs absorb about 50% of their total water absorption capacity. The superplasticizer content was 0.2 and 1.5% solids in ratio to cement consumption for the mixtures of 35 MPa and 60 MPa, respectively.

2.3 *Mixing Procedure*

Due to the high water absorption of the RCA, a specific methodology was adopted for the mixing process: the total water was divided into two equal parts and the addition of the parts was performed at different times of the mixture.

The mixing procedure was carried out in the following steps: first, all the aggregates were mixed for 1 min; then, 50% of the total water was added and the mixing continued for another 1 min; after, the cement was added and the materials were mixed for another 1 min; finally, the superplasticizer and the second half of the water were added, and all materials were mixed for 8 min.

2.4 *Test Methods*

The fresh state property was determined through slump tests [16]. For mechanical characterization the compressive strength, elastic modulus and tensile splitting strength at 28 days were determined [17, 18]. Total water absorption, voids index and density were performed on cylindrical samples (100 mm of diameter and 200 mm of height), according to NBR 9778 [19]. Autogenous shrinkage tests were performed on sealed prismatic specimens (75 × 75 × 285 mm, room temperature of 23.0 ± 2.0 °C and relative humidity of 50 ± 4%).

3 Results and Analysis

3.1 Workability and Physical Properties of RAC

Table 3 presents the means values and the coefficient of variations of slump, total water absorption, voids index and density.

The slump test results indicate that RCA do not negatively influence the workability of the recycled concrete mixtures. Note that the four mixtures presented slump values ranging between 160 and 195 mm.

The physical properties of RACs were more influenced by the use of RCA. In this case, the total water absorption was higher for recycled concrete mixtures, as for C35-D-C1 was 3.5% in comparison with the reference mixture with 3.0%. For the high strength class, the recycled concrete presented 1.4% of total water absorption while the reference mixture exhibit 1.1% on this property.

The voids index is a property directly related to total water absorption. Thus, the RACs presented higher voids index than the reference mixtures, for both strength classes. Additionally, concrete mixtures with only natural aggregates presented higher density than RACs.

Table 3 Workability and physical properties of concrete mixtures

Mixture	Slump (mm)	Total water absorption (%)	Voids index (%)	Density (kg/m^3)
C35-NAT	175	3.0 (±1.4%)	7.0 (±1.2%)	2303 (±0.3%)
C35-D-C1	195	3.5 (±2.2%)	7.8 (±2.3%)	2247 (±0.1%)
C60-NAT	165	1.1 (±1.7%)	2.7 (±1.3%)	2411 (±0.6%)
C60-D-C1	160	1.4 (±1.3%)	3.3 (±1.5%)	2376 (± 0.2%)

Table 4 Mechanical properties of concrete mixtures

Mixture	$f_{c,28}$ (MPa)	$\varepsilon_{c,28}$ ($\mu\varepsilon$)	$E_{c,28}$ (GPa)	$f_{t,28}$ (MPa)	$f_{t,28}/f_{c,28}$ (%)
C35-NAT	34.2 (±2.4%)	2898 (±3.2%)	21.3 (±2.1%)	2.7 (±1.7%)	7.9
C35-D-C1	33.5 (±2.3%)	2941 (±2.7%)	20.9 (±2.7%)	2.6 (±2.2%)	7.8
C60-NAT	60.1 (±1.5%)	2665 (±1.7%)	29.1 (±3.2%)	3.9 (±3.2%)	6.5
C60-D-C1	59.7 (±0.5%)	2691 (±2.1%)	29.5 (±1.9%)	4.1 (±3.0%)	6.9

3.2 Mechanical Behavior of RAC

Table 4 presents the means values and the coefficient of variations of 28-day compressive strength ($f_{c,28}$), strain at maximum stress ($\varepsilon_{c,28}$), Elastic modulus ($E_{c,28}$), 28-day splitting tensile strength ($f_{t,28}$) and the relation between 28-day splitting tensile strength and 28-day compressive strength and ($f_{t,28}/f_{c,28}$).

The compressive strength results obtained for RACs (C35 with 33.5 MPa and C60 with 59.7 MPa) were very similar to the natural concretes (C35 with 34.2 MPa and C60 with 60.1 MPa). In the normal strength class, the concretes had a maximum difference of 4% compared to the predicted value of 35 MPa. In the high strength class, the maximum difference was only 0.5% in relation to the desired value of 60 MPa. This fact shows that the use of a scientific mix-design that considers the intrinsic properties of RCA combined with the accurate consideration of the water absorption during the mixing process allowed positive results to be obtained.

The elastic modulus results of C35 class were 21.3 GPa for the natural concrete and 20.9 GPa for the RAC. In the high strength class, the value obtained for the natural concrete was 29.1 GPa while for the RAC was 29.5 GPa.

The results of 28-day splitting tensile strength show that, for high strength concretes, the RAC presented better tensile strength (with 6.9 MPa) than the natural concrete with 6.5 MPa. For the normal strength class, the splitting tensile strength was very similar for both mixtures.

3.3 Autogenous Shrinkage of RAC

The whole experimental curves (up to 350 days of test) of autogenous shrinkage are summarized in Fig. 2.

Figure 2 highlights that, as expected, the strains caused by autogenous shrinkage were more intense in the first three days of test: this is more evident by the analysis of the graphs proposed in Fig. 3 which shows the autogenous shrinkage curves for the first six days of the test. These deformations were intensified due to the use of initial high strength cement that promoted acceleration in hydration reactions of cement.

Moreover, from Fig. 2, it can be highlighted that no significant increment autogenous of shrinkage is registered after 90 days of test.

Comparing the behavior of the two strength classes, it is noted that the high strength mixtures presented higher values of autogenous shrinkage deformation than the normal strength class mixtures (see Fig. 4). In fact, the higher cement consumption combined with low water-cement ratio causes an increase in this type of deformation. Thus, the risk of cracking in high strength concrete structures is higher, especially in the early ages.

For both strength classes, the mixtures containing recycled aggregates presented lower autogenous shrinkage values than the mixtures containing natural aggregates after 365 days of test (for C35 class: 134 $\mu\varepsilon$ for C35-D-C1 and 142 $\mu\varepsilon$ for C35-NAT;

Fig. 2 Autogenous shrinkage curves up to 350 days of test

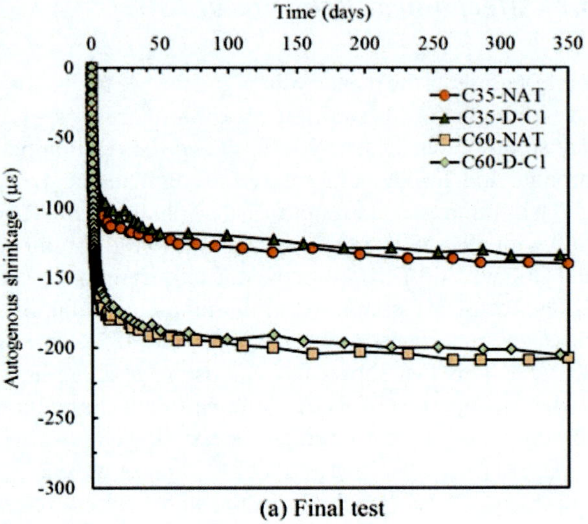

Fig. 3 Autogenous shrinkage curves up to 6 days of test

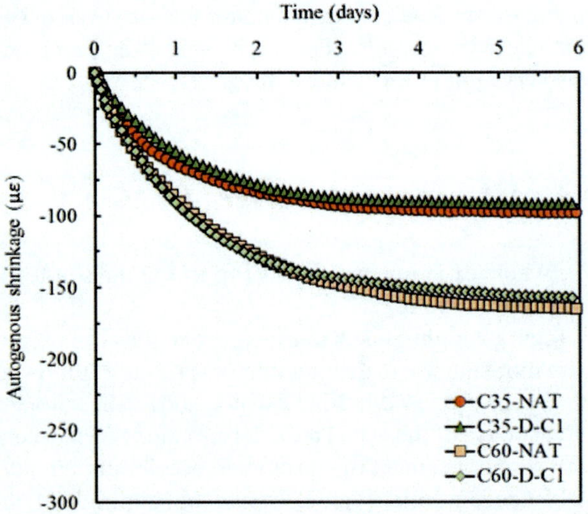

for C60 class: 205 με for C60-D-C1 and 210 με for C60-NAT). This behavior can be explained by the large amount of pores present in the recycled aggregates that function as small reservoirs of water, gradually releasing this stored water for cement hydration reactions. This is called internal curing of concrete.

Therefore, the recycled aggregate promoted a decrease in the deformations caused by cement hydration reactions, contributing positively to the reduction of cracking effects mainly in the early concrete ages.

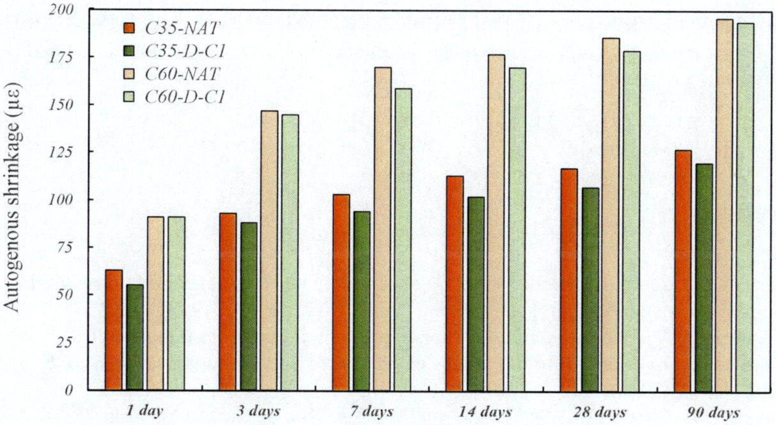

Fig. 4 Autogenous shrinkage values for RAC and NAT mixtures up to 90 days

4 Conclusion

This study analyzed the influence of RCA in the physical and mechanical behavior and in the autogenous shrinkage of concretes with normal and high strength. The following consideration can be remarked:

- The recycled concrete aggregate did not affect the compressive strength of recycled concrete in comparison with natural concrete for both strength classes. The mix-design method adopted (i.e. CPM) proved to be appropriated for the composition of recycled concrete mixtures with the same mechanical behavior as natural concretes.
- The results of elastic modulus and splitting tensile strength were not influenced by the presence of recycled aggregates for both C35 and C60 strength classes. In fact, these properties are strongly related to the compressive strength of concrete, so the mechanical behavior was similar between the two mixtures for each strength class.
- The physical properties were more influenced by the use of RCA. In fact, higher porosity associated with the recycled aggregate of grains caused an increase in water absorption and a reduction in density of the RAC in comparison to natural concrete for both strength classes.
- The autogenous shrinkage of recycled concrete mixtures was lower than the natural mixtures for both normal and high strength classes. The presence of a highly porous aggregate reduced the shrinkage of recycled concrete by promoting the internal curing of the mixtures, i.e. the RCA were able to release water stored inside gradually for the cement hydration reactions.

Finally, the results of this study reinforce that RCA can be used to obtain concrete mixtures with similarly behavior to concretes containing only natural aggregates, provided that an adequate mix-design method is adopted for these mixtures.

Acknowledgements The study is part of SUPERCONCRETE Project (H2020-MSCA-RISE-2014, n. 645704): the Authors wish to acknowledge the financial contribution of the EU-funded Horizon 2020 Programme.

References

1. Manzi, S., Mazzotti, C., Bignozzi, M.C.: Short and long-term behavior of structural concrete with recycled concrete aggregate. Cement Concr. Compos. **37**, 312–318 (2013)
2. Gholampour, A., Ozbakkaloglu, T.: Time-dependent and long-term mechanical properties of concretes incorporating different grades of coarse recycled concrete aggregates. Eng. Struct. **157**, 224–234 (2018)
3. Rangel, C.S., Toledo Filho, R.D., Amario, M., Pepe, M., de Castro Polisseni, G., de Andrade, G.P.: Generalized quality control parameter for heterogenous recycled concrete aggregates: a pilot scale case study. J. Clean. Prod. **208**, 589–601 (2019)
4. Wu, L., Farzadnia, N., Shi, C., Zhang, Z., Wang, H.: Autogenous shrinkage of high-performance concrete: A review. Constr. Build. Mater. **149**, 62–75 (2017)
5. Liu, J., Shi, C., Ma, X., Khayat, K.H., Zhang, J., Wang, D.: An overview on the effect of internal curing on shrinkage of high-performance cement-based materials. Constr. Build. Mater. **146**, 702–712 (2017)
6. Jensen, O.M., Lura, P.: Techniques and materials for internal water curing of concrete. Mater. Struct. **39**(9), 817–825 (2006)
7. Medjigbodo, S., Bendimerad, A.Z., Rozière, E., Loukili, A.: How do recycled concrete aggregates modify the shrinkage and self-healing properties? Cement Concr. Compos. **86**, 72–86 (2018)
8. Maruyama, I., & Sato, R.: A trial of reducing autogenous shrinkage by recycled aggregate. In: Proceedings of 4th international seminar on self-desiccation and its importance in concrete technology, pp. 264–270 (2005)
9. NBR NM 53: Coarse aggregate—determination of the bulk specific gravity, apparent specific gravity and water absorption. ABNT (2009)
10. NBR NM 52: Fine aggregate—determination of the bulk specific gravity and apparent specific gravity. ABNT (2009)
11. NBR NM 30: Fine aggregate—test method for water ab-sorption. ABNT (2001)
12. NBR 5733: High early strength Portland cement—Specification. ABNT (1991)
13. De Larrard, F.: Concrete mixture proportioning: a scientific approach. CRC Press (1999)
14. Amario, M., Rangel, C.S., Pepe, M., Toledo Filho, R.D.: Optimization of normal and high strength recycled aggregate concrete mixtures by using packing model. Cement Concr. Compos. **84**, 83–92 (2017)
15. Pepe, M., Toledo Filho, R.D., Koenders, E.A., Martinelli, E.: A novel mix design methodology for recycled aggregate concrete. Constr. Build. Mater. **122**, 362–372 (2016)
16. NBR NM 67: Concrete—slump test for determination of the consistency. ABNT (1998)
17. NBR 5739: Concrete—compression test of cylindric specimens—method of test. ABNT (2007)
18. NBR 7222: Concrete and mortar—determination of the tension strength by diametrical compression of cylindrical test specimens. ABNT (2011)
19. NBR 9778: Hardened mortar and concrete—determination of absorption, voids and specific gravity. ABNT (2005)

Circular Economic Modelling—Barriers and Challenges Throughout the Value Circle

Birgitte Holt Andersen, Giovanni Salvetti, and Anastasija Komkova

Abstract The transition from a linear to a circular economy is a process that has already started throughout our societies as one important strategy to reduce CO_2 emissions and by decoupling the use of natural resources, materials and fossil fuels from economic activity. Using examples from the Construction Industry, this paper argues that key premises must be present for circularity to happen. The business case at each level of the value circle must be viable. New building standards backed up by adequate policy measures, e.g. green public procurement, CO_2 and virgin materials taxation is likely to be required. Finally, securing future supply of secondary raw materials (SRM) must be demonstrated and adequate supply chains of SRM to emerge.

Keywords Circular economic modelling · Transition to circular economy · Barriers and opportunities · Secondary raw material (SRM) · Geopolymer cement · Construction materials

1 Background

1.1 Construction Sector

The construction sector as a whole account for 36% of global energy use and 39% of energy related CO_2 emissions [1]. Around a third of the emission derives from the production of concrete, amounting to 5–8% of CO_2 global emissions [2, 3]. Additionally, the sector is the most material resource consuming sector with an

B. H. Andersen (✉) · G. Salvetti
Institute for Applied Economics and Health Research (ApHER), Ewaldsgade 3, 2200 Copenhagen, Denmark
e-mail: bgha@appliedeconomics.dk

A. Komkova
ETH Zurich, Chair of Sustainable Construction, Stefano-Franscini-Platz 5, 8093 Zurich, Switzerland

V. M. C. F. Cunha et al. (eds.), *Proceedings of the 3rd RILEM Spring Convention and Conference (RSCC 2020)*, RILEM Bookseries 35,
https://doi.org/10.1007/978-3-030-76543-9_12

Fig. 1 Content of OPC concrete compared with Geo-polymer concrete. *Source* DurSaam

annual average of about 5.4 billion tonnes of raw materials, almost 4 times higher than the second most resource consuming sector; agriculture and fishing [4].

The most common cement is Ordinary Portland Cement (OPC), which is produced by calcination of limestone, which then is heating up to 1400 °C to produce the clinker [5]. To produce one tonne of OPC results in the emission of 0.82 tonnes CO_2 [6].

On the positive side, the construction sector represents a huge potential for circularity, in particular since many of the virgin materials currently used in building materials can be replaced by a wide range of secondary raw materials (SRMs), ranging from by-products or waste streams of industrial production to the use of recycled end-of-life materials from construction and demolition waste (CDW). Figure 1 shows the ingredients required to produce 1 m^3 of traditional concrete compared with 1 m^3 of a green type of concrete, the so-called geopolymer concrete.

1.2 On-going Research in Circularity Related to Developing Green Cements

The present research will take advantage of three on-going research projects: URBCON,[1] DuRSAAM[2] and WOOL2LOOP[3] as the basis to develop a circular economic model. All three projects aim at developing an alternative to OPC. From an ecological point of view, the geopolymer cement has two distinct advantages to Portland cement: (a) lower energy requirements, (b) it can be produced almost entirely using SRMs, such as mineral wool, mining slags, fly ash, etc. It has been assumed that geopolymer cement has the potential to replace at least 20% of current cement applications, mainly for precast applications, e.g. façade panels, floorings, ceiling elements and other more specialized applications where geo-polymer cement

[1] Interreg NWE project: 'By-products for sustainable concrete in the urban environment', www.nweurope.eu/urbcon.

[2] Marie Curie PhD training network, http://www.dursaam.ugent.be.

[3] Horizon2020 project: Geopolymer technology for the development of mineral wool waste value chains. https://www.wool2loop.eu/en/.

has absolute technical performance advantages compared to OPC. Furthermore, comparing geopolymer cement with OPC the potential reduction of CO_2 emissions range between 26 and 80% [6].

While the focus of the Wool2Loop project is to collect and reuse mineral wool waste (MWW) to produce specific geo-polymer products, the focus of URBCON is to use local waste to produce 'urban concrete', a circularity concept closely associated with relevant policies of the cities involved. Finally, DuRSSAM is a collaborative PhD framework aiming at delivering world-leading training in this multidisciplinary field through 13 PhDs in interrelated aspects of Geopolymer concrete, fibre reinforced high-performance concrete, and textile-reinforced mortar, as well as sustainability assessment.

1.3 The Aim of This Paper

The aim of this paper is to report on progress in developing a circular economic model is to (1) understand the premises for which a transition to a more circular economy within the building material sector can take place by analysing the business case at each step of the value chain from sourcing of SRM; to production of geopolymer cement; design and manufacturing of the building materials; use phase and end of life, (2) identify the main hindrances facing such transition, including technical, legislative, standardisation, market type barriers and user acceptance (3) assess security of future supply of suitable SRMs as competing demands for critical SRMs can be expected in the future, combined with scarcity of some SRMs such as fly ash (due to fewer coal power plants in the future) and blast furnace slags (due to reduced steel production in Europe). The first aim and the third aim will be the primary focus of this paper.

1.4 Approach

The current research includes literature review, data collection including statistics, model developing, interviews with industry and other stakeholders. As the research projects are still on-going, the results presented in this paper are preliminary. The technical work related to the development of the geo-polymer cement is not yet finalised, wherefore the results of LCA and LCCA are also not yet available.

2 Results Related to Circular Economic Modelling

In the following we will present the preliminary findings regarding the premises for which this specific area of the construction industry can turn circular by substituting virgin materials with industrial waste streams resulting also in a considerable reduction in CO_2 emissions. Likewise, the paper presents a first estimation of availability of key secondary raw materials (SRM) critical to secure future supply for production of geopolymer cement.

2.1 The Business Case at Each Step of the Value Circle

The current research has taken a pragmatic approach in assessing and describing the business cases along the value chain in order to understand the premises for which a transition to a circular economy can take place.

Sourcing SRM

The first step of the value circle is sourcing of suitable Secondary Raw Materials (SRMs) for the production of geopolymer cement. Examples of suitable SRMs are depicted in Fig. 2. Each SRM representing a distinct sourcing value chain.

Construction and Demolition Waste (CDW) accounts for more than one third of all waste in Europe, in the order of 850 million tons a year. That means on average each of us produce more than 1½ tons of CDW on a yearly basis [7]. The CDW is usually made up by Minerals (concrete, natural stones, plaster/gypsum, foamed clay bricks, glass, mineral wool); Organic materials (wood, asphalt, various plastics and PVC); and Metals: (iron, steel copper). According to the Waste Framework Directive (2008/98/EC), article 11.2 demands Member States to achieve a minimum of 70% recycling of non-hazardous CDW by weight in 2020. Currently, the recycling of CDW stands at around 50%, with huge variations across the different member states. CDW management protocols has been introduced by the EC as non- binding guidelines to facilitate higher recycling rates. *Pre-demolition audit (PDA) is an activity organized*

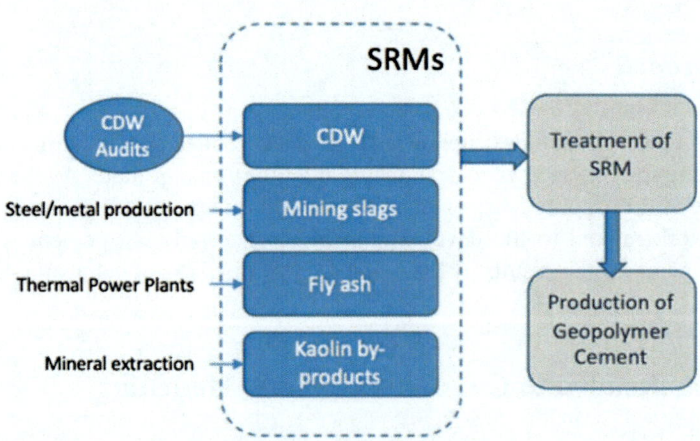

Fig. 2 Examples of waste streams feeding geopolymer cement

by the owner of the building or infrastructure resulting in the inventory of materials and components arising from the future demolition, deconstruction or refurbishment projects, and their management and recovery options. It was originally initiated in order to identify hazardous materials, such as lead, PCB and asbestos, prior to demolition. PDAs are already compulsory in some Member States, such as Belgium, The Netherlands, France, Sweden, Finland, Austria, Bulgaria, and Czech Republic [8].

So, who owns the waste or the CDW? Traditionally, a waste has been seen as a liability and associated with a certain deposit or landfill cost, which typically lies in the range of 100EUR-200EUR per ton. As waste is gradually becoming an asset representing a given value, it is interesting to understand who owns the waste or the SRM. In case of CDW, the building itself is obviously owned by the property owner, but at the moment a contract is signed for demolition it is the demolition company that owns the CDW. Hence, the business case is viable, if you can turn a negative cost (a landfill fee) for a given waste into an income generating material.

Treatment of SRM

The recycling market has emerged over the past decades as a results of increased regulated waste management. Recycled steel, iron, glass, paper, asphalt, aggregates markets are well established. According to EuRIC,[4] the recycling industry consisting of some 5500 companies, generate a turnover of 95BEUR. In the case of 'new' waste streams, such as mineral wool waste or certain type of mining slags, new SRM value chains and new recycling companies are likely to emerge.

In case of mineral wool waste (MWW), certain investments are needed in order to make the waste useful. The main source of MWW is from demolition. At the demolition site, the MWW has to be separated from other types of demolition waste. Due to the low density of the wool, some kind of processing at the demolition site is needed to reduce the volume of the MWW prior to transporting. Before the MWW can be used for geopolymer cement production is has to be grinded into very fine particles. The milling process could be done either by the demolition company as an intermediate supplier or by the final user of the material, e.g. the cement company. The cost of an industrial milling machine with a capacity of 15 tons/hour is 50 K EUR.[5]

The costs of the MWW powder can be added up as follows:

$$\begin{aligned}\text{Cost price of MWW powder} &= [\text{Labour Cost of separation}/T] \\ &+ [\text{Transport cost of MWW powder}/T] + [\text{Milling machine CAPEX}/T] \\ &+ [\text{Milling machine OPEX}/T]\end{aligned}$$

If the market price of the MWW is higher than the cost price, or the deposit costs cost of MWW is higher than the Cost price of MWW, the business case is valid.

---

4 The European Recycling Industries Confederation.

5 According to the Wool2Loop project.

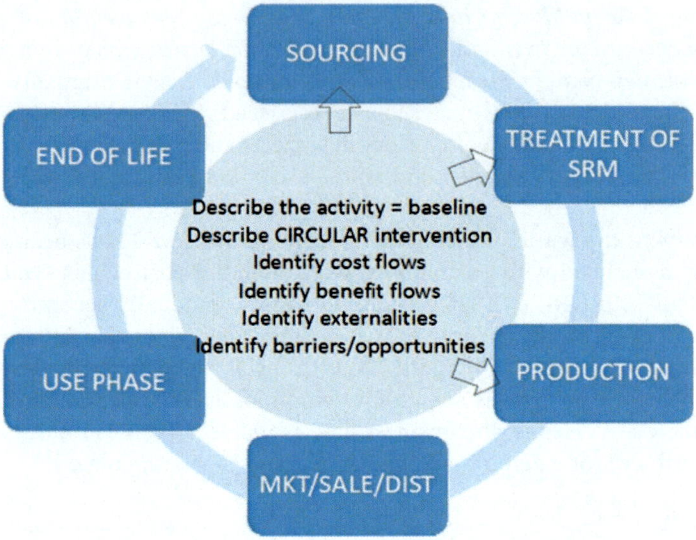

Fig. 3 Conceptual framework to assess the potential for economic circularity

Production of Geopolymer cement

At the level of production of geopolymer cement, at least two scenarios can be considered. The production can either take place at existing production facilities of cement or construction materials. Alternatively, new players can establish themselves to specialise in the production of geopolymer cement. For existing production facilities, only marginal investments will be required to accommodate the use of CRM for the production of geopolymer cement. Additional cost flows for established players concerning the production of geopolymer cement and products based on geopolymer cement, can be grouped as follows: CAPEX and OPEX of production facilities, training of staff for handling and additional health and safety procedures to handle the activator while preparing the geopolymer cement mix. However, the total investment due to using geopolymer cements is not expected to vary considerable compared to other normal alterations in the production flow.[6] This aspect will be further investigated through the pilot phase of both Wool2Loop and URBCON project.

The market perspectives, the use phase and end of life considerations is still ongoing research.

Based on the conceptual circularity model and excel model is under development to capture and analyse all the data and narratives being collected from the various stakeholders in the value circle in order the access the economic viability (Fig. 3).

[6]According to industrial partners in Wool2Loop and URBCON.

2.2 Main Hindrances for Turning Circular

Globally, concrete is the most widely used construction material [9] and demand for sustainable concrete is gradually increasing. However, there are several barriers preventing the transition towards circular economy in the construction sector.

While the technology associated with geopolymer production is evolving and are promising, certain technological barriers remains, but currently examined within the ongoing research projects, (Wool2Loop, URBCON and DuRSAAM). The developing geoploymer technology is associated with lack of long-term data on concrete durability in real world settings [10]. Another barrier for adapting geopolymers at the industrial scale is the absence of relevant regulatory standards for geopolymer concrete.

The main market hindrances are associated with lack of incentives fostering the concrete industry to adapt to sustainable technologies. As the prices of the virgin materials remain relatively low, the concrete producers tend to use conventional materials to remain competitive in the market. On the other hand, the application of the industrial by-products and waste materials in the concrete production can minimise the use of the finite natural resources and create new business opportunities as pressures from various stakeholders increase as public opinion demands actions to reduce CO_2 footprint and deliver against the UN sustainability goals.

However, the perception and reaction to the geopolymer cement representing a green solution to the construction industry remains uncertain and does potentially constitute a main barrier. Key stakeholders include construction companies, architects, construction engineers, certification agencies, but also Public procurement policies are seen as potential driver for public construction investments. Costs, performance, durability, LCA and LCCA will form part of the value proposition.

2.3 Availability of Secondary Raw Materials (SRM)

In order to secure supply of the alternative raw materials it is important to investigate stocks and flows of those materials that can partially or completely substitute virgin raw materials in concrete production. In the context of producing geopolymer cement we have made a preliminary mapping and assessment of order of magnitude availability of key SRMs. These includes Mineral Wool Waste (MWW), Bio Fly Ash (BFA), Ground granulated blast furnace slags, and Kaolin by-products (Fig. 4).

As there is no publicly available data on the volume of mineral wool produced in Europe, the previously developed models suggest that MWW amounts to approximately 2.5 mt in year 2020, based on assumption that MWW constitute 0.2% of the total CDW generated in Europe [11]. Due to the increase in the use of insulation over the last decades, the availability of MWW is expected to reach 4mt in year 2040 and assuming shorter building refurbishment trends, it can rise up to 10 mt.[7]

[7]Based on internal communication within Wool2Loop project.

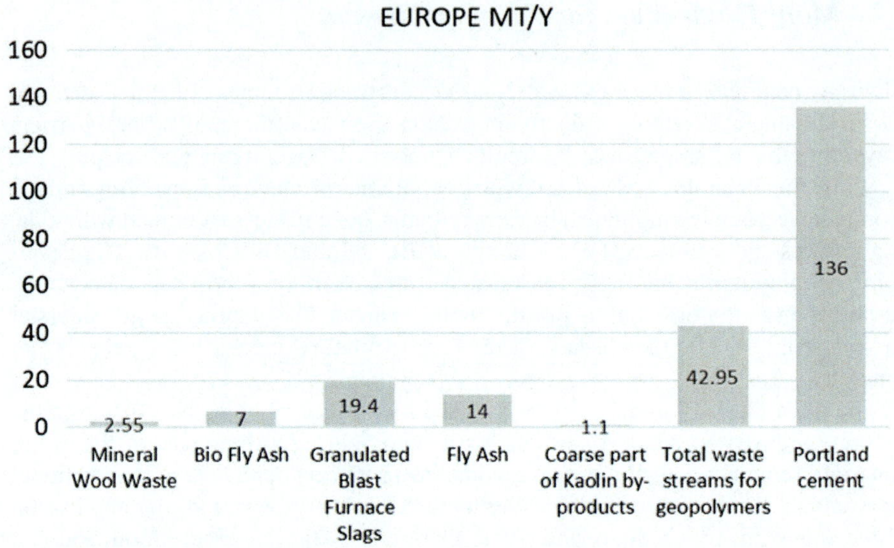

Fig. 4 Waste streams available for geopolymer binders

Fly ash is produced during coal combustion at thermal plants, and the European production of fly ash was approximately 35 mt in 2008, with 13.6 mt was used in concrete production [12]. It has been estimated that 1 ton of fly ash can produce approximately 3 m^3 geopolymer concrete [13]. In the future, however, fly ash from coal fired power plants will decrease as coal will be out phased as EU are shifting towards greener energy sources. Instead fly ash from bio-based thermal power plants will increase. The stockpiled fly ash can be used as alternative reserves, where stock size should be estimated.

The calcined natural clays such as metakaolin are often used in geopolymer concrete production as a precursor, however the coarse kaolin by-products available at a rate of 1.1 mt can be used as an alternative precursor.[8]

Blast furnace slag is a by-product of steel industry. In Europe, the total production of blast furnace slag was 24.6 mt in year 2016, with granulated blast furnace slag was approximately 78.9% or 19.4 mt [14]. Approximately 60% of blast furnace slag has been applied in cement production/concrete addition and about 24% for road construction [15].

Within the concrete production process, the above-mentioned industrial by-products and waste materials are activated with alkaline solutions. Activators, such as sodium silicate, potassium silicate, sodium hydroxide and potassium hydroxide, typically used in geopolymer concrete production have a high carbon footprint as manufacturing processes of these chemicals are energy demanding. Activators therefore account for the major part of environmental impact of geopolymer concrete [16]. In geopolymer mixtures, the inorganic activators can be partially substituted

[8] According to industrial partners in Urbcon project.

with waste materials. For example, rice husk ash, which is a by-product from the rice industry, can be used to minimize sodium silicate content in the geopolymer concrete [17]. The alternative materials include urban and industrial glass waste that can be used as an alkaline activator for blast furnace slag [18].

The main learnings from this preliminary exercise is that in the order of 43 tons of suitable SRM are available to produce geopolymer cement on a yearly basis, more than enough to replace up to 20% of the current OPC applications.

3 Conclusions

According to the statistics there is still a long way to a full circularity circle within the construction industry. Ideally, full material circularity is reached when only recycled materials are used. In principle all LCAs will be zero in a perfect circular economy.

The good news is that there exist technological and economic potential in using the SRMs in geopolymer cement production and that the future supply of relevant SRMs are plentiful. There exist opportunities to create a business case for geopolymer cement and we have identified potential stakeholders that can benefit from the emerging markets.

3.1 Future Research

- Complete data collection and assessment of costs flows and benefit flows to document the business case at each stage of the value circle
- Further stakeholder interactions to understand willingness to change and acceptance of new building standards
- Further literature reviews into the subject of pricing of SRMs
- The economic aspects of the activators

Acknowledgements Thanks to the projects Wool2Loop and URBCON and all partners in contributing with knowledge, insights as well as hard and soft data.

References

1. UN: Global Status Report, 2017. https://www.worldgbc.org/news-media/global-status-report-2017. Last accessed 2019/12/20
2. Lehne J, Preston F (2018) Making Concrete Change. Innovation in Low-Carbon Cement and Concrete. The Royal Institute of International Affairs (UK), Chatham House
3. Habert, G., Ouellet-Plamondon, G.: Recent update on the environmental impact of geopolymers. RILEM Techn. Lett. **1** (2016)

4. BIO Intelligence Service: Sectoral resource maps. European Commission, Prepared in response to an Information Hub request (2013)
5. Gartner, E.: Industrially interesting approaches to "low-CO_2" cements. Cem. Concr. Res. **34**(9), 1489–1498 (2004)
6. Turner, L.K., Collins, F.G.: Carbon dioxide equivalent (CO_2-e) emissions: a comparison between geopolymer and OPC cement concrete. Constr. Build. Mater. **43**, 125–130 (2013)
7. Eurostat: Waste Generation (2016). https://ec.europa.eu/eurostat/statistics-explained/index.php/Waste_statistics. Last accessed 2019/12/19
8. European Commission: EU Construction and Demolition Waste Protocol and Guidelines. https://ec.europa.eu/growth/content/eu-construction-and-demolition-waste-protocol-0_en. Last accessed 2019/12/20
9. Meyer, C.: The greening of the concrete industry. Cement Concr. Compos. **31**(8), 601–605 (2009)
10. Duxson, P., Van Deventer, J.S.J.: 17—Commercialization of geopolymers for construction—opportunities and obstacles. In: Provis, J.L., van Deventer, J.S.J. (eds.) Geopolymers. Woodhead Publishing, pp 379–400 (2009)
11. Väntsi, O., Kärki, T.: Mineral wool waste in Europe: a review of mineral wool waste quantity, quality, and current recycling methods. J. Mater. Cycles Waste Manage. **16**(1), 62–72 (2014)
12. Feuerborn, H.-J.: Coal combustion products in Europe–an update on production and utilization, standardisation and regulation. In: World Coal Ash Conference (2011)
13. Lloyd, N., Rangan, V.: Geopolymer concrete with fly ash. In: Proceedings of the Second International Conference on Sustainable Construction Materials and Technologies, UWM Center for By-Products Utilization (2010)
14. EUROSLAG: Statistics 2016—Euroslag. https://www.euroslag.com/products/statistics/statistics-2016/. Last accessed 2019/12/06
15. Coppola, L., Buoso, A., Coffetti, D., Kara, P., Lorenzi, S.: Electric arc furnace granulated slag for sustainable concrete. Constr. Build. Mater. **123**, 115–119 (2016)
16. Habert, G., De. Lacaillerie, J.D.E., Roussel, N.: An environmental evaluation of geopolymer based concrete production: reviewing current research trends. J. Clean. Prod. **19**(11), 1229–1238 (2011)
17. Kamseu, E., Moungam, L.M., Cannio, M., Bilong, N., Chaysuwan, D., Melo, U.C., Leonelli, C.: Substitution of sodium silicate with rice husk ash-NaOH solution in metakaolin based geopolymer cement concerning reduction in global warming. J. Clean. Prod. **142**, 3050–3060 (2017)
18. Puertas, F.,Torres-Carrasco, M.: Use of glass waste as an activator in the preparation of alkali-activated slag. Mechanical strength and paste characterisation. Cem. Concr. Res. 57:95–104 (2014)

Development of Sustainable Perspective of Carbon Fibers Recycling and Reusing for Construction Materials

R. Napolitano, P. Vitale, C. Menna, and D. Asprone

Abstract The development of sustainable methods is a demanding challenge posed by Global civilizations for improving the life quality in many ways. A key role in achieving sustainability goals could be the implementation of the new building materials models for the future environmental requirements. The presented process deals with the recycling of carbon fibers wastes deriving from automotive industry and the reusing within a building system, increasing in value the approach of circular economy and providing a solution to many negative economic and environmental impacts. To quantify potential benefits, in this work the behavior of cement-matrix composites containing short carbon fibers is described. An experimental campaign, starting from a specific surface treatment of the "pre peg" composite material sheets and an accurate study of the dimensional distribution of the carbon fibers, is presented. The aim is the evaluation of the influence of the percentage and length of fibers usage on the mechanical performance of cement based carbon fibers-reinforced mortars, following different approaches of mixture. The results show a good increase in stiffness and in flexural strength up to 11% and 20% respectively, for the samples characterized by fibers in percentage of 6.4% by volume of mixture and of 2–5 mm length and by a reduced amount of cement. A strong connection between the mixture workability and fibers size is observed. Finally, the assessment of the entire life cycle of the product is discussed for quantifying effective environmental impacts.

Keywords Circular economy · Carbon fibers recycling · Sustainable building material · Life cycle assessment

1 Introduction

Construction is one of the worlds' leading industrial sector, becoming pivotal also in terms of environmental burdens, for example, on a global scale, around 39% of GHG (Green House Gases) emissions and approximately 40% of primary energy used

R. Napolitano (✉) · P. Vitale · C. Menna · D. Asprone
Department of Structures for Engineering and Architecture, University of Naples Federico II, Via Claudio, 21, 80125 Naples, Italy
e-mail: rosanna.napolitano@unina.it

V. M. C. F. Cunha et al. (eds.), *Proceedings of the 3rd RILEM Spring Convention and Conference (RSCC 2020)*, RILEM Bookseries 35,
https://doi.org/10.1007/978-3-030-76543-9_13

refers to this sector, in Europe the 30% of the wastes generated come from construction and demolition activities [1]. Inside this sector, cement production embodies a central role, by representing 8% of global CO_2 emissions [2]. For this reason, there is a need to find out different solutions to promote sustainability in this sector as of the production process of cement.

Furthermore, in the last decades, the total amount of carbon fibers reinforced polymers (CFRP) production is strongly increased, mainly due to its physical and mechanical properties, for this reason, different industrial sectors, mainly aerospace and automotive, employ CFRP in a wide number of applications [3].

However, at the same time, the amount of CFRP wastes' volume augmented effectively reaching around 4 million tons in 2017 worldwide [4], and there is the necessity to manage them in the more sustainable way in order to mitigate their impacts. Actually, despite the fact that the European Community has indicated a drastic reduction in the amount of waste to be sent to landfill [5, 6], in Italy, there are still some lacks about the most appropriates waste managements for carbon fiber.

Many efforts have been done for recovering and recycling of carbon fiber, on this purpose, one of the first study was carried out by Pannkoke et al. [7] showed that recycling of CFRP was feasible working at low temperature, less than 30° C, useful to harden the resin and make prepregs cuttable, conserving a big share of the original properties.

Recycling and re-manufacturing processes are growing up by numbers and by systems, Oliveux et al. [8] and Pimenta [9] reviewed different type of recycling technologies for CFRP, from mechanical treatment, pyrolysis, fluidized bed and chemical treatment, demonstrating the technical feasibility of re-introducing the recycled parts in the market for the non-critical structural applications. Saccani et al. [10] investigated the possibility to recycle scraps of CFRP, coated in epoxy or other matrix, without any previous chemical or high temperature treatment, and then used to increase flexural strength and toughness of different mortars.

Unfortunately, most of these kind of wastes have been landfilled, losing resources and money, because even though there are several recycling systems, or in general recovering technics available at lab scale, there is not any possibility to find the same efficiency on a commercial scale. To comply with this problem, other researchers tried to mix carbon fiber with cementitious or concrete mortar in order to improve their mechanical properties. For example Badanoiu and Holmgren [11] studied how solve the problem of the organic compounds of resins on the surface of fibers which decrease the bond between the cement and the carbon fibers. Others like Garcés et al. [12], Mastali et al. [13], Mastali and Dalvand [14], and Ogi et al. [15] focused their activity on the improvement of mechanical properties like strength, flexural strength, impact resistance, compressive strength, tensile strength and modulus of elasticity by mixing carbon fiber with cement in concrete and by testing several samplings and specimens.

On this line, this study suggests a model in line with the circular economy by enhancing a feasible industrial symbiosis between two sections, in a win–win solution. On the one hand, there is a techno-textile company using CFRP, with the necessity to improve the sustainability related to waste management, ensuring natural resources optimization, recycling or recovering of energy from waste, and, at the same time reducing the cost for this service. On the other hand, there is the need to move forward to a more sustainable cement production, mitigating environmental and economic impacts during the production process.

The aim of this paper is to quantify the environmental performances of an ordinary cement mortar compared with two experimental mortars. The first one, in which a precise amount of carbon fibers (obtained from CFRP prepregs as waste of an Italian Company) substitute a share of sand, and the second one, in which a discrete amount of carbon fibers can substitute a share of cement. To do this, an attributional and comparative LCA (Life Cycle Assessment) [16, 17] has been carried out on different specimens on a lab scale, which have been also tested and compared on mechanical performances, in the specific flexural and compressive strength.

2 Materials and Methods

The experimental campaign, conducted at Laboratory of the Department of Structures for Engineering and Architecture, University of Naples "Federico II", aims at optimizing the mechanical performances of cementitious material reinforced with short carbon fibers deriving from CFRP from automotive industry, by producing a new type of material that could be reused within a building system.

In order to study the effects of short carbon fibers on the behavior of cement composites, flexural and compression tests were carried out on different groups with and without carbon fibers. For the case of cement composite with fiber, it was reinforced by 0.7–1 mm, 1–2 mm, 2– mm fibers. The variable percentages of fiber content, chosen for this investigation were 4, 6, 8 and 12% by volume of mixture, with varying the approaches and the workability of mixture.

2.1 Carbon Fibers

The carbon fibers used in this research work was obtained from the Italian Company, market leader in the production of carbon fibers components for the automotive industry. In particular, the material used by the company is "pre-preg", i.e. carbon fibers sheets pre-impregnated with a pre-catalyzed resin system. Table 1 shows the properties of carbon fibers.

Table 1 Properties of carbon fibers

Tensile strength	1061.9 MPa	EN ISO 527-4
Elastic modulus	200–600 GPa	
Density	1750 kg/m^3	

2.2 Recycling Process

As mentioned above the recycling process aims to produce a more performing mortar in terms of mechanical properties, meaning an increased value of ductility, tensile strength and durability. The addition of carbon fibers in the mortar lends a reduction of cracks caused by plastic shrinkage due to evaporation of the water into the mixing.

After the wastes collection from industry storage, the carbon fibers have been cooked at a temperature of about 90 °C for 30 min that allows the polymerization of the pre-preg sheets, in order to harden the sticky resin and consequently enabling the workability of the fibers. The second phase involves the shredding of CFRP fibers, a pre-treatment operation consisting in reducing the size of material into fragments by using a mechanical grinder. Then, by using a series of sieves superimposed with opening gradually decreasing from top to bottom, the measurement of grain sized has been detected for enhancing the physical and chemical properties of the materials. Thanks to this operation, the length of fibers fragments is defined, in order to make the choice of fibers dimension mixed inside the cement composite more reasonable and defined.

2.3 Dimensional Distribution

The characterization of the fibers has been implemented only for the substitution of mixtures. Starting from the images reproducing the distribution of the fibers sieved, the data processing provided by a Matlab application and reported here as distributions of length and diameter diagrams. An important result is the aspect ratio, i.e. the ratio between the expected value of the fibers length and diameter, in tabular form (see Table 2). As the aspect ratio increases, the adherence between the materials increases and consequently the performance of the fiber-mixture improves, since the fibers subjected to tensile forces are more difficult to slip off.

Table 2 Dimensional distribution results

Fiber	Length		Diameter		Aspect ratio
	Expected value	σ	Expected value	σ	
0.425–0.8	2.84	0.30	0.53	0.39	5.33
0.8–1	3.77	0.29	0.92	0.38	4.11
1–2	4.40	0.28	1.61	0.47	2.73
2–5	8.72	0.32	2.79	0.34	3.12

Table 3 Properties of mortar specimens in substitution of sand and cement

Label	Mortar type	Amount of fibers by volume (%)	Expected value length of fibers (mm)	Length of fibers (mm)
Sand substitution				
S-R32.5	Standard	–	–	–
SS-R32.5_4%_0.425-0.85	Standard	4	2.84	0.425–0.8
SS-R32.5_4%_0.85–1	Standard	4	3.77	0.8–1
SS-R32.5_4%_1-2	Standard	4	4.40	1–2
SS-R32.5_4%_2-5	Standard	4	8.72	2–5
SS-R32.5_8%_2-5	Standard	8	8.72	2–5
Cement substitution				
S-R32.5	Standard	–	–	
SC-R32.5_6.4%_1-2	Standard	6.4	4.40	1–2
SC-R32.5_6.4%_2-5	Standard	6.4	8.72	2–5
SC-32.5_12.8%_2-5	Standard	12.8	8.72	2–5

2.4 *Fiber Reinforced Mortar Sample*

In this study, ordinary mortar was used as reference base material. The study follows different criteria for the specimens' production: (a) workability; (b) the fibers percentage; (c) the fibers length; (d) substitution of cement and or sand. In the criterion (d) the amount of cement and sand is reduced to the extent which corresponds to the same amount of fiber added. In each cast, elements of plain mortar were poured and tested to find out the compressive and flexural strength. The amount of sand, cement and carbon fiber were dry mixed manually in order to ensure a uniform distribution of cement and fiber in the mixture. The amount of water was carefully added to the dry mix and afterwards components were mixed thoroughly. Mix proportions of mortar–fiber mixtures are given in Table 3, where the label SS or SC refers to mixture approach (d), taking account the sand or cement substitution respectively; R32.5 is the strength class of cement, according to UNI EN 197-1; the percentage refers to the amount of fibers in the sample and the last numbers stand for sieve size.

2.5 *Mechanical Properties Optimization*

The specimens were tested under flexural and compression loading conditions, according to UNI EN 196-1 [18] protocol and the resulting average flexural and compression strength were calculated. The results are summarized in Tables 4 and 5.

Table 4 Flexural response of composites with the variation of fibre length and content

Label	Reference flexural value (MPa)	Average flexural strength (MPa)	Flexural Δ (%)
Sand substitution			
S-R32.5	–	7.53	–
SS-R32.5_4%_0.425-0.85	7.53	7.38	−2
SS-R32.5_4%_0.85-1	7.53	8.65	15
SS-R32.5_4%_1-2	7.53	7.94	5
SS-R32.5_4%_2-5	7.53	8.88	18
SS-R32.5_8%_2-5	7.53	8.68	15
Cement substitution			
S-R32.5	–	7.53	–
SC-R32.5_6.4%_1-2	7.53	8.45	12
SC-R32.5_6.4%_2-5	7,53	8,36	11
SC-32.5_12.8%_2-5	7,53	6,47	−14

Table 5 Compression response of composites with the variation of fibre length and content

Label	Reference compression value (MPa)	Average compression strength (MPa)	Compression Δ (%)
Sand substitution			
S-R32.5	–	38.0	–
SS-R32.5_4%_0.425-0.85	38.0	40.4	6%
SS-R32.5_4%_0.85-1	38.0	42.3	11%
SS-R32.5_4%_1-2	38,0	39,6	4%
SS-R32.5_4%_2-5	38.0	41.7	10%
SS-R32.5_8%_2-5	38.0	38.55	1%
Cement substitution			
S-R32.5	–	38.0	–
SC-R32.5_6.4%_1-2	38.0	35.76	−6%
SC-R32.5_6.4%_2-5	38.0	37.68	−1%
SC-32.5_12.8%_2-5	38.0	27.76	−27%

The best performances are showed by SC-R32.5_6.4%_2-5 series and are referred to a restrained increasing in strength but reduced amount of cement, in order to achieve a more sustainable product. In fact the optimal result in the case of mixture containing fibers of 6% by volume with fiber lengths of 2–5 mm in substitution of cement opens the opportunity of sustainable goals development, through an approach of circular economy for reducing negative environmental impacts. The relative load–deflection curves are illustrated in Fig. 1 compared to the control specimens without fibers.

In graph (Fig. 2), the flexural strength variance in percentage is reported with respect to corresponding reference value, for the mixes characterized by cement substitution.

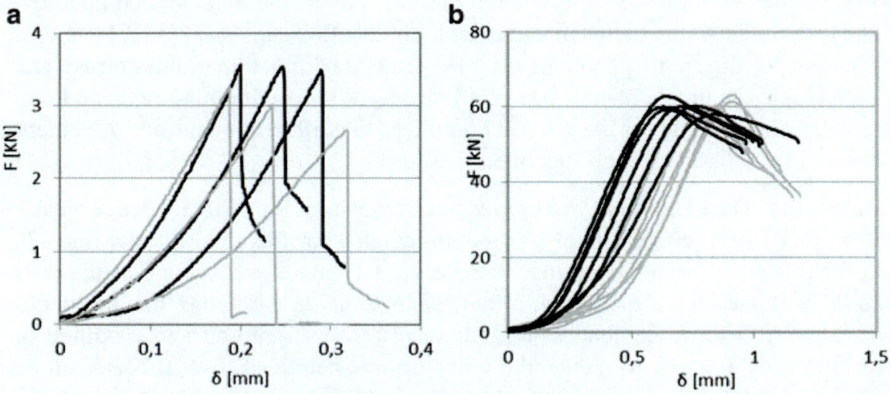

Fig. 1 **a** Load–deflection relationship in flexural regime (grey lines: control specimens, black lines: fiber content 6.4%); **b** Load–deflection relationship in compression regime (grey lines: control specimens, black lines: fiber content 6.4%)

Fig. 2 Flexural strength variance of different mixes with respect to corresponding reference value

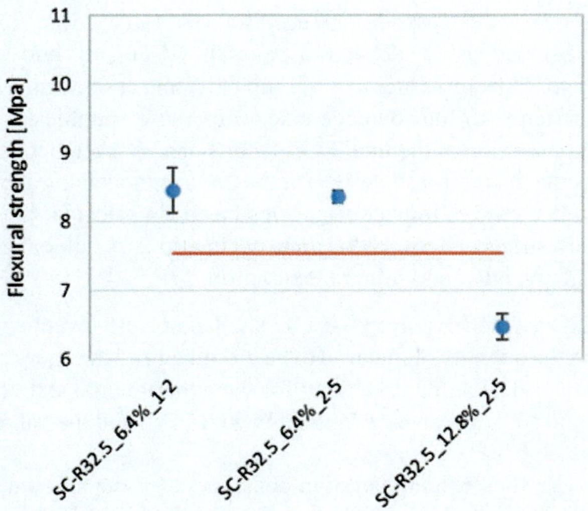

3 Life Cycle Assessment

3.1 Goal and Scope Definition

System under analysis and system boundaries. The objective of the study is to quantify the mechanical and environmental performances of an ordinary cement mortar compared with two experimental mortars. An attributional and comparative LCA (Life Cycle Assessment) in agreement with the ISO standard [16, 17] has been carried out for different specimens on a lab scale. The functional unit corresponds to each single specimen, having size of 40 mm × 40 mm × 160 mm, realized in the laboratory and suitable for the tests. It comprises the different quantities of cement, water, sand and fibers in each specimen.

Data quality. The LCA study was carried out by means of SimaPro 8.0.5, a dedicated tool provided by Pré-Consultant [19] recommended for this kind of analysis. The data for this study derive from the Ecoinvent v.3.1 database [20], one of the most reliable database on the market with thousands items. In particular, the production processes referring to cement, sand and water has been chosen and modified on research needs, in order to reduce the range of uncertainties [21–24]. For example, all the entry related to several types of transportations out of the system boundaries of this study has been deleted, to make it as close as possible to the real case.

LCIA Methodology. This stage requires the use of a mathematic model to convert values for all types of emissions, energy, heat dissipation, noise etc. into impact categories to provide essential information for a decision-making process. IMPACT 2002+ is the methodology used for this study [25], a recognized tool used already in other studies [26, 27]. It transposes the LCI results into 15 midpoint categories, then it is possible, by reducing the complexity and at same time losing scientific information, rearrange to four damage categories more suitable for a non-scientific public. The presentation of the final LCA results can be limited to few impact categories, chosen as the most representative for the environmental burdens to summarize the scenario, in this case, 6 impact categories has been selected, like: respiratory inorganics, RI, terrestrial ecotoxicity, TE, land occupation, LO, global warming, GW, non-renewable energy, NRE and mineral extraction, ME.

Life Cycle Inventory—LCI. The Life Cycle Inventory (LCI) in this study is based on the quantity of material to be produced and the energy necessary to grind the CFRP fibers. The amount, defined for the experimental test, in terms of water, cement and sand are suggested by UNI EN 196-1 [18], and the values has been reported in Table 6.

Furthermore, the system considers also the amount of electricity needed for the grinding machine to form regular size of CFRP. A technical data sheet has been used to calculate the energy consumptions of the machinery. The internal sizes of the machinery are 0.6 m × 0.4 m × 0.25 m, and so the internal volume is 0.06 m^3, assuming that it regularly works with a load of CFRP around the 10% of the total

Table 6 Series of sampling with substitution of sand and cement

	Cement (g)	Water (g)	w/c [−]	Sand (g)	Fibers [g]	Fibers percentage (%)
Sand substitution						
S-R32.5	450	225	0.5	1350	–	–
SS-R32.5_4%_0.425-0.85	450	225	0.5	1296	36	4
SS-R32.5_4%_1-2	450	225	0.5	1296	36	4
SS-R32.5_4%_0.85-1	450	225	0.5	1296	36	4
SS-R32.5_4%_2-5	450	225	0.5	1296	36	4
SS-R32.5_8%_2-5	450	225	0.5	1242	72	8
Cement substitution						
S-R32.5	450	225	0.5	1350	–	–
SC-R32.5_6.4%_1-2	421.2	225	0.53	1350	36	6.4
SC-R32.5_6.4%_2-5	421.2	225	0.53	1350	36	6.4
SC-R32.5_12.8%_2-5	394.4	225	0.57	1350	72	12.8

volume, then, the total working volume is 0.006 m^3 and a total weight of 0.8–0.9 kg. The amount of carbon fibers necessary for a specimens ranges from 36 to 72 g, so, it is reasonable and conservative assume 5% as of the share of energy consumptions accountable to the shredded CFRP needed for the mixture of a sample.

The grinding machine has a nominal power of roughly 400 W, then the time to grind a total load is 30 min, the supposed time needed only for the fibers necessary for the quantity of one specimens has been calculated as:

$$t_1 = t_0(\mathrm{Vol}_1{:}\mathrm{Vol}_1) \tag{1}$$

The time is 0.025 h, therefore the energy consumptions are less than 0.01 kWh.

The life cycle inventories for cement, sand and water come out from the database Ecoinvent 3.1, the items selected are "Cement, Portland production" for cement, "Silica sand" for the amount of sand and "Tap water production" for the use of water. All the production sites are located in Campania region, south of Italy, and all the processes are similar to those selected in the database.

3.2 *Life Cycle Impact Assessment—LCIA*

Based on the data coming out from the LCI, it was possible to calculate the impact. As mentioned above the midpoint or impact categories selected are: respiratory inorganics, RI, terrestrial ecotoxicity, TE, land occupation, LO, global warming, GW, non-renewable energy, NRE and mineral extraction, ME.

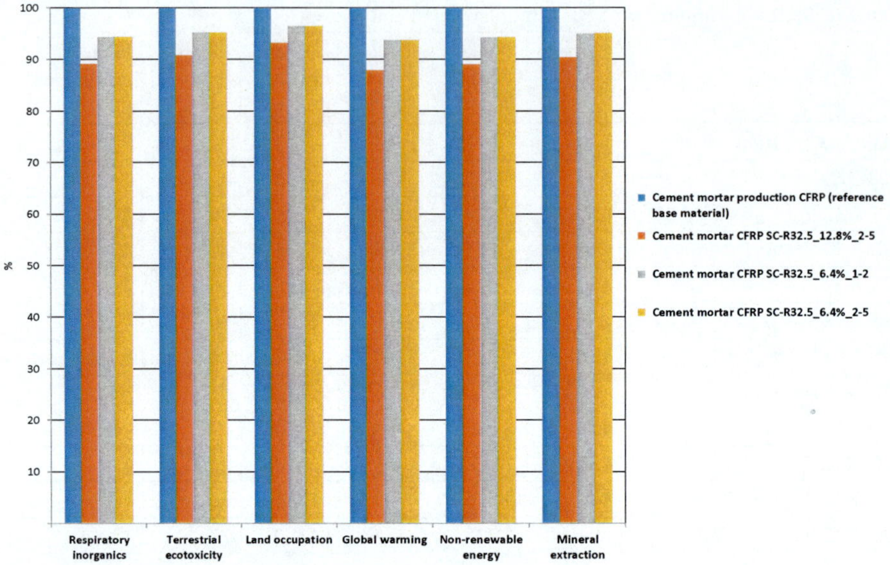

Fig. 3 Characterization of impacts for cement substitution

Figure 3 shows the results of the characterization stage that compare the baseline cement mortar and those with a substitution of CFRP fibers instead of cement. It should be noted that there are substantial environmental benefits, gradually, depending on the amount of cement that has been replaced. In fact, mortars with a substitution rate of 6.4% when compared to baseline mortar presents environmental benefits around 5%, while mortar with a substitution rate of 12.8% shows benefits for around 10%.

4 Results and Discussion

The experimental results have revealed that the addition of the recycled short carbon fibers gives the cement composite strategic properties concerning the mechanical and environmental aspects. From a mechanical point of view, the enhancement of cementitious behavior is linked to the ductility and increase in strength. Referring to the results of cement composites with addition of fibers in substitution of sand summarized in Table 3, the best performances have been obtained in the case of fiber content 4% and fiber length 2–5 mm. This mix showed an increase with respect to reference material in flexural and compression strength of 18% and 10% respectively, a significant increase in stiffness quantified in 27%, and a greater ductility. Concerning the mixes with approach in substitution of cement, the optimal result has been achieved by cement paste having carbon fibers in the amount of 6.4% by volume of mixture and length of 2–5 mm. About this, the Table 4 records a greater improvement in flexural behavior, as the increase in the slope of the load branch shows by the curves in Fig. 1a, against a drop in compression strength, due evidently to the lower amount of cement (see Table 5).

Therefore, considering SC-R32.5_6.4%_2-5 sample as the optimal solution on the one hand a good mechanical optimization is guaranteed and on the other a sustainable perspective is reproduced, as demonstrated by the life cycle assessment. In fact, the results of the life cycle assessment support those of the experimental tests on mechanical performance. Specifically, as already observed in Fig. 3, when specimens replacing cement were analyzed, the benefits were higher, especially for the SC-R32.5_12.8%_2-5 series, saving even more than 10% for most of midpoint categories analyzed. In fact, for GW from 0.394 kg of CO_2 eq of the baseline to 0.346 of the experimental mix, NRE from about 1.88 MJ primary of the baseline to about 1.67 MJ primary of the experimental mix and finally RI from 0.000146 kg of PM2.5 eq to about 0.00013 of the experimental mix.

5 Conclusions

The aim of this paper was to quantify the environmental performances of an ordinary cement mortar compared to two experimental mortars. Through such circular economy application, the traditional production model based on the use of natural resources and oriented towards maximizing gaining is overcome and any product is perceived as a benefit for economic and environmental improving. To this end, this work has developed an integrative criterion of recycling of short carbon fibers deriving from CFRP from automotive industry and reusing as a new type of cementitious material for construction industry. The results of this work have revealed clearly that the addition of a certain amount of short carbon fibers increased the mechanical performances of ordinary cement mortar, specifically in terms of flexural strength and ductility. Furthermore, the experimental results have been validated significantly by an attributional Life Cycle Assessment, in order to present a valid tool embracing different critical aspects related to the built environment posed by Global civilizations and leading to sustainable development paths.

References

1. United Nations Environmental Program (UNEP), Note 12: Towards a Life Cycle Sustainability Assessment (2011)
2. Andrew, R.M.: Global CO_2 emissions from cement production. Earth Syst. Sci. Data Discuss. 1–52 (2017)
3. Meng, F., McKechnie, J., Turner, T.A., Pickering, S.J.: Energy and environmental assessment and reuse of fluidised bed recycled carbon fibres. Compos. Part A Appl. Sci. Manuf. **100**, 206–214 (2017)

4. Sciencepractice, B.O.F.: Бюллетень науки и практики — Bulletin of Science and Practice научный журнал (scientific journal). Бюллетень Науки И Практики — Bull Sci. Pract. **6**(3466), 46–53 (2016)
5. The Council of the European Union: Off. J. Eur. Commun. **10** (1999)
6. European Parliament and Council: Directive 2008/98/EC of the European Parliament and of the Council of 19 November 2008 on waste and repealing certain directives (Waste framework). LexUriServ. do, pp. 3–30 (2008)
7. Pannkoke, K., Oethe, M., Busse, J.: Efficient prepreg recycling at low. Cryogenics (Guildf) **38**(1), 155–159 (1998)
8. Oliveux, G., Dandy, L.O., Leeke, G.A.: Current status of recycling of fibre reinforced polymers: review of technologies, reuse and resulting properties. Prog. Mater. Sci. **72**, 61–99 (2015)
9. Pimenta, S., Pinho, S.T.: Recycling carbon fibre reinforced polymers for structural applications: technology review and market outlook. Waste Manag. **31**(2), 378–392 (2011)
10. Saccani, A., Manzi, S., Lancellotti, I., Lipparini, L.: Composites obtained by recycling carbon fibre/epoxy composite wastes in building materials. Constr. Build. Mater. **204**, 296–302 (2019)
11. Badanoiu, A., Holmgren, J.: 2 Cementitious Composites Reinforced with Continuous Carbon Fibres.PDF, vol. 25, pp. 387–394 (2003)
12. Garcés, P., Fraile, J., Vilaplana-Ortego, E., Cazorla-Amorós, D., Alcocel, E.G., Andión, L.G.: Effect of carbon fibres on the mechanical properties and corrosion levels of reinforced portland cement mortars. Cem. Concr. Res. **35**(2), 324–331 (2005)
13. Mastali, M., Dalvand, A., Sattarifard, A.: The impact resistance and mechanical properties of the reinforced self-compacting concrete incorporating recycled CFRP fiber with different lengths and dosages. Compos. Part B Eng. **112**, 74–92 (2017)
14. Mastali, M., Dalvand, A.: The impact resistance and mechanical properties of self-compacting concrete reinforced with recycled CFRP pieces. Compos. Part B Eng. **92**, 360–376 (2016)
15. Ogi, K., Shinoda, T., Mizui, M.: Strength in concrete reinforced with recycled CFRP pieces. Compos. Part A Appl. Sci. Manuf. **36**(7), 893–902 (2005)
16. ISO—International Standard Organization: ISO 14044. Environmental Management—Life Cycle Assessment—Principles and Framework. Geneva, Switzerland (2006)
17. ISO: ISO 14044 Enviro Mgmt LCA Requirements and Guidelines, vol. 1 (2006)
18. UNI EN 196-1 (2016)
19. Goedkoop, M., Oele, M.: "SimaPro 6—Introduction to LCA with SimaPro," ... to LCA with SimaPro. Pré Consult. Rep. version, Oct 2004, pp. 1–78 (2004)
20. Wedema, B.P., et al.: Data quality guideline for the ecoinvent database version 3. Ecoinvent report 1 (v3), Swiss Cent. Life Cycle Invent. **3**(1) (2013)
21. Clavreul, J., Guyonnet, D., Christensen, T.H.: Quantifying uncertainty in LCA-modelling of waste management systems. Waste Manag. **32**(12), 2482–2495 (2012)
22. Groen, E.A., Heijungs, R., Bokkers, E.M., de Boer, I.: Sensitivity analysis in life cycle assessment. In: Proceedings of 9th International Conference on LCA Food, San Fransisco, USA, 8–10 Oct 2014, pp 482–488 (2014)
23. Häfliger, I.F., et al.: Buildings environmental impacts' sensitivity related to LCA modelling choices of construction materials. J. Clean. Prod. **156**, 805–816 (2017)
24. Hoxha, E., Habert, G., Lasvaux, S., Chevalier, J., Le. Roy, R.: Influence of construction material uncertainties on residential building LCA reliability. J. Clean. Prod. **144**, 33–47 (2017)
25. Jolliet, O., et al.: IMPACT 2002+: A New Life Cycle Impact Assessment Methodology. Int. J. Life Cycle Assess. **8**(6), 324–330 (2003)
26. Blengini, G.A., Di. Carlo, T.: The changing role of life cycle phases, subsystems and materials in the LCA of low energy buildings. Energy Build. **42**(6), 869–880 (2010)
27. Rosado, L.P., Vitale, P., Penteado, C.S.G., Arena, U.: Life cycle assessment of construction and demolition waste management in a large area of São Paulo State, Brazil. Waste Manag. (2019)

Optimization of Alkali-Activated Mineral Wool Mixture for Panel Production

Majda Pavlin, Ana Frankovič, Barbara Horvat, and Vilma Ducman

Abstract A significant amount of mineral wool waste is generated during the construction and demolition of buildings. At the moment, most of this material ends up in a landfill without further utilization. Alkali activation is one technology already recognized to produce low carbon dioxide binders and other products/materials from several industrial by-products, and could also be employed to recycle mineral wool waste. This study shows the result of milling and homogenization of different types of wool waste and their subsequent use as precursors in the alkali-activation process. Two different types of mineral wool waste were taken from the mining company Termit (Slovenia). Stone and glass wool were milled, pulverized and sieved below 63 μm. After homogenization, different alkali-activated pastes were prepared using two different alkali activators (NaOH and/or Na–water glass). The compressive and flexural strength of each alkali-activated material was measured, showing higher values for glass wool in comparison to stone wool. In addition, different curing temperatures were assessed (room temperature and 40 °C). The compressive and flexural strength of glass wool after three days at 40 °C was 34.7 and 8.7 MPa respectively, compared to values of 29.1 and 9.3 MPa for stone wool. After 3 days at room temperature, the strengths were not measurable, however, after 28 days the respective compressive and flexural strength were 29.9 and 14.4 MPa in the case of glass wool, and 40.6 and 14.9 MPa for stone wool.

Keywords Mineral wool · Construction and demolition waste · Recycling · Alkali-activated material · Mechanical properties · Circular economy

M. Pavlin · A. Frankovič · B. Horvat · V. Ducman (✉)
Slovenian National Building and Civil Engineering Institute, Dimičeva ulica 2, 1000 Ljubljana, Slovenia
e-mail: vilma.ducman@zag.si

M. Pavlin
e-mail: majda.pavlin@zag.si

V. M. C. F. Cunha et al. (eds.), *Proceedings of the 3rd RILEM Spring Convention and Conference (RSCC 2020)*, RILEM Bookseries 35,
https://doi.org/10.1007/978-3-030-76543-9_14

1 Introduction

Mineral wool (stone and glass wool) is the most common insulating building material worldwide. Due to the massive production of mineral wool, a huge amount of construction and demolition waste has been generated. Mineral wool waste has a low density (20–200 kg/m^3) and contains formaldehyde-based components that could leach into the environment after disposing to landfill. Due to its low weight and density, it consumes a large amount of space. In addition, sorting and separation, logistics and the lack of economically feasible uses for mineral wool waste impede its recycling. This study, which is part of the H2020 project Wool2Loop [8] aims to upgrade construction and demolition wool waste into a valuable raw material which could be used in new products and for applications based on alkali activation technology. By combining smart demolition practices with on-site analysis, the costs of the mineral wool waste can be reduced to a level which would enable reuse. Alkali activation technology is an important method to utilise mineral wool waste as a raw material to convert into new materials. A careful mix design enables the production of various value-added products with different properties (e.g. hollow core slabs, fibre-reinforced panels, acoustic panels, pavement slabs, facade panels, etc.).

Alkali activation technology can be used as an alternative to Portland cement production to prepare a low carbon dioxide binder and other products/materials from several industrial by-products, including mineral wool waste. Moreover, concrete or ceramic-like materials originating from mineral wool generate 50–80% less CO_2 emissions than regular concrete [1, 6]. Alkali activated materials are generated through a combination of an alkali source (typically sodium or potassium silicate or hydroxide) and an aluminosilicate component (blast-furnace slag, fly ash, calcined clay, etc.) [7]. During alkali activation, a structurally disordered, highly cross-linked aluminosilicate gel is formed [5]. Mineral wool waste could be a potential precursor for alkali activation due to its consistent chemical and physical composition, high content of Si in amorphous phase [4, 9].

The goal of the present work is the development of mixtures based on wool waste by applying alkali-activation technology. Glass and stone wool, which were used as precursors, were ground, homogenized and activated using sodium hydroxide and/or sodium silicate. The microstructure before and after grinding, as well as after alkali activation was analysed, and the effect of different curing temperatures on mechanical properties was assessed.

2 Experimental Part

Different types of mineral wool waste (glass and stone wool) were obtained from an open waste dump belonging to the company Termit d.d.. Samples were first separated, shaken by hand to remove particles such as pebbles and wood, and cut into small pieces. About 1 kg of wool waste was placed in a classic concrete mixer and mixed

Table 1 AAMs mix design and liquid to solid ratio (calculated according to precursor) for glass and stone wool waste

	GW1	GW2	GW3	GW4	SW1	SW2	SW3	SW4
Mineral wool waste (g)	150	150	150	150	150	150	150	150
NaOH (g)	12	3	/	6	12	3	/	6
Na_2SiO_3 (g)	/	60	96	91	/	60	129	117
H_2O (g)	65	22	/	/	98	42	/	/
L/S	0.43	0.38	0.37	0.35	0.65	0.51	0.5	0.45

for 2 h. The sample was homogenized and dried at 105 °C for 24 h in a drying oven then sieved below 63 μm. The residue was ground again in the homogenizer, sieved and a very small amount left for disposal.

Alkali-activated materials (AAM) were prepared using glass (labelled as GW) or stone (labelled as SW) wool sieved below 63 μm, and alkali activators NaOH (Donau Chemie Ätznatron Schuppen, EINECS 215-785-5) and/or Na-water glass (received from mining company Termit, with mass percentage of Na_2O 12.8%, and mass percentage of SiO_2 29.2%). NaOH and Na-water glass were prepared in different ratios, stirred until the liquid became clear (when NaOH was added into Na-water glass), and poured into the sample whilst constantly mixing.

The paste mixtures (precursors and activators) were moulded into prisms of 80 × 20 × 20 mm^3. Compressive and flexural strength were measured using a compressive and flexural strength testing machine (ToniTechnik ToniNORM) after three days cured at 40 °C and after 1, 3, 7, 14 and 28 days at room temperature (Table 1).

XRD analysis of glass and stone wool samples was performed using an Empyrean PANalytical X-ray Diffractometer (Cu X-Ray source) between 4 and 70° at intervals of 0.0263°, under cleanroom conditions in powder sample holders.

For XRF analysis (Thermo Scientific ARL Perform'X Sequential XRF) powder samples were first treated for two hours at 950 °C in a furnace (Nabertherm B 150) to remove organic compounds and carbonate and were then mixed with Fluxana (FX-X50-2, Lithium tetraborate 50%/Lithium metaborate 50%) in a ratio of 1:10 in order to lower the melting temperature. XRF analysis was performed with software OXAS on melted disks, while data was characterized using the software UniQuant 5.

The specific surface area (BET) of the milled glass and stone wools was determined by nitrogen adsorption at 77 K over a relative pressure range of 0.05–0.3 (Micrometrics ASAP 2020, Micrometrics, Norcross, GA, USA). Before BET analysis samples were heated at 105 °C for 24 h and degassed to 10^{-3} Torr (Micrometrics Flowprep equipment, Micrometrics, Norcross, GA, USA).

Scanning electron microscopy (SEM; Jeol JSM-IT500) with a tungsten filament cathode as an electron source was used to investigate samples sputtered with gold under high vacuum conditions. Samples were dried at 40 °C in a vacuum chamber before observation.

Table 2 Chemical composition of glass and stone wool, loss of ignition at 550 °C determined by the gravimetric method and BET analysis for glass and stone wool

Chemical component (wt%)	Glass wool (GW)	Stone wool (SW)
Na_2O	12.1	3.9
SiO_2	54.0	43.7
Al_2O_3	8.7	16.0
MgO	7.9	11.6
Fe_2O_3	3.0	5.3
CaO	12.5	16.8
LOI 550 °C (%)	5.38	4.62
BET (m^2/g)	0.4948	0.3722

3 Results and Discussion

3.1 *Chemical and Mineralogical Analysis*

The data showing loss of ignition at 550 °C, and the specific surface area of the different types of mineral wool is presented in Table 2. Ignition loss is higher in glass wool, suggesting that glass wool contains more organic species [2]. BET analysis demonstrates similar results for glass and stone wool; however, glass wool has a bigger specific surface area. XRF analysis shows the presence of different oxides in glass and stone wool (Table 2). Mineral wools are mainly composed of silicon, aluminium, calcium, magnesium and iron. The glass wool contains 54.0 wt % of SiO_2 and 12.1 wt % of Na_2O, which is slightly lower than in typical glass wool, but 8.7 wt % of Al_2O_3 and 7.9 wt % of MgO which is a little higher than in typical glass wool collected from Termit d.d. [3]. In contrast, the stone wool has a composition as expected. In this study, it is assumed that the glass wool may not be pure glass wool but contains some proportion of stone wool. During the collection of waste glass wool, it is hard to separate glass and stone wool completely, and the final material could be a mixture of both.

XRD analysis of precursors is shown in Fig. 1. As expected, the curves show that both types of wool are almost completely amorphous. Traces of minerals are present in the mineral wools: quartz, calcite and dolomite in the glass wool (Fig. 1a), and quartz and dolomite in the stone wool (Fig. 1b).

3.2 *Microstructural Analysis*

SEM pictures before (a, b) and after (c, d) grinding the glass and stone wool are presented in Fig. 2. Samples are composed of different fibres which vary in size and diameter, cylindrical rods and irregular random-shape particles. Although all material was sieved below 63 μm, some particles are bigger. Some particles with a

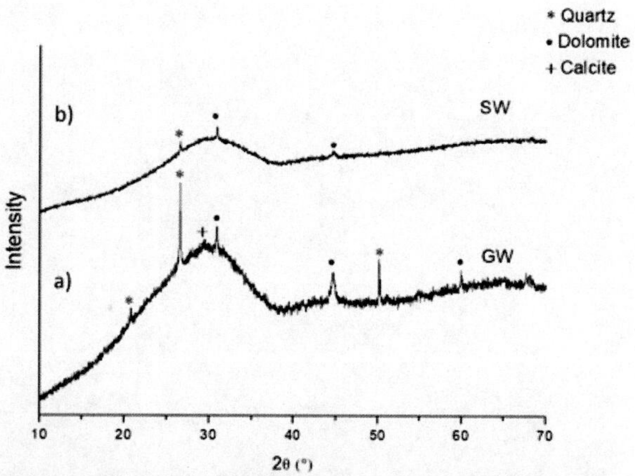

Fig. 1 XRD of precursors glass wool and stone wool

Fig. 2 SEM micrographs before pulverizing the glass (**a**) and rock wool (**b**). After grinding, samples were sieved to below 63 μm. **c** Represents glass wool with different elongated particles, while in **d** many random-shape particles are also observed

Fig. 3 SEM micrographs of the matrix from alkali activated glass wool

diameter smaller than 63 μm may enter perpendicularly during the sieving process (Fig. 2 c, d).

The SEM of mineral wool waste after alkali activation is presented in Fig. 3 (glass wool) and 4 (stone wool). The AAMs differ according to which activator was used. When NaOH was used almost no matrix is formed (GW1 and SW1). Na-water glass as an activator gave a porous structure with both mineral wools (GW2-GW4 and SW2-SW4). However, fewer pores are present in the samples GW4 and SW4. It can be seen that the initial fibres of glass wool (Fig. 3) are better incorporated into the matrix compared to stone wool fibres (Fig. 4), no matter which activator was used.

3.3 Mechanical Analysis

The glass and stone wool were alkali-activated using NaOH, Na-water glass or a combination of the two. The results for compressive and flexural strength of the different AAMs are presented in Figs. 5 and 6. All samples were cured for three days at 40 °C. The results show better compressive strengths for the mixtures prepared with glass wool, compared to those using stone wool. However, the strength is dependent on the type of activator. If using only NaOH (GW1 and SW1), the results for compressive strength are below 11 MPa for glass wool and below 9 MPa for stone

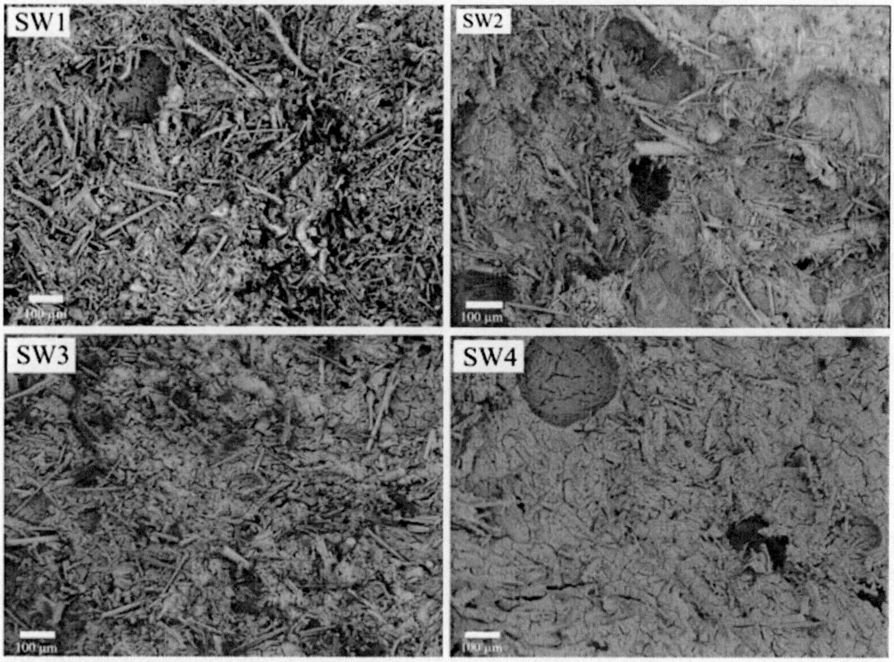

Fig. 4 SEM micrographs of the matrix from alkali activated stone wool

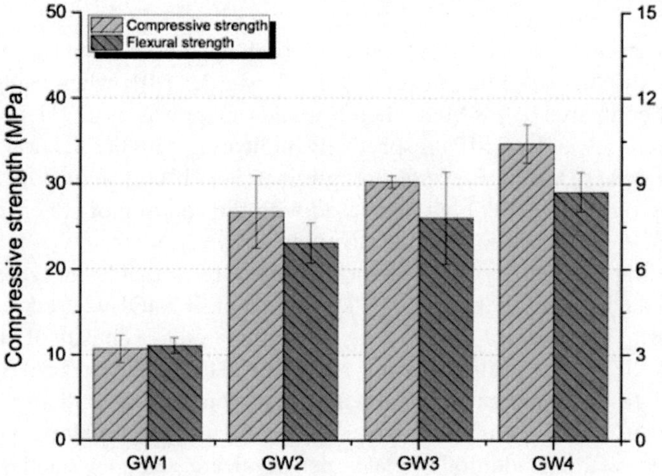

Fig. 5 The results of compressive and flexural strength for different alkali-activated glass wool materials after 3 days of drying at 40 °C

wool. Using Na-water glass as an activator or a combination of both activators (NaOH and Na-water glass) significantly improves the compressive strength, giving values

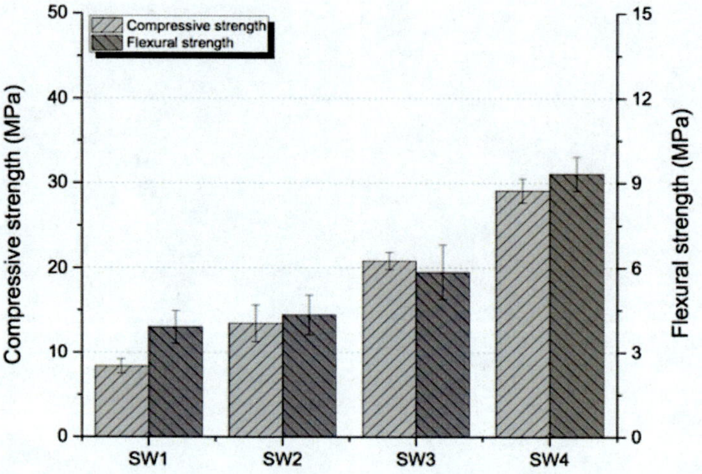

Fig. 6 The results of compressive and flexural strength for different alkali-activated stone wool materials after 3 days of drying at 40 °C

above 30 MPa for glass wool (GW2-GW4, Fig. 5) and 20 MPa for stone wool (SW2-SW4, Fig. 6). The addition of water slightly decreases the compressive strength in the case of glass wool (26.7 MPa, GW2 in Fig. 5), while compressive strength reduces significantly the stone wool (below 15 MPa, SW2 in Fig. 6). The highest compressive strengths were achieved in mixtures GW4 and SW4, eliciting values of 34.7 and 29.1 MPa, respectively.

The flexural strength of all tested mixtures was below 10 MPa when cured at 40 °C for three days. In glass wool, the flexural strength was 3.3 MPa when using NaOH as an activator compared to 7.8 MPa when Na-water glass was used; for stone wool these values were 3.9 and 5.9 MPa respectively, indicating a higher flexural strength when Na-water glass is used. Flexural strength increases if both activators are used simultaneously (around 9 MPa) but decreased with the addition of water (6.9 MPa for glass and 4.3 for stone wool), similar to compressive strength.

Data regarding the density of different mixtures are available in Table 3. The density of AAMs depends on the type of alkali activator. If NaOH is used (GW1 and SW1), the density is between 1.2 and 1.6 kg m^{-3} compared to a density of more than 1.8 kg m^{-3} in the AAMs prepared using Na-water glass (GW3 and SW3). Using both activators (a combination of NaOH and Na-water glass) for alkali activation results in a slightly lower density compared to the Na-water glass alone (GW4 and SW4). However, with the addition of water, density decreased (GW2 and SW2).

Since manufacturing AAMs at room temperature would be favourable for the project Wool2Loop (to avoid the energy required for curing at elevated temperatures), in the next step, the most promising mixtures of glass (GW4) and stone wool (SW4) were chosen. Compressive and flexural strength was measured after 1, 3, 7, 14 and 28 days of drying at room temperature. The results are shown in Fig. 7. After three days of curing at room temperature, samples were still too soft to be demoulded.

Table 3 The density of each prepared sample calculated after three days at 40 °C

Samples	Density (kg m^{-3})
GW1	1.52
GW2	1.57
GW3	1.86
GW4	1.83
SW1	1.23
SW2	1.52
SW3	1.87
SW4	1.83

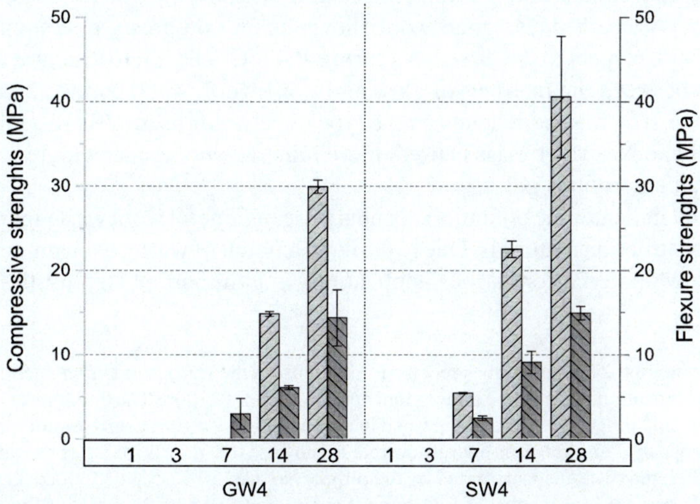

Fig. 7 The compressive and flexural strengths of alkali activated GW4 and SW4 measured after 1, 3, 7, 14 and 28 days, dried at room temperature

After one week, the compressive strength for stone wool was below 10 MPa. In the case of glass wool, it was not possible to measure the compressive strength after one week due to the fragility of the material. Curing at room temperature for 28 days (Fig. 7) results in better compressive strength for stone wool (40.5 MPa) compared to glass wool (30.0 MPa), whereas when samples were cured at 40 °C for three days (Figs. 5 and 6) better compressive strengths were observed for glass wool (34.7 MPa) compared to stone wool (29.1 MPa). The flexural strength after one week at room temperature ranged between 2 and 3 MPa for both types of AAM. After 28 days of curing, the flexural strength ranged between 14 and 15 MPa, which is better than the flexural strength of around 9 MPa which was achieved after drying the AAMs for three days at 40 °C.

4 Conclusions

Glass and stone wool waste were used for the preparation of AAMs. Both types of wool were milled and sieved under the same conditions, and the effects of different activators on the final compressive and flexural strengths has been studied. NaOH as an alkali activator results in low compressive and flexural strength, whilst using Na-water glass, or a combination of both activators significantly improves the strength properties. Compressive and flexural strength also depends on the curing temperature, whereby a higher temperature (40 °C) results in faster drying and better compressive and flexural strengths after three days compared to curing at room temperature. However, the compressive strengths of glass wool after three days at the higher temperature are comparable to those measured after drying for 28 days at room temperature. After 28 days, stone wool shows better compressive strength versus glass wool with respect to the three days curing at 40 °C. The microstructure indicates differences between the matrices of glass and stone wool, whereby the glass wool is more reactive. The matrix depends on the type of activator used. Where a mixture of both NaOH and Na-water glass is used, more mineral wool reacts with the activator and forms an amorphous gel.

This study indicates the potential secondary use of mineral wool waste for different building/construction materials. Due to the huge amount of waste, using mineral wool waste as a raw material for various applications is an important step for the circular economy.

Acknowledgements This project has received funding from the European Union's Horizon 2020 research and innovation programme under grant agreement No. 821000. The opinions expressed in this document only reflect the authors' view and in no way reflect the European Commission's opinions. The European Commission is not responsible for any use that may be made of the information it contains. The study was also supported by the project No. C3330-17-529032 "Raziskovalci-2.0-ZAG-529032" granted by Ministry of Education, Science and Sport of Republic of Slovenia. The investment is co-financed by the Republic of Slovenia, Ministry of Education, Science and Sport and the European Regional Development Fund.

References

1. Habert, G., Ouellet-Plamondon, C.: Recent update on the environmental impact of geopolymers. RILEM Techn Lett **1**, 17–23 (2016). https://doi.org/10.21809/rilemtechlett.v1.6
2. Heiri, O., Lotter, A., Lemcke, G.: Loss on ignition as a method for estimating organic and carbonate content in sediments: reproducibility and comparability of results. J Paleolimnol. **25** (2001). https://doi.org/10.1023/A:1008119611481
3. Horvat, B., Češnovar, M., Pavlin, A., Ducman, V.: Upcycling with alkali activation technology (2018). https://doi.org/10.18690/978-961-286-211-4.22
4. Kinnunen, P., Yliniemi, J., Talling, B., Illikainen, M.: Rockwool waste in fly ash geopolymer composites (full text available at: http://rdcu.be/ncgH). J. Mater. Cycles Waste Manag. (2016). https://doi.org/10.1007/s10163-016-0514-z

5. Provis, J. (2014). Geopolymers and other alkali activated materials: why, how, and what? Mater. Struct. 47. https://doi.org/10.1617/s11527-013-0211-5
6. Provis, J., Deventer, J.: Alkali activated materials: state-of-the-art report, RILEM TC 224-AAM (2014). https://doi.org/10.1007/978-94-007-7672-2
7. Provis, J., & Deventer, J. (2019). Geopolym. Other Alkali-Activated Mater. https://doi.org/10.1016/B978-0-08-100773-0.00016-2
8. WOOL2LOOP (2019). Retrieved 3 Jan 2020, from https://www.wool2loop.eu/en/
9. Yliniemi, J., Kinnunen, P., Karinkanta, P., Illikainen, M.: Utilization of mineral wools as alkali-activated material precursor. Materials **9**, 312 (2016). https://doi.org/10.3390/ma9050312

Suitability of Different Stabilizing Agents in Alkali-Activated Fly-Ash Based Foams

Katja Traven, Mark Češnovar, and Vilma Ducman

Abstract Alkali-activated foams (AAFs) are inorganic materials produced from aluminosilicate sources, such as fly-ash and metallurgical slag or clay, which contain air voids in their matrices. One possible route to the pore-forming process is chemical foaming through the use of a blowing agent such as hydrogen peroxide. Gaseous products (e.g. O_2) are formed during this process and then become trapped in a solidified structure during the hardening stage. In order to avoid the collapse of pores during this process various stabilizing agents (surfactants), such as sodium oleate, sodium dodecyl sulfate (both anionic surfactants), and Triton X 100 (nonionic surfactant), are also added to the mixture. In the present study, the AAFs were formed using fly-ash from a Slovenian thermal plant and H_2O_2 as a foaming agent. The air voids were stabilized through the addition of three different surfactants: sodium oleate, sodium dodecyl sulfate, and Triton. The effects of different quantities of foaming and stabilizing agents and different types of stabilizing agents on the mechanical properties and microstructure of foams were investigated.

Keywords Alkali-activated foams · Fly-ash · Chemical foaming · Stabilizing agent · Mechanical properties

1 Introduction

Alkali-activated materials, also called geopolymers or inorganic polymers, are important materials which could provide a suitable alternative to ordinary Portland cement

K. Traven (✉) · M. Češnovar · V. Ducman
Slovenian National Building and Civil Engineering Institute (ZAG), Dimičeva 12, 1000 Ljubljana, Slovenia
e-mail: katja.traven@zag.si

V. Ducman
e-mail: vilma.ducman@zag.si

M. Češnovar
International Postgraduate School Jožef Stefan, Jamova 39, 1000 Ljubljana, Slovenia

V. M. C. F. Cunha et al. (eds.), *Proceedings of the 3rd RILEM Spring Convention and Conference (RSCC 2020)*, RILEM Bookseries 35,
https://doi.org/10.1007/978-3-030-76543-9_15

composites and some other industrial materials, primarily due to their excellent characteristics and the reduced CO_2 footprint of their production process compared to conventional products [1]. Recent research in the field of alkali-activated materials has led to the development of several new or improved types of alkali-activated products, among them foams (AAFs) sometimes also called alkali-activated lightweight concrete [2–4], based on the use of different waste materials such as fly-ashes, clays, volcanic ashes, and various slags [5, 6]. It is worth noting that fly-ash based alkali-activated foams possess relatively good mechanical and thermo-acoustic properties, their production is cost efficient due to the use of waste materials and their environmental footprint is favourable compared to similar products which might be produced at significantly higher temperatures (such as foamed glass or ceramic foams). They can therefore be used in the building sector, e.g. in acoustic panels and lightweight pre-fabricated components for thermal insulation, as well as for certain other industrial applications [7–9].

There are several approaches to the formation of AAFs: (i) direct foaming, (ii) replica method, (iii) sacrificial filler method, (iv) additive manufacturing, and (v) other methods [10]. The most used technique is the direct foaming method, whereby air voids can be introduced into the slurry either mechanically, where pre-made foam is physically mixed into the alkali-activated material, or chemically using blowing agents such as hydrogen peroxide (H_2O_2), sodium perborate or aluminium (Al), forming gaseous products during the process [11].

$$2H_2O_2(l) \rightarrow 2H_2O(l) + O_2(g) \tag{1}$$

As shown in Eq. (1), one of the reaction products in the case of hydrogen peroxide is O_2, which is released into the slurry. The resulting structure is a highly-porous lightweight fire-resistant material with acceptable mechanical properties and good thermal or acoustic properties, which can be used in construction applications.

In order to avoid pore collapse due to the release of pressure inside the matrix the use of a stabilizing agent (SA) or surfactant is recommended. Such additives include sodium oleate and sodium dodecyl sulfate (both anionic surfactants), and Triton X 100 (a nonionic surfactant), all of which decrease the surface tension of the air/slurry system and therefore stabilize the wet foam by reducing the coalescence of bubbles. Moreover, their presence enables a better control of the following parameters: (i) cell size, (ii) size distribution and (iii) the ratio between open and closed cells [10]. The influence of different SAs on the mechanical and microstructural characteristics of AAFs has been investigated by various authors. Esmaily and Nuranian [12] used three different SAs in the same system (alkyl ether sulphate, oleic acid and sodium lauryl sulphate) and showed that specimens produced using sodium lauryl sulphate had a more homogeneous porous structure and higher strength than samples obtained using the other two SAs. Cilla et al. [13] as well as Korat et al. [14] tested various amounts of SAs in the same system, resulting in AAFs with different average cell size and open porosity values.

The aim of the present paper was to compare the effect of different quantities of foaming and stabilizing agents as well as the addition of different types of stabilizing agents on the mechanical properties and microstructure of the hardened AAFs. The alkali activation and foaming were formed using fly-ash from a Slovenian thermal plant in Na_2SiO_3 and NaOH solutions. A chemical method was used to create the foaming process with H_2O_2 (1.0 or 2.0 mass %) and the air voids were stabilized through the addition of three different stabilizing agents (sodium oleate, sodium dodecyl sulfate, and triton; 1.0 and 2.0 mass %). The microstructure of foams was investigated by means of SEM analysis.

2 Experimental

2.1 *Materials and AAFs Preparation*

Fly-ash (FA) was obtained from a thermal plant (Slovenia). Full characterization of this precursor has already been investigated by Traven et al. [15], where the presence of quartz, akermanite-gehlenite, hematite, magnesioferrite, mullite and anhydrite was confirmed, as well as over 70% of amorphous phase by means of XRD. The presence of alumosilicates was also shown through chemical analysis by means of XRF. Water glass (sodium silicate Crystal 0112 produced by Tennants distribution, Ltd., SiO_2:Na_2O = 1.97 mass %; 54.2 mass % aqueous solution) and NaOH (produced by Donau Chemie, 41.7 mass % water solution) were used as activators. 30% hydrogen peroxide solution (produced by Carlo Erba Reagents) was used as a foaming agent and three different surfactants, as described in Table 1 and in Fig. 1, used as stabilizing agents.

For this study 12 different samples of AAFs were prepared by mixing the FA (precursor) with Na_2SiO_3 and NaOH (activators) with the same solid/liquid ratio of 3.5 for all mixtures but varying the quantity of foaming agent (1.0 and 2.0 mass %) as well as the type and quantity of stabilizing agent (sodium oleate, sodium dodecyl sulfate or triton; 1.0 and 2.0 mass %). The contents and designations of the mixtures are shown and explained in Table 2. The fresh foamed pastes were poured into 20 × 20 × 80 mm^3 moulds and cured at 70 °C for 3 days. The density of all AAFs was determined by weighing the individual foams and dividing the thus determined

Table 1 Key information regarding the stabilizing agents used

Name	Internal abbreviation	Producer	Type	M (g/mol)	Appearance
Sodium dodecyl sulfate	SDS	Acros organics	Anionic	288.38	White powder
Sodium oleate	OL	Honeywell	Anionic	304,44	White powder
Triton X 100	T	Fisher scientific	Non-ionic	av. 625	Clear colourless liquid

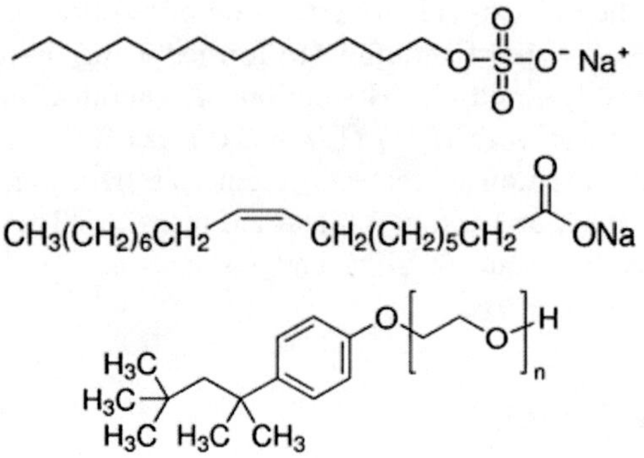

(a)

(b)

(c)

Fig. 1 Structural formulas of the following stabilizing agents: sodium dodecyl sulfate (**a**), sodium oleate (**b**) and Triton X 100 (**c**)

Table 2 List of compositions together with the sample designations[a] of different AAFs prepared for the study (in mass %)

Sample designation	FA	Na_2SiO_3	NaOH (aq)	H_2O_2	SDS	OL	T
H1SDS1	0.73	0.22	0.03	0.01	0.01	/	/
H1OL1					/	0.01	/
H1T1					/	/	0.01
H1SDS2					0.02	/	/
H1OL2					/	0.02	/
H1T2					/	/	0.02
H2SDS1				0.02	0.01	/	/
H2OL1					/	0.01	/
H2T1					/	/	0.01
H2SDS2					0.02	/	/
H2OL2					/	0.02	/
H2T2					/	/	0.02

[a]Sample designations explanation: H stands for hydrogen peroxide followed by its concentration followed by the stabilizing agent abbreviation followed by SA' concentration

weights by the corresponding dimensions of the specimens (i.e. so called geometrical density).

2.2 *Characterization Methods and Instruments*

Mechanical strength (compressive strength) was determined at an age of 3 days using Toninorm test equipment (Toni Technik, Germany), using a force application rate of 0.005 kN/s. Microstructural analysis of cross sections of the hardened AAFs was performed using a JEOL JSM-IT500 LV (back-scattered electrons image mode) SEM in a low vacuum.

3 Results and Discussion

The influence of the three selected stabilizing agents was investigated by applying two different concentrations (1.0 and 2.0 mass %) of each to two different amounts of the selected foaming agent (H_2O_2). The resulting density and compressive strength measurements are presented in Table 3 and Fig. 2. The measured density decreases as the amount of foaming agent increases, but the effect of a change in the amount of SA added varies according to the type of SA used. As the amount of SDS increases, the density decreases, in the case of OL the density stays almost the same and in the case of T the density slightly increases (when the 1.0 mass % of the foaming agent is added to the mixture). The trend for SDS supports the findings made by Korat et al. [14]. Since the trend is not exactly the same when the amount of H_2O_2 is 2.0 mass % general conclusions for OL and T could not be made solely on the basis of this study.

As expected, the mechanical strengths correlate with the sample densities. For instance, the compressive strength of the sample H1SDS1 with the highest density of 0.83 g/cm^3 is 3.47 MPa but it decreases significantly in the case of sample H2SDS2 (to 0.15 MPa with a density of 0.37 g/cm^3). The foams with 2.0 mass % of H_2O_2 added are in all cases very fragile with a very low compressive strength, as is evident from Table 3 and the macro-images in Table 4, and were therefore not suitable for further investigation by means of SEM. With regard to the effects of the type of SA added on the mechanical strength, the best performances are shown by SDS when 1.0 mass % of SA is added and by T, when higher amounts of SA are added to the mixture (i.e. 2.0 mass %).

Table 4 shows the macroscopic captures of the surface pore structure when 1.0 or 2.0 mass % of hydrogen peroxide is used for foaming and different types and

Table 3 The measured compressive strengths and densities (together with the corresponding standard deviations) of the studied AAFs

Sample designation	Compressive strength (MPa)	Density (g/cm^3)
H1SDS1	3.47 (0.44)	0.83 (0.04)
H1OL1	2.70 (0.58)	0.81 (0.03)
H1T1	1.67 0.31)	0.70 (0.01)
H1SDS2	1.25 (0.15)	0.63 (0.01)
H1OL2	2.53 (0.67)	0.80 (0.04)
H1T2	2.54 (0.26)	0.73 (0.02)
H2SDS1	0.45 (0.15)	0.44 (0.04)
H2OL1	0.27 (0.12)	0.46 (0.04)
H2T1	0.76 (0.26)	0.53 (0.02)
H2SDS2	0.15 (0.08)	0.37 (0.01)
H2OL2	0.71 (0.38)	0.51 (0.13)
H2T2	0.42 (0.10)	0.49 (0.02)

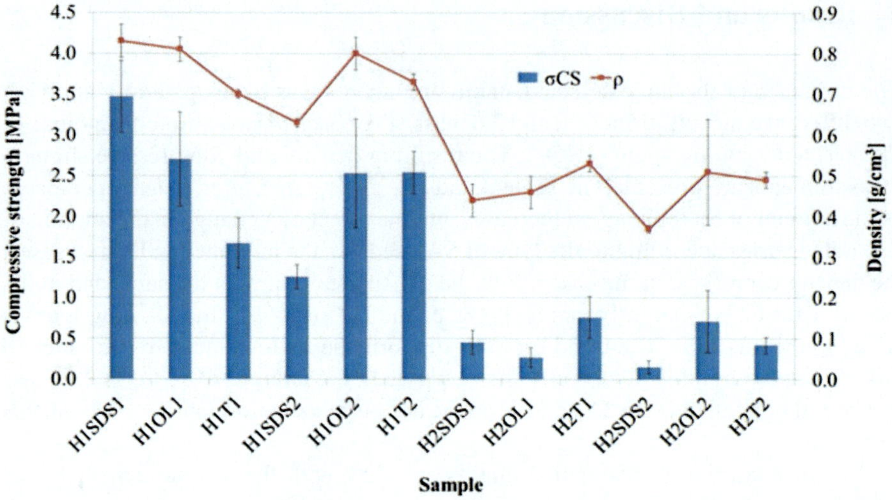

Fig. 2 Comparative graphic presentation of the compressive strength (σ_{CS}) and density (ρ) results shown in Table 3

amounts of stabilizing agent are used. The increased addition of foaming agent led to an increase in pore size and macro-pore distribution, particularly where OL and T were used as stabilizing agents. All samples also showed almost identical collapsing of pores when a higher amount of foaming agent was added. This is most evident in sample H1T1, where the upper side of the material is visibly curved due to the pore collapse.

Secondly, when comparing the effects of adding different amounts of SA on the final structure of the AAFs we can conclude that in the case of SDS addition foam samples appeared to be comparable at macro level. We could draw similar conclusions for the sample pairs H1T1, H1T2 and H2OL1, H2OL2. Contrarily there is an observable difference in the sample pair H1OL1 and H1OL2, where the addition of higher amounts of SA resulted in a greater number of smaller pores (also confirmed by SEM pictures in Fig. 3). The difference is also noticeable in the samples H2T1 and H2T2, where the higher amount of added SA prevents the pores collapsing, meaning that the sample H2T2 is not curved on the upper side of the AFF.

Last but not least, comparison of the AAFs with different types of SA reveals the biggest difference among them. In the samples where SDS and OL are used as SAs the pores are smaller and more uniformly distributed, whereas in the case of the addition of T the pores are bigger and more varied in size. Overall we can conclude that both the type and amount of SA could significantly affect the final structure of the materials.

The findings described above are supported by the SEM microstructure analysis of the selected AAF samples, which was made in order to more precisely define the specific contrast between pore size and their distribution (Fig. 3). Low magnification of the inner microstructure consists of clearly visible micro-pores, meso-pores, and

Table 4 Macro-images of all 12 sample cross-sections

		1.0 mass % of H_2O_2	2.0 mass % of H_2O_2
SDS	1.0 mass %		
	2.0 mass %		
OL	1.0 mass %		
	2.0 mass %		
T	1.0 mass %		
	2.0 mass %		

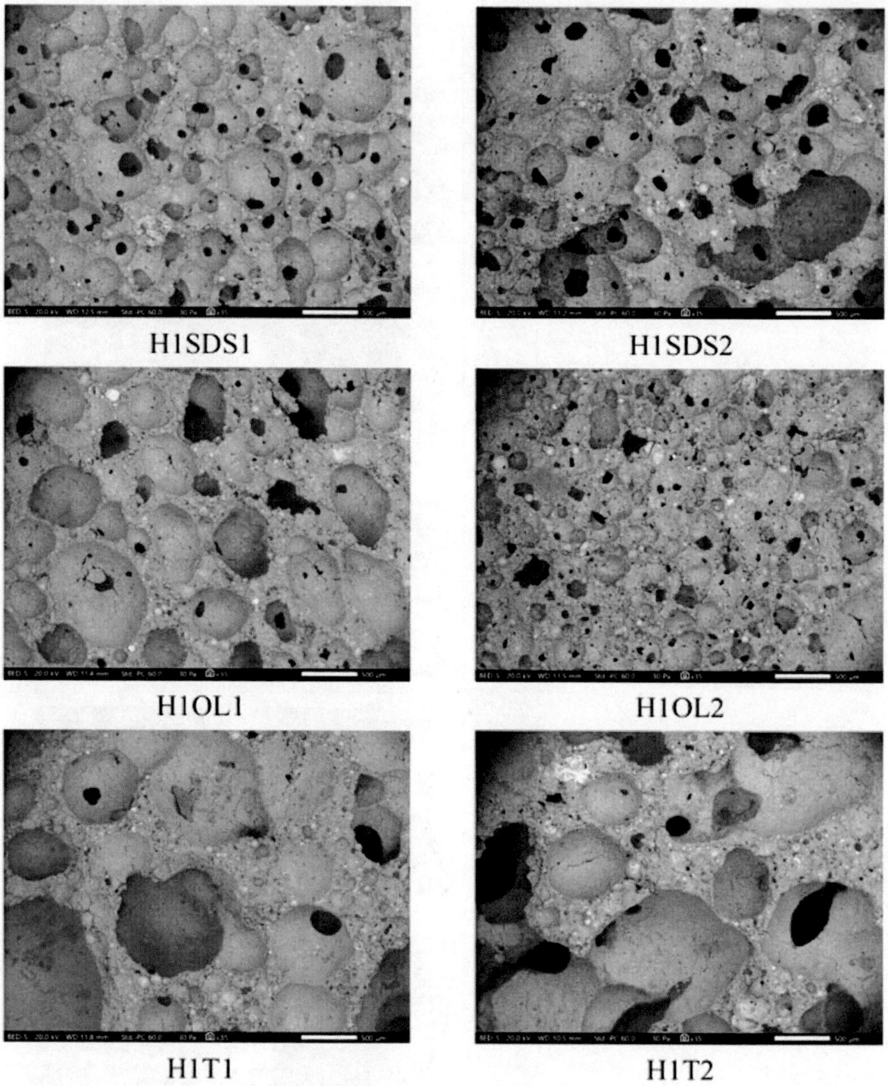

Fig. 3 SEM images of the selected AAFs at 35× magnification

cracks. Pore structures are round shaped for all sample types regardless of the SA type, whereas the average pore size differs significantly among them (from the smallest average pore size of 200 μm for sample H1OL2 to the biggest average pore size of 700 μm for sample H1T2). In all samples the spherically-shaped remains of unreacted FA precursor are also observed. To add, the effect of pore percolation is also observed in all selected AAFs.

4 Conclusions

The present study investigated the usability of different types and amounts of stabilizing agents in an alkali-activated foaming process. The alkali activation and foaming were performed using fly-ash from a Slovenian thermal plant (Šoštanj) in Na_2SiO_3 and NaOH solutions. The influence of sodium dodecyl sulfate (SDS), sodium oleate (OL) and triton (T) as selected stabilizing agents was investigated by applying two different concentrations (1.0 and 2.0 mass %) of each to two different amounts of selected foaming agent (H_2O_2). Assessment of density and mechanical strength measurements as well as SEM analysis allows the following conclusions to be made:

- The material density decreases with an increased amount of foaming agent, whereas the effect of the amount of SA added varies according to the type. The densities of AAFs range between 0.37 and 0.83 g/cm^3.
- The compressive strengths correlate with the sample densities and are between 0.15 and 3.47 MPa.
- SEM analysis revealed that the pore structures are round shaped for all sample types regardless of the SA type whilst the average pore size differs significantly among them (from the smallest average pore size of 200 μm for sample H1OL2 to the biggest average pore size of 700 μm for sample H1T2).
- The overall conclusion is that the compressive strength as well as the microstructure of AAFs is significantly affected by the quantity of stabilizing and foaming agents as well as by the type of stabilizing agent used. The best performance was achieved using the oleate as a stabilizing agent (sample H1OL2 with 1.0 mass % of H_2O_2 and 1.0 mass % of OL), giving the smallest average pore size together with a satisfactory compressive strength (i.e. 2.53 MPa).

Acknowledgements The authors wish to thank the Slovenian Research Agency (ARRS) for the project grant J2 9197.

References

1. Provis, J.L., Bernal, S.A.: Geopolymers and related alkali-activated materials. Annu. Rev. Mater. Res. **44**, 299–327 (2014)
2. Zhang, Z.H., Provis, J.L., Reid, A., Wang, H.: Geopolymer foam concrete: an emerging material for sustainable construction. Constr. Build. Mater. **56**, 113–127 (2014)
3. Masi, G., Rickard, W.D.A., Vickers, L., Bignozzi, M.C., van Riessen, A.: A comparison between different foaming methods for the synthesis of light weight geopolymers. Ceram. Int. **40**(9), 13891–13902 (2014)
4. Cilla, M.S., Morelli, M.R., Colombo, P.: Effect of process parameters on the physical properties of porous geopolymers obtained by gelcasting. Ceram. Int. **40**(8), 13585–13590 (2014)
5. Abdollahnejad, Z., Pacheco-Torgal, F., Felix, T., Tahri, W., Aguiar, J.B.: Mix design, properties and cost analysis of fly ash-based geopolymer foam. Constr. Build. Mater. **80**, 18–30 (2015)

6. Zhang, R.F., Feng, J.J., Cheng, X.D., Gong, L.L., Li, Y., Zhang, H.P.: Porous thermal insulation materials derived from fly ash using a foaming and slip casting method. Energy Build. **81**, 262–267 (2014)
7. Novais, R.M., Buruberri, L.H., Ascensao, G., Seabra, M.P., Labrincha, J.A.: Porous biomass fly ash-based geopolymers with tailored thermal conductivity. J. Clean. Prod. **119**, 99–107 (2016)
8. Lach, M., Korniejenko, K., Mikula, J.: Thermal insulation and thermally resistant materials made of geopolymer foams. Ecol. New Build. Mater. Prod. **151**, 410–416 (2016)
9. Rickard, W.D.A., Vickers, L., van Riessen, A.: Performance of fibre reinforced, low density metakaolin geopolymers under simulated fire conditions. Appl. Clay Sci. **73**, 71–77 (2013)
10. Bai, C.Y., Colombo, P.: Processing, properties and applications of highly porous geopolymers: a review. Ceram. Int. **44**(14), 16103–16118 (2018)
11. Ducman, V., Korat, L.: Characterization of geopolymer fly-ash based foams obtained with the addition of Al powder or H_2O_2 as foaming agents. Mater. Charact. **113**, 207–213 (2016)
12. Esmaily, H., Nuranian, H.: Non-autoclaved high strength cellular concrete from alkali activated slag. Constr. Build. Mater. **26**(1), 200–206 (2012)
13. Cilla, M.S., Colombo, P., Morelli, M.R.: Geopolymer foams by gelcasting. Ceram. Int. **40**(4), 5723–5730 (2014)
14. Korat, L., Ducman, V.: The influence of the stabilizing agent SDS on porosity development in alkali-activated fly-ash based foams. Cement Concr. Compos. **80**, 168–174 (2017)
15. Traven, K., Češnovar, M., Škapin, S., Ducman, V.: Evaluation of fly ash based alkali-activated foams at room and elevated temperatures. In: Bogataj, M., Kravanja, Z., Novak-Pintarič, Z. (eds.) CONFERENCE 2019, International Conference on Technologies & Business Models for Circular Economy, Portorož, Slovenia (2019) (article in press)

Influence of Specific SCM on Microstructure and Early Strength of Sustainable Cement Blends

O. Rudic, J. Juhart, J. Tritthart, and M. Krüger

Abstract Blended binders combining different hydraulically active materials and inert fillers, have a high potential to improve sustainability of cement and concrete. In this study a portion of ordinary Portland cement (OPC) up to 40 wt.% was replaced by supplementary cementitious material (SCM): two limestone powders (LSP) of different fineness and ground granulated blast furnace slag (GGBFS). Their influence on microstructure and strength development and of blended formulations is presented. While strength development was assessed by compressive strength measurements up to three months, the effects on microstructure were investigated by Mercury intrusion porosimetry and air permeability measurements. The results show that refinement of pore size distribution and altered capillary porosity is strongly influenced by both, the ratio of water to total binder content (w/b) and of water to hydraulically effective portion of binder (w/b_{hy}) related to the effects of combined SCMs. Significantly higher strength gains (between 1st to 28th and 1st to 91st day) of blended mixes were found due to joint contribution of latent-hydraulic reaction of GGBFS and strength accelerating "filler-effect" of inert LSP. Sole influence of inert fillers was found to be significantly higher in case of lower w/b (water content per volume) due to increased packing density.

Keywords Cement replacement materials · Blended cement mix design · Microstructure · Compressive strength development

1 Introduction

The worldwide production of Portland cement (PC) of approximately 4.2 billion tons per year, with 1-ton production of cement clinker being responsible for nearly 0.87 ton

O. Rudic (✉) · J. Juhart · J. Tritthart · M. Krüger
Institute of Technology and Testing of Building Materials (IMBT-TVFA), Graz University of Technology, Inffeldgasse 24, 8010 Graz, Austria
e-mail: ognjen.rudic@tugraz.at

J. Juhart
e-mail: joachim.juhart@tugraz.at

V. M. C. F. Cunha et al. (eds.), *Proceedings of the 3rd RILEM Spring Convention and Conference (RSCC 2020)*, RILEM Bookseries 35,
https://doi.org/10.1007/978-3-030-76543-9_16

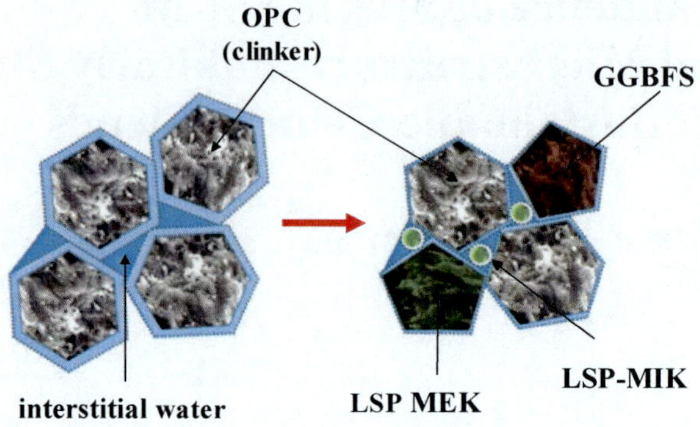

Fig. 1 Ordinary OPC paste (left) and optimized paste (right) with blended cements (OPC, GGBFS, MEK, MIK). Blended cements can have enhanced packing density at equal flowability as well as beneficial effects for functional properties and environmental impact

of carbon dioxide exhaust in atmosphere, presents a serious environmental issue [1, 2]. In order to satisfy the increasing worldwide demand of cement and concrete based materials and to reduce CO_2 emissions at the same time, many studies recommend substituting PC by SCMs to reduce environmental impact of the binder material itself but also to improve concrete performance and durability [3–21].

The principles of mix-design of eco-efficient blended binders for sustainable concrete were pointed out inter alia by [1, 10, 19]. The mix-design approach of this study is illustrated in Fig. 1 according to [10]. OPC is blended with SCMs of different fineness, different particle size distribution, mean diameter respectively and then compared to pure OPC paste. Cement clinker—which represents 95 wt.% of OPC—is primarily responsible for GWP (global warming potential) and PEt (total primary energy consumption) of normal concrete [20, 21]. A blend is optimized in terms of its environmental impact when OPC with its high GWP and PEt is partly substituted by properly selected SCMs, which on the rule have lower environmental loads. The SCMs used within this study can be distinguished in very fine micro-fillers (MIK) and coarser eco-fillers (MEK). An appropriate combination can (i) optimize packing density in a way that lower the water demand of the blend for achieving a specific flowability or (ii) increase the specific surface area (SSA) of the mix which may affect early strength and (iii) be beneficial regarding durability [10]. MEK can be hydraulically active materials like GGBFS as well as inert fillers like LSP-MEK used in this study. MIK can also be hydraulically active materials (not used in this study) or inert materials like the used LSP-MIK. An optimized mix of OPC/MEK/MIK according to Fig. 1 requires less water for achieving a specific flowability than ordinary OPC paste (of 100% OPC), mainly due to the physical filler effect of MIK but also the alteration of particle interaction. Moreover, addition of superplasticizers (SP) reduces the water demand for achieving a specific workability. As SP have a high environmental impact, they must be used sparingly in an eco-optimization process. For a systematic comparison of different blends in this study binder mixes were made with two given, nearly constant w/b-ratios as pointed out below, where b denotes to the sum of all powders (<125 μm) of the blend.

2 Materials and Mix Design

The mineral raw constituents show notable differences in particle size and specific surface characteristics (Table 1). The most prominent difference is the d_{50} values of MEK and finer MIK. Considering chemical composition of OPC and SCM materials, notable difference is in the calcite content (Table 1). Mix design of blended cement pastes and mortars was prepared by mixing CEM I 52.5R (OPC) with varied quantities of GGBFS and LSP (MIK and MEK) and with two different water contents (Table 2). Note that with (nearly) constant w/b-ratio, the w/cl-ratio (i.e. water to clinker ratio, calculated with the assumption of 95 wt.% of clinker in OPC according to EN 197-1 [22]) and the w/b_{hy} ratio (i.e. water to clinker + GGBFS ratio) vary. The main aim during mix design was to achieve very similar flow characteristics of all fresh mixes (Table 2), which was achieved by adding SP (0.07–0.15 wt.%/binder) to mixes with low water content per volume (w/b = 0.36–0.45).

Table 1 Properties of investigated mineral raw materials

Type	ρ (g/cm^3)	d_{50} (μm)	Blaine (mm^2/g)	Calcite (wt.%)
CEM I	3.12	7.1	4270	3.6
GGBFS	2.85	10.9	3046	1.2
MEK	2.70	5.0	4785	≥ 95.0
MIK	2.70	1.2	(10,560)[a]	≥ 97.0
QUARTZ SAND	2.63	984	/	/

[a]Out of the reliable measuremnt range of Blaine method

Table 2 Mix design for the reference cement and cement blends (w/b—water to binder ratio, w/cl—water to clinker ratio and w/b_{hy}—water to hydraulically binder ratio)

Sample	CEM I (wt.%)	GGBFS (wt.%)	MEK (wt.%)	MIK (wt.%)	w/b	w/cl	w/b_{hy}	Flow (mm)
A1	100				0.40	0.42	0.42	178
A2	100				0.60	0.63	0.63	183
B1	70	30			0.40	0.60	0.42	171
B2	70	30			0.60	0.90	0.62	200
J1	60	30		10	0.45	0.79	0.52	191
J2	60	30		10	0.60	1.05	0.69	201
G1	60		32.5	7.5	0.36	0.63	0.63	183
G2	60		32.5	7.5	0.60	1.05	1.05	200
K1	60	22.5	10	7.5	0.45	0.79	0.57	195
K2	60	22.5	10	7.5	0.60	1.05	0.75	201
M1	60		40		0.45	0.79	0.79	169
M2	60		40		0.60	1.05	1.05	200

2.1 Sample Preparation

The different paste mixes were placed in cylindrical plastic tubes (diameter of 70 mm, height of 150 mm) and rotated for 24 h to prevent bleeding and segregation. After that the samples were packed in plastic bags and stored for another 27 days at 20 °C and 65% of relative humidity. After the 28-day curing period the paste samples were unpacked and dried at 105 °C for 24 h in order to stop further hydration process. Then they were again sealed in plastic bags and stored at ambient room temperature. The microstructure was investigated at a sample age of about 3 months.

In addition to pure paste mixes mortar samples were prepared as well. The volume of the paste was kept constant for all mortar samples (43.5 vol.%) regardless to the water content. For the remaining 56.5 vol.%. of the mix quartz sand with a maximum aggregate size of 4 mm was used (Table 1). Two types of mortar samples were prepared; firstly, prisms with standard dimensions of 16 × 4 × 4 cm to be used for compressive strength (CS) determination and secondly, with dimensions of 36 × 14 × 10 cm for air-permeability investigation. All mortar specimens were cured for the first 24 h in molds covered with plastic sheet at 20–23 °C and then stored until 28th day in the same way as the paste samples.

2.2 Testing and Evaluation Methods

Porous structure modification of hardened paste blends was investigated by mercury intrusion porosimetry (MIP) with Pascal 440 Thermo-Scientific. Mercury (density 13.53 g/cm^3, surface tension 0.485 N/m and contact angle 130°) was used for calculating the pore sizes by applying the Washburn equation [23]. Skeletal density (i.e. solid density) was measured using Pycnomatic ATC (Thermo-Scientific) instrument with Helium as a working gas medium. In this study, pore size distribution (PSD) obtained by MIP method is divided into three main groups: gel- and mesopores, middle capillary pores and larger capillary pores with pore radius in the ranges of 0–0.05 μm, 0.05–0.1 μm and 0.1–100 μm, respectively. The intrudable porosity and total pore volume were evaluated as well as critical pore radius (R_{cr}) according to [24] with two specimens investigated for each mix.

Gas diffusion of mortar samples with blended cements was assessed by non-destructive technique of air-permeability measurement by a permeaTORR-device [25] in accordance to SIA 262/1:2013 [26]. The air-permeability measurement was performed on the upper surface of the samples (not affected by mould walls) with 5 measurements per each sample. Compressive strength development of reference and mortar samples with blended cements were assessed was measured according to EN 196-1 [27] over time up to 91 days.

3 Results and Discussion

3.1 Microstructure Assessment

According to MIP porosity assessment the substitution of OPC by SCMs leads to noticeable changes of microstructural properties by evident change of pore structure (Table 3, Fig. 2). Overall, mixes with higher w/b-ratios (mixes A2 to M2), which corresponds to higher water content per volume of the paste, possess higher porosity (i.e. total intrudable porosity, total pore volume and R_{cr}) than those with lower w/b (mixes A1 to M1). At (nearly) constant w/b-ratios (w/b = 0.36–0.45; Fig. 2a vs. w/b

Table 3 Porosity parameters measured by He-pycnometry and MIP

Sample	Skeletal density (g/cm^3)	Intrudable porosity (%)	Total pore volume (mm^3/g)	R_{cr} (μm)
A1	2.336	25.5	167	0.096
A2	2.122	36.0	265	0.108
B1	2.174	26.9	174	0.072
B2	2.115	38.6	327	0.162
J1	2.089	30.8	212	0.111
J2	2.034	41.2	356	0.156
G1	2.313	29.0	180	0.119
G2	2.024	44.0	369	0.188
K1	2.183	30.0	223	0.135
K2	2.173	43.1	321	0.161
M1	2.225	31.6	189	0.135
M2	2.116	41.0	334	0.144

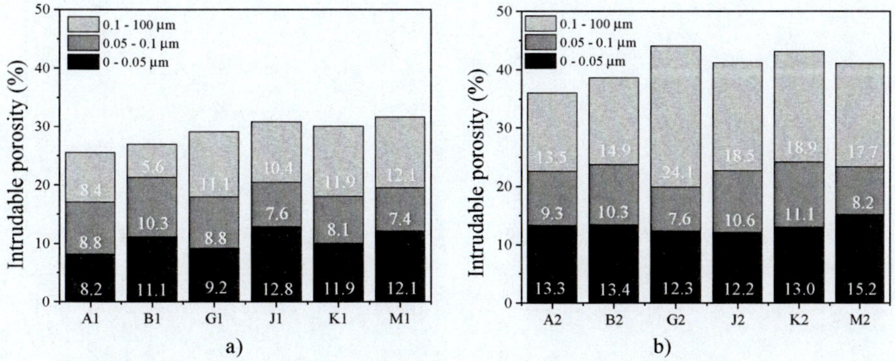

Fig. 2 Fractional analysis of pore size distrbution for hardened pastes with **a** w/b = 0.36–0.45; lower water quantity per volume and **b** w/b = 0.6; higher water quantity per volume

= 0.6; Fig. 2b), however, there are noteworthy differences depending on the content of GGBFS and LSP (MEK and MIK). Mixes with sole GGBFS as OPC substitute (B1 and B2) show only a slight increase of overall intrudable porosity in comparison to reference pastes (A1, A2), which is in accordance with findings of [28]. With a longer period of hydration of the GGBFS-samples, this relation could change in favour of GGBFS versus pure OPC, which however was not investigated in this study. At low w/b-ratio of 0.4, a lower portion of coarser capillary pores (0.1–100 μm) and lower R_{cr} of GGBFS-mix B1was observed compared to reference A1 (Fig. 2; Table 3).

With an increasing proportion of LSP at (nearly) constant w/b, the total intrudable porosity increases, Fig. 2. The majority of the mixes (J1 vs. B1, J2 vs. B2, K1 vs. J1; G2 and M2 vs. A2) also show an increasing portion of coarser capillary pores (0.1–100 μm) with more LSP. But there is no systematic trend over all mixes, as for example, G1 and M1 show similar or even lower proportion of middle capillary pores (0.05–0.1 μm) than comparable pure OPC paste A1, e.g. mix G1 with LSP (w/b = 0.36, w/cl = 0.63) shows significantly lower intrudable porosity than reference mix A2 (w/b = 0.6; w/cl = 0.63), although w/cl-ratio is the same. This is due to the filler effect of the LSP and the overall reduction of clinker (resp. reactive powder) content in the mix. Further confirmation of described LSPs influence on overall intrudable porosity increase can be seen when OPC replacement ratio was increased by adding MIK and MEK to mixes together with GGBFS (as can be seen by comparing J and K paste blends with B paste blends in Fig. 2) regardless to water content. The observed increase occurred due to pore size distribution refinement predominantly in capillary porosity range (Fig. 2).

By plotting the intrudable porosity versus w/b_{hy} and distinguishing between mixes with and without GGBFS (Fig. 3a), the influence of solely inert and inert/latent hydraulically binders on porous structure modification can be estimated. The obtained simplified linear correlation label very good mutual relationship of intrudable porosity with w/b_{hy} with $R^2 = 0.927$ and $R^2 = 0.817$ for mortar mixes with and

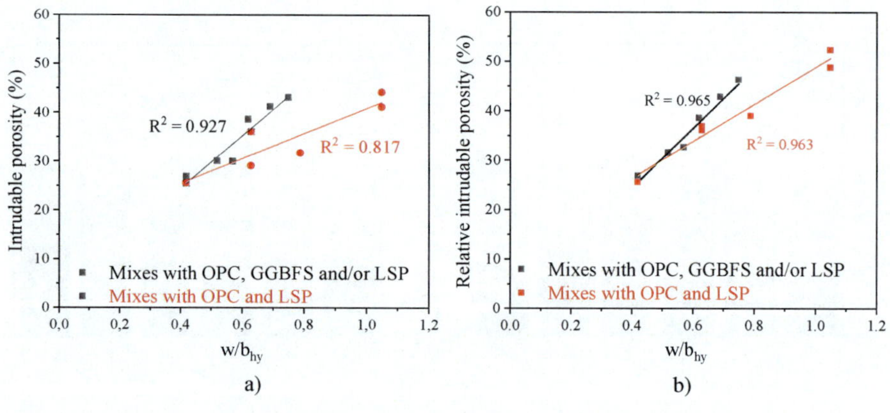

Fig. 3 Correlation of **a** intrudable porosity versus w/b_{hy}, **b** relative intrudable porosity versus w/b_{hy} with a distinction between mixes with solely inert and hydraulic and inert SCMs

without GGBFS, respectively. It has to be noted that the degree of hydration of all mixes was assumed the same, which does not hold true.

However, if the LSP is assumed as a pure inert filler material that does not have a significant effect on total porosity, the porosity measured by MIP can be assigned to the volumetric reactive powder/water fraction and the corresponding hydration products. The relationship between intrudable porosity relative to the calculated water and reactive binder's volumetric fractions ("relative intrudable porosity") and w/b_{hy} is shown in Fig. 3b. Performing analysis in such manner, the correlation between the calculated relative porosity and w/b_{hy} ratio is better than 0.96. Thus, it can be seen that the intrudable porosity development is dominated by the volume of hydraulically reactive powder fractions and water during hydration process. Additionally, mixes with inert filers show lower relative intrudable porosity at the constant w/b_{hy} ratio. The above analysis may indicate that the intrudable porosity of an individual mix composition may be specified (or predicted) as a function of the hydrated or bound water (representable as w/b_{hy}) taking into account the volumetric contribution of hydraulically binders optimized by volumetric content of inert filers. The functional relationship between porosity and such mix parameters will be the subject of further, more detailed investigations.

3.2 *Air-Permeabilty Assessment*

As expected, air permeability coefficients kT of blended mortars are lower for the mixes with lower w/b (w/b = 0.36–0.45, mixes A1 to M1, except of B1) compared to the ones with higher w/b (see Fig. 4). All blended mortar samples with lower w/b fall into the classes of covercrete permeability "very low" (kT = 0.001–0.01 × 10^{-16} m^2) or "low" (kT = 0.01–0.1 × 10^{-16} m^2) according to Swiss standard SIA 262-1:2013 and suggest possible good durability.

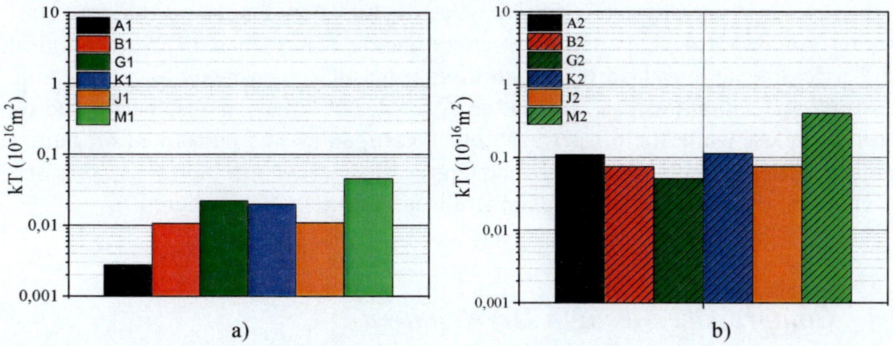

Fig. 4 Results of air permeability assessment for **a** lower (w/b = 0.36–0.45) and **b** higher (w/b = 0.6) water content

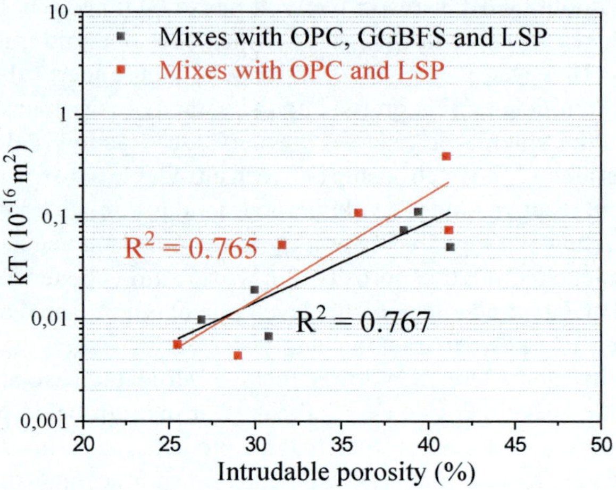

Fig. 5 Corelation between air-permeability coefficient (kT) and total intrudable porosity with a distinction between mixes with solely inert and hydraulic and inert SCMs

Mortars with higher w/b are on the limit to class "moderate air permeability" (kT = 0.1–1 × 10^{-16} m^2), mix M2 with LSP MEK clearly falls into this higher permeability class. It thus should be characterized as more pronounced to corrosion processes. Similar to MIP pore structure assessment (Sect. 3.1), it was found that air-permeability strongly depends on the type of SCM used. Namely, the highest kT values were detected in case of M1 sample, with LSP MEK followed by G1 (LSP MEK plus MIK) and K1 (GGBFS, MEK plus MIK) samples, while J1 and B1 has very similar air-permeability values.

A linear correlation between kT and total intruded porosity was found with a reasonable correlation coefficient (Fig. 5) for mixes separated in those with and without GGBFS. However, general dependency of mortar samples air permeability on paste samples intruded porosity obtained by MIP technique is not followed by appropriate correlation coefficient as in case of microstructural characterization (Fig. 3). There are two reasons for the observed discrepancy.

First, a different physical characteristics of fluid flow employed during examination (air flow for air permeability investigation versus mercury penetration at MIP assessment) may lead to certain divergence of air permeability versus total porosity correlation analysis. On the other hand, MIP analysis was performed on paste samples, while the air permeability investigation was performed on mortar samples. Consequently, impact of sand aggregate employed in mortar samples and its effect on porous structure development is not adequately considered.

3.3 Compressive Strength Development

Figure 6 highlights the different increase of CS achieved up to 1st day, between 1st and 28th day and 1st day and to 91th day. It indicates that mortar mixes with a portion

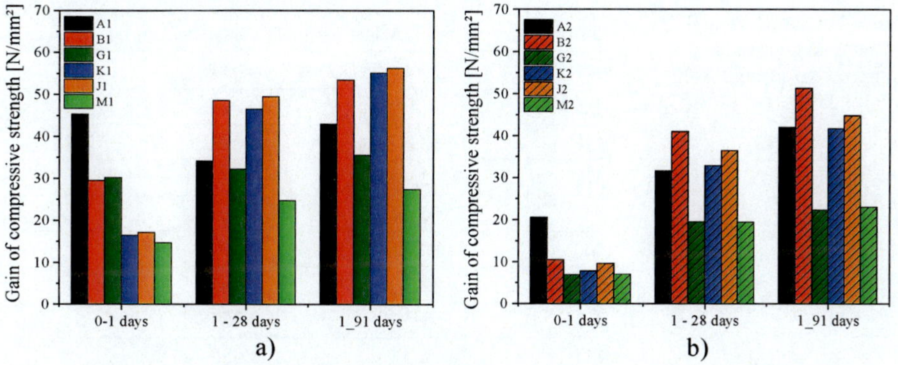

Fig. 6 Compressive strength gain in 3 time periods of **a** w/b = 0.36–0.45 and **b** w/b = 0.6

of GGBFS, namely B, J and K possess higher gain of CS in time range between 1st and 28th as well as 91st day than reference mortar mixes with pure OPC (mixes A) in case of both w/b groups. On the one hand pure OPC reference mixes achieved much higher absolute CS values than blended cement mixes after 1st day, which is attributed to the influence of dilution of OPC in blended mixes (Fig. 6). The influence of OPC replacement by combined SCMs on CS was highlighted by normalizing CS per mass of hydraulically binder (sum of OPC and GGBFS) per unit of mix volume, Fig. 7.

Figure 7 shows normalized CS over time of all mixes, while Table 4 shows comparison of strength values at 28th day of hydration. The highest effectiveness when CS is related to the hydraulically active binder content (OPC + GGBFS)—i.e. a normalized CS—value per mass of hydraulically effective binder per volume (Fig. 7)—can be observed in case of LSP-MIK-blended mixes G1 and K1 at low w/b-ratios (Fig. 7a). Up to 7 days of hardening, also mix M1 with LSP-MEK was more effective than pure OPC. This fact indicates a higher activity of OPC and GGBFS particles during

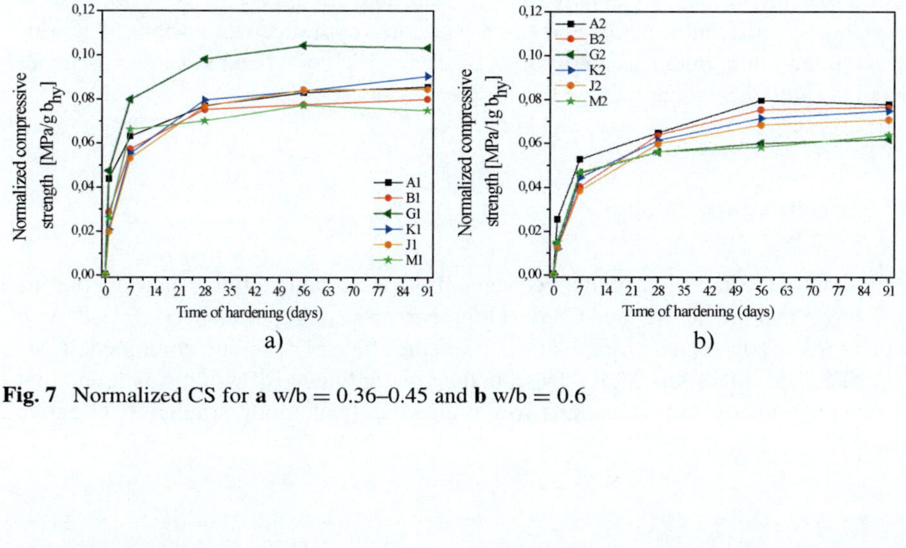

Fig. 7 Normalized CS for **a** w/b = 0.36–0.45 and **b** w/b = 0.6

Table 4 Compressive strength and normalized compressive strength values at 28th day

Sample	CS (MPa)	CS (MPa/g of h_{by})
A1	79.4	0.077
B1	77.9	0.075
G1	62.3	0.098
K1	62.9	0.079
J1	66.6	0.076
M1	39.3	0.070
A2	52.1	0.065
B2	51.5	0.063
G2	26.4	0.056
K2	40.6	0.061
J2	46.0	0.059
M2	26.4	0.050

hydration in case of LSP being present at relatively low w/b-ratios, i.e. low water content per volume respectively. Our results correspond to analogous conclusions reported in literature [29–31].

Acceleration of the hydration reaction and therefore better utilization possibilities of hydraulically active binders (OPC and GGBFS) by LSP was observed also when micro—and nano-limestone was incorporated leading to improvement of environmental and mechanical properties [10, 32]. It was not detected when coarser LSP with a d_{50} of > 4 μm were added to replace OPC [10, 33].

Regarding the outlined results, we can also conclude that for LSP-MIK with a d_{50} value of 1.2 μm, strength accelerating effects (due to nucleation of CSH growth or formation of new phases [9]) outweigh dilution, while for the coarser LSP-MEK, the dilution effect may predominate. The influence of inert fillers (MEK and MIK) appears to be significantly higher in case of lower w/b than higher water content per volume (Fig. 6a vs. 6b and Fig. 7a vs. 7b). The observed CS-development (especially aside from the first 24 h of hardening) and normalized CS assessment implies the existence of different kinetics and microstructural rearrangements (both of porous structure and mineralogical/amorphous phases) which must be the subject of more detailed future investigation.

4 Conclusions

Properties of the investigated blended cements in pastes and mortars regarding porous structure, air-permeability and CS development are strongly influenced by both, w/b ratio or water content per volume of paste and the effect of different, combined SCMs (GGBFS, LSP MEK and MIK). Despite the great influence of w/b it was found that intrudable porosity correlates quite well with w/b_{hy} (water to hydraulically effective powders in the binder). However, if we neglect that the inert filler has an effect on porosity and we assign the measured porosity to the volume fraction of hydraulic binder and water, correlation becomes even better. Considering these results, well known porosity assumptions based on w/b ratio like that of T.C. Powers are not directly applicable, but might be modified accordingly. In addition, a good correlation of air permeability with intrudable porosity was found.

As expected, the strength development of mortar samples with blended cements depends strongly on its w/b value. However, the proportion of SCMs in the blends also influences this development. A higher strength gain can be observed between the 1st and 91th day on blends of OPC, GGBFS and LSP compared with pure OPC. The effect is more pronounced at lower w/b than at higher w/b. At higher water content, the dilution effect of inert LSP overweight the strength increasing "filler" effect of LSP (i.e. nucleation effect, etc.), whereas the latter is more noticeable at low w/b. Addition of finely grinded LSP appears to accelerate early strength development of blended cements by better utilization of hydraulically active components (OPC and GGBFS) during hydration. Strength was related to the content of hydraulically effective powders (OPC + GGBFS), which revealed that mixes with LSP show higher binder—efficiency than mixes with pure OPC. In particular, LSP–MIK accelerates the hydration of OPC and GGBFS, so that particularly large increases in strength per g/l binder can be achieved. Optimized combinations of OPC, GGBFS and LSP thus represent promising future blends with appropriate functional performance and low environmental impact. The obtained results indicate higher potential of ternary (quaternary) blended cement formulation over binary mixes, with further need of mix design optimization (related to packing density).

Acknowledgements The authors would like to thank the Austrian society for construction technology and the Research Promotion Agency FFG for the funding under grant no. ÖBV 2017-05-2. Furthermore, we thank Dr. Neven Ukrainczyk from the Institute of Construction and Building Materials of TU Darmstadt for the MIP analyses and fruitful discussion.

References

1. Scrivener, K., John, V., Gartner, E.: Eco-efficient cements: Potential, economically viable solutions for a low-CO_2, Cement- based materials industry. UNEP (2016)
2. USGS: Cement—Mineral Commodity Summaries (2017)
3. Snellings, R.: Assessing, understanding and unlocking cement replacement materials. RILEM Tech. Lett. **1**, 50–55 (2016)
4. Scrivener, K., Martirena, F., Bishnoi, S., Maity, S.: Calcined clay limestone cements (LC^3). Cem. Concr. Res. **114**, 49–56 (2018)
5. Wang, D.H., Shi, C.J., Farzadnia, N., Shi, Z.G., Jia, H.F., Ou, Z.H.: A review on use of limestone powder in concrete: Mechanism, hydration and microstructures. Constr. Build. Mater. **181**, 659–672 (2018)
6. Meddah, M., Lambachiya, M., Dhir, R.: Potential use of binary and composite limestone cements in concrete production. Constr. Build. Mater. **58**, 193–205 (2014)

7. Vance, K., Aguayo, M., Oey, T., Sant, G., Neithalath, N.: Hydration and strength development in ternary Portland cement blends containing limestone and fly ash or metakaolin. Cem. Concr. Compos. **39**, 93–103 (2013)
8. Deschner, F., Winnefeld, F., Lothenbach, B., Seufert, S., Schwesig, P., Dittrich, S., Neunhoeffer, F.G., Neubauer, J.: Hydration of Portland cement with high replacement by siliceous fly ash. Cem. Concr. Res. **42**(10), 1389–1400 (2012)
9. Scrivener, K., Lothenbach, B., Belie, N., Gruyaert, E., Skibsted, R., Snellings, R.: TC 238-SCM: hydration and microstructure of concrete with SCMs—state of the art on methods to determine degree of reaction of SCMs. Mater. Struct. **48**(4), 835–862 (2015)
10. Juhart, J., David, G.A., Saade, M., Baldermann, C., Passer, A., Mittermayr, F.: Functional and environmental performance optimization of Portland cement-based materials by combined mineral fillers. Cem. Concr. Res. **122**, 157–178 (2019)
11. Li, W.G., Huang, Z.Y., Zu, T.Y., Shi, C.J., Duan, W.H., Shah, S.P.: Influence of nanolimestone on the hydration, mechanical strength, and autogenous shrinkage of Ultrahigh-performance concrete. J. Mater. Civil Eng. **28**(1), 04015068 (2016)
12. Mohammadi, J., South, W.: Effect of up to 12% substitution of clinker with limestone on commercial grade concrete containing cement replacement materials. Constr. Build. Mater. **115**, 555–564 (2016)
13. Aashay, A., Gaurav, S., Narayanan, N.: Ternary blends containing slag and interground/blended limestone: hydration, strength, and pore structure. Constr. Build. Mater. **102**, 113–124 (2016)
14. Puerta-Falla, G., Balonis, M., Le. Saout, G., Falzone, G., Kumar, A., Zhang, C.: The influence of the aluminous source on enhancing limestone reactivity in cementitious materials. J. Am. Ceram. Soc. **98**, 4076–4086 (2015)
15. Sanjuán, M.A., Piñeiro, A., Rodríguez, O.: Ground granulated blast furnace slag efficiency coefficient (k value) in concrete, applications and limits. Mater. Constr. **61**, 303–313 (2011)
16. Sisomphon, K., Franke, L.: Carbonation rates of concretes containing high volume of pozzolanic materials. Cem. Concr. Res. **37**, 1647–1653 (2007)
17. Sanjuán, M.A., Estévez, E., Argiz, C., del Barrio, D.: Effect of curing time on granulated blast-furnace slag cement mortars carbonation. Cem. Concr. Compos. **90**, 257–265 (2018)
18. Lyu, K., She, W., Miao, C., Chang, H., Gu, Yue.: Quantitative characterization of pore morphology in hardened cement paste via SEM-BSE image analysis. Constr. Build. Mater. **202**, 589–602 (2019)
19. Proske, T., Rezvani, M., Palm, S., Müller, C., Graubner, C.A.: Concretes made of efficient multi-composite cements with slag and limestone. Cem. Concr. Compos. **89**, 107–119 (2018)
20. Gartner, E., Hirao, H.: A review of alternative approaches to the reduction of CO_2 emissions associated with the manufacture of the binder phase in concrete. Cem. Concr. Res. **78**, 126–142 (2015)
21. Glavind, M., et al.: Guidelines for green concrete structures. Int. Feder. Struct. Concrete FIB Bull. **67** (2012)
22. EN 197-1: Cement—Part 1: Composition, specifications and conformity criteria for common cements (2011)
23. Washburn, E.: Note on a method of determining the distribution of pore sizes in a porous material. Proc. Natl. Acad. Sci. USA **7**(4), 115–116 (1921)
24. Scrivener, K., Snellings, R., Lothenbach, B.: A Practical Guide to Microstructural Analysis of Cementitious Materials, 2nd edn. CRC Press, New York (2017)
25. Torrent, R.J.: A two-chamber vacuum cell for measuring the coefficient of permeability to air of the concrete cover on site. Mater. Struct. **25**(6), 358–365 (1992)
26. Swiss Standard SIA 262/1:2013, Concrete Construction—Complementary Specifications
27. EN 196-1: Methods of testing cement. Determination of strength
28. Namoulniara, K., Turcry, Ph., Aït-Mokhtar, A.: Measurement of CO_2 effective diffusion coefficient of cementitious materials. Eur. J. Environ. Civil Eng. **20**(10), 1183–1196 (2016)
29. Martin, C., Philippe, L., Erick, R.: Efficiency of mineral admixtures in mortars Quantification of the physical and chemical effects of fine admixtures in relation with compressive strength. Cem. Concr. Res. **36**, 264–277 (2006)

30. Adu-Amankwah, S., Zajac, M., Stabler, C., Lothenbach, B., Black, L.: Influence of limestone on the hydration of ternary slag cements. Cem. Concr. Res. **100**, 96–109 (2017)
31. Berodier, E., Scrivener, K.: Understanding the filler effect on the nucleation and growth of C-S-H. J. Am. Ceram. Soc. **97**, 3764–3773 (2014)
32. Sato, T., Beaudoin, J.: Effect of nano-$CaCO_3$ on hydration of cement containing cement replacement materials. Adv. Cem. Res. **23**(1), 1–11 (2011)
33. De, W.K., Ben, H.M., Le, S.G., Kjellsen, K.O., Justnes, H., Lothenbach, B.: Hydration mechanisms of ternary Portland cements containing limestone powder. Cem. Concr. Res. **41**, 279–291 (2011)

Affecting Factors in Rehabilitating Water Distribution Networks

Rahimi A. Rahman, Noor Suraya Romali, Siti Sarah Sufian, and Mazlan Abu Seman

Abstract High levels of non-revenue water (NRW) reflect that an area's water distribution networks (WDN) are losing vast volumes of clean water. Therefore, reducing NRW is crucial for sustainable water management. NRW levels in most developing countries are high, ranging between 35 and 50%. While one of the significant causes of NRW is difficulties in rehabilitating old piping networks, studies on factors that are influencing WDN rehabilitation in practice is limited. This study aims to identify affecting factors for WDN rehabilitation. To achieve that objective, a series of individual interviews with industry practitioners that manage WDN are analyzed using the thematic analysis. The major findings from the analysis are: (1) internal and external factors are influencing WDN rehabilitation; (2) internal factors are related to cost, location, and design; and (3) external factors are related to local authorities and surrounding communities. This research adds to the body of knowledge by providing a set of affecting factors for rehabilitating WDN, which can assist researchers and practitioners in developing strategies to reduce NRW for achieving sustainable water management.

Keywords Sustainable development · Water distribution networks · Non-revenue water · Rehabilitation

1 Introduction

Reducing water losses is key to sustainable water management but challenging. Rapid population growth, income growth, and urbanization, in combination with a fixed supply of total renewable water resources, are pressuring numerous nations

R. A. Rahman (✉) · N. S. Romali · S. S. Sufian · M. A. Seman
Faculty of Civil Engineering Technology, Universiti Malaysia Pahang, Lebuhraya Tun Razak, 26300 Kuantan, Pahang, Malaysia
e-mail: arahimirahman@ump.edu.my

R. A. Rahman
Earth Resources and Sustainability Centre, Universiti Malaysia Pahang, Lebuhraya Tun Razak, 26300 Kuantan, Pahang, Malaysia

V. M. C. F. Cunha et al. (eds.), *Proceedings of the 3rd RILEM Spring Convention and Conference (RSCC 2020)*, RILEM Bookseries 35,
https://doi.org/10.1007/978-3-030-76543-9_17

in identifying renewable water resources. On the other hand, high levels of non-revenue water (NRW) reflect vast volumes of water are lost through leaks (real losses) and drinking water not being invoiced to customers (apparent losses) and unbilled authorized consumption [1]. The NRW levels in developing countries, including Malaysia, are high, ranging from 35 to 50% of the water produced [2]. Therefore, the water distribution sector needs to improve the way it uses its water resources significantly, especially in NRW, to attain sustainable water management.

Reducing NRW is compelling to save much money, and significant volumes of water can be used for more productive purposes [3]. One of the approaches for reducing NRW involves repairing and replacing (i.e., rehabilitating) old pipelines of water distribution networks (WDN). However, despite some proposed strategies for NRW management and reduction, most WDN continue to experience high levels of water losses [1, 2, 4, 5]. Furthermore, the study on NRW and the rehabilitation of WDN is limited and yet to be given adequate attention. Therefore, having additional insights into the factors influencing the success of rehabilitating WDN is vital.

This study aims to identify the affecting factors for rehabilitating WDN. To achieve that objective, this study analyzes a series of interviews with industry practitioners that have experience and knowledge in managing WDN. This study contributes to the current body of knowledge by providing a list of factors that are influencing the success of rehabilitating WDN. Researchers and practitioners from both the public and private sectors can use the findings to develop appropriate strategies for enhancing WDN rehabilitation. These results can be utilized in reducing NRW, achieving sustainable water management, and the United Nation's Sustainable Development Goals of responsible consumption and production.

2 Background

2.1 Causes of Non-revenue Water

Water utilities are facing major challenges due to a high level of NRW [1]. NRW is calculated as the difference between the water produced (also known as the system input volume) and the water delivered (also known as the revenue water) as a proportion of the water produced [1]. NRW has three components: apparent losses, real losses, and non-revenue authorized consumption [3]. Apparent losses occur due to unauthorized use, personnel errors, management, and operational errors, and data-handling errors. These losses cost-utility revenue and distort data on customer consumption patterns. Real losses comprise leakage from system elements and overflows of storage tanks. These losses are also caused by poor operations and maintenance activities, lack of active leakage control, poor quality of underground assets, and mainly contain visible and invisible leakage. Non-revenue authorized consumptions include water used by the utility for operational purposes, water used for firefighting, and water provided for free to certain consumer groups [3].

Due to the high level of NRW worldwide, authorities in water companies seek the most effective NRW reduction activities. Despite some proposed approaches for NRW management and reduction in recent years, most WDN continues to experience high levels of water losses [3]. Experience has shown that the essential contributing factor for water shortage is related to real losses, which is either due to pipe corrosion or leakage from the distribution network [4–6]. Hence, the water distribution system needs to be monitored closely by replacing the existing water infrastructure to minimize physical losses. Reducing non-revenue water in the water services industry is not only crucial for saving water but also essential for securing its future efficiency and development.

2.2 Strategies for Reducing Non-revenue Water

One of the approaches for reducing NRW includes rehabilitation of WDN. Rehabilitation strategies need to be in place to ensure that the WDN continues to operate efficiently and economically within defined operating requirements over an extended period [4]. Prior studies are proposing strategies for rehabilitation and expansion of WDN using several modeling approaches such as the Markov model, Monte Carlo simulation, and PALM+ system [4–6]. On the other hand, other studies have collected data from the literature, industrial experience, and direct data and feedback from industry practitioners to identify key variables in the construction management field of research. Some examples of this type of study with similar methodology include the identification of best practices for: waste management [7], energy efficiency [8], supply chain management [9], and water management in the hospitality sector [10]. Other examples of using this type of methodology include the identification of key challenges for: effective application of anti-corruption measures in infrastructure projects [11]; construction management for rural transit projects [12]; and design-construction interfaces of large building construction projects [13]. Also, this methodology has been used to identify parameters for: public–private-partnerships in developing a developed country [14]; public-housing projects in developing countries [15]; and project success for companies in Brazil [16].

2.3 Positioning of This Study

However, in developing countries, the study on the rehabilitation of WDN is yet to be given adequate attention, and the kinds of literature are limited. Due to urbanization and rapid economic development, the provision of quality water and sewerage services has become more challenging. For example, in 2015, Malaysia's NRW was at 35.5%, and the main factors that contribute to that are physical losses due to the old network of pipes and poor quality by contractors, especially in new development areas, commercial losses, and also lack knowledge and expertise in NRW

[17]. Hence, understanding the affecting factors of WDN rehabilitation is essential to assist in the development of global NRW management.

3 Methodology

3.1 Data Collection

Data on the affecting factors for WDN rehabilitation are collected by interviewing practitioners that have experience and knowledge in managing WDN to maintain a level of quality of the interviewees (i.e., purposeful sampling). Also, this approach has been used to identify success factors in other construction management topics, including design-build public sector projects [18] and highway projects [19]. Individuals from different water distribution organizations are interviewed to ensure the data is comprehensive. Interviews allow industry practitioners to provide implicit knowledge of their situation. Also, open-ended questions encourage participants to contribute as much detailed information as desired while enabling investigators to ask probing questions as a means to follow-up. The open-ended questions are: (1) What are the challenges for rehabilitating WDN? And (2) What are the factors that are affecting the success of rehabilitating WDN? Participants addressed the questions while the investigators took notes and provide follow-up questions. For verification purposes, the notes are summarized and sent to the interviewees. The interviews will be completed by following the principle of saturation.

3.2 Data Analysis

The interview data are analyzed using thematic analysis to formulate lists of affecting factors for WDN rehabilitation. This approach is selected because it can assist in making sense of qualitative data [20]. Other construction management topics that use this method to analyze qualitative data include identifying problems in construction projects [21], attributes of change agents in construction companies [22], and parameters for highway construction projects [23]. The analysis involves coding the interview data, organizing the codes into concepts, forming categories of related concepts, elaborating patterns and linkages between categories, and developing a theme, subthemes, and codes that explain the data. Developing the themes, subthemes, and codes identify the affecting factors for WDN rehabilitation.

4 Results and Discussion

Figure 1 summarizes the affecting factors for rehabilitating WDN from the analysis of individual interviews with industry practitioners of water distribution operators. The factors are themed into two main categories, internal and external factors. Internal factors can be associated with variables that are usually actionable from by water distribution operators (ex. having a good design process for rehabilitating old pipes). On the other hand, external factors relate to those that are often uncontrollable by water distribution operators (ex. customer not paying water bills). The factors are further discussed in the subsequent subsections.

4.1 Internal Factors

Cost is one of the internal factors that are influencing the success of rehabilitating WDN. Similar to other rehabilitating other types of infrastructures, the cost of rehabilitating old pipes is high. To illustrate this situation, the cost can be associated with capital expenditures rather than repairs and maintenance expenses. As the objectives of capital improvements relate to increasing the value of an organization's assets, water distribution operators are expected to make decisions based on the return of

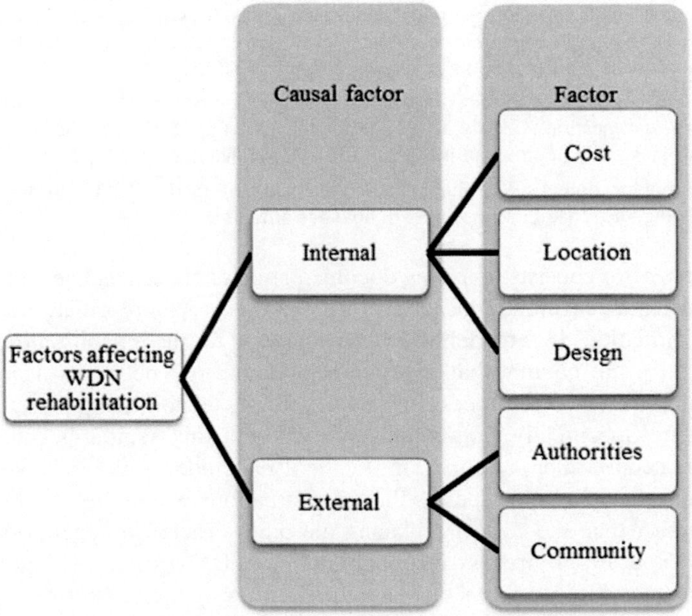

Fig. 1 Affecting factors in rehabilitating water distribution networks (WDN)

investment (ROI). Therefore, operators tend to rehabilitate old pipes partially while targeting an effective reduction of water loss. Also, operators need to balance between the proportionate savings and the costs for rehabilitating larger areas. In other words, water distribution operators need to identify strategies for optimizing the process of rehabilitating WDN. Example of the responses from the summary of interviews that illustrate this factor include:

> To change old pipes, we cannot rehabilitate in small scales, for example, 6 m or 12 m. To make it cost-effective, around 500 m to 1 km needs to be rehabilitated to ensure the new pipes can last longer. Rehabilitating on this scale involves much money.
>
> The cost to rehabilitate old pipes is quite high. So, we can only rehabilitate in smalls scales. So, the critical locations are prioritized first to make sure water loss is reduced. Also, if the process involves rehabilitating a big area, it will take time and money to complete.
>
> The cost of pipes is high. So, the provisions are under capital expenditure, where the return of investment (ROI) will be taken into account.

Location involves the positioning of the WDN' pipes. Specifically, water distribution operators are having difficulties in rehabilitating pipes that are positioned underneath a built facility. While WDN might be designed without any facility on top of them, economic developments are resulting in buildings and roads constructed on top of the WDN's alignment. Therefore, in an attempt to reduce complications in rehabilitating WDN in both the near and distant future, proper urban planning that ensures minimal facilities or obstacles are constructed above this type of network is necessary. Samples of the response include:

> The alignment of pipes is located underneath main roads. Most pipes were designed 20 years ago. The country's economic development results in existing pipes to be located at the alignment of roads. This makes the pipe replacement work difficult.
>
> Existing location of pipes are constructed with another facility such as houses and roads. Before this, the location is easily accessible for rehabilitating. But, after the time being, houses or roads are built on top of it. So, it is difficult to rehabilitate the pipes.
>
> If the existing location is still available and no one had built facilities at the location from the time being, the rehabilitating work will not have difficulties.

Design. This factor consists of having documentation on the current design of WDN as most systems are developed decades before. Without proper information, designers are having difficulties in remodeling existing systems for the rehabilitation process. Also, designers are having challenges in remodeling old networks according to current standards when the process involves explicitly rehabilitating partial sections of the system. Specifically, non-compliance with existing standards can result in problems related to water pressure. This factor also includes deciding between rehabilitating and replacing old pipes as there are situations where the cost of rehabilitating is higher than the cost of replacing old pipes. Therefore, operators need to establish a good design process for rehabilitating WDN. Responses that illustrate from the summary of interviews this factor are:

> There are also problems in the design process. Designers need to observe the existing system before designing the new pipelines to avoid water pressure problems. Modeling or hydraulic analysis needs to be done to ensure the new system works properly.

Loss of pressure can also be a factor in pipe replacement.

In the case of repeated water loss, the cost of repairing is higher than the replacement of the pipe.

Types of pipes can also affect the rehabilitation process. Usually, old pipes do not adhere to current standards. Choosing an inappropriate type of pipe can have a negative impact during the rehabilitation process.

Problems in identifying the location of old pipelines because most pipes that are being replaced are over 30 years old. So, data and records involving these old pipes are not complete.

4.2 External Factors

Authorities, specifically approval from authorities is one of the external factors that influence the process of rehabilitating WDN. To start rehabilitating old pipes, water distribution operators need approvals from certain government authorities such as administration bodies that control the construction and maintenance of public infrastructures and local governments. Also, operators need to acquire approval from other utility institutions. Conversely, contractors and subcontractors require approvals from the operators. While administrating construction and maintenance of public infrastructure works are more than necessary, the approval process is suggested to take quite some time (sometimes up to a year). Therefore, having an appropriate process for administrating this type of works (including rehabilitating old pipes) is vital. Samples of responses from the summary of interviews that illustrate this success factor include:

Constraints to get the permit from the government authorities and local government authorities to replace the old pipes.

Constraints due to laws of local government authorities, public work departments, and utility companies to get approval for the new pipeline installation.

There is one time when they apply for a permit, and they take more than one year to give it. So, it causes the work for pipe replacement to be delayed and slow.

Community. On the other hand, communities that surround or use the WDN (i.e., the public) also plays a role in the successful rehabilitation of WDN. As discussed in the previous subsection, water distribution operators rehabilitate old pipes based on their ROI because the work requires a significant amount of monies. However, these operators are facing problems in allocating sufficient funds because users are not paying their water bills. Also, dissenting behavior from the public, such as complaining through local authorities and social media from additional traffic congestion, noise, and dirt from the rehabilitation process of WDN, can result in the stoppage of the rehabilitation works. Similarly, public complaints to local authorities are slowing down and even idling the construction of highway projects [19]. Therefore, awareness from the public on the importance of supporting the rehabilitation of WDN is crucial. Examples that illustrate these results include:

People need to pay water bills to rehabilitate the pipes.

The first challenge relates to the local community. In detail, there are complaints from the public, such as dirty roads resulted from the rehabilitation process. Although it can sometimes be true, the local community should understand the process and provide cooperation during the process. Sometimes, the public shares information through social media. The public should understand that after the rehabilitation process is completed, then the roads will be fixed.

5 Conclusion

This study identifies the affecting factors for WDN rehabilitation by analyzing individual interview data with seven professionals from water distribution operators using the thematic analysis. The major findings include:

- The affecting factors in rehabilitating WDN is associated with internal and external factors.
- The internal factors are related to cost, location, and design.
- The external factors are related to both local authorities and surrounding communities.

These findings highlight the need for water distribution operators to develop strategies to optimize the process of rehabilitating WDN because it involves high costs, hard to reach locations, and complicated designs. Conversely, the local government can improve the approval process and provide awareness to the public on the importance of WDN rehabilitation work. Research and industry practitioners can use these findings to develop strategies to ensure the success of rehabilitating WDN. The key theoretical contribution of this research is by providing a set of factors that influence WDN rehabilitation in a developing country.

Acknowledgements The authors wish to thank Universiti Malaysia Pahang for supporting this study through the financial grant on "Sustainable water: Approaches for rehabilitating water distribution network to reduce non-revenue water," as well as the project managers that agreed to participate in this work.

References

1. See, K.F., Ma, X.: Does non-revenue water affect Malaysia's water services industry productivity? Utilities Policy **54**(2018), 125–131 (2018)
2. Bakri, B., Arai, Y., Inakazu, T., Koizumi, A., Yoda, H., Pallu, S.: Selection and concentration of pipeline mains for rehabilitation and expansion of water distribution network. Procedia Environ. Sci. **28**, 732–742 (2015)
3. Tabesh, M., Roozbahani, A., Roghani, B., Faghihi, N.R., Heydarzadesh, R.: Water Resour. Manage. **32**, 3647–3670 (2018)
4. Engelhardt, M.O., Skipworth, P.J., Savic, D.A., Saul, A.J., Walters, G.A.: Rehabilitation strategies for water distribution networks: a literature review with a UK perspective. Urban Water **2**, 153–170 (2000)

5. Rogers, D., Calvo, B.: Defining the rehabilitation needs of water networks. Procedia Eng. **119**, 182–188 (2015)
6. Aşchilean, I., Badea, G., Giurca, I., Naghiu, G.S., Iloaie, F.G.: Choosing the optimal technology to rehabilitate the pipes in water distribution systems using the AHP method. Energy Procedia **112**, 19–26 (2017)
7. Gálvez-Martos, J.L., Styles, D., Schoenberger, H., Zeschmar-Lahl, B.: Construction and demolition waste best management practice in Europe. Resour. Conserv. Recycl. **136**, 166–178 (2018)
8. Galvez-Martos, J.-L., Styles, D., Schoenberger, H.: Identified best environmental management practices to improve the energy performance of the retail trade sector in Europe. Energy Policy **63**, 982–994 (2013)
9. Styles, D., Schoenberger, H., Galvez-Martos, J.-L.: Environmental improvement of product supply chains: proposed best practice techniques, quantitative indicators and benchmarks of excellence for retailers. J. Environ. Manag. **110**, 135–150 (2012)
10. Styles, D., Schoenberger, H., Galvez-Martos, J.L.: Water management in the European hospitality sector: Best practice, performance benchmarks and improvement potential. Tour. Manage. **46**, 187–202 (2015)
11. Owusu, E., Chan, A.P.: Barriers affecting effective application of anticorruption measures in infrastructure projects: Disparities between developed and developing countries. J. Manag. Eng. **35**(1), 04018056 (2019)
12. Tran, D.Q., Hallowell, M.R., Molenaar, K.R.: Construction management challenges and best practices for rural transit projects. J. Manag. Eng. **31**(5), 04014072 (2014)
13. Sha'ar, K.Z., Assaf, S.A., Bambang, T., Babsail, M., Fattah, A.A.E.: Design–construction interface problems in large building construction projects. Int. J. Constr. Manag. **17**(3), 238–250 (2017)
14. Osei-Kyei, R., Chan, A.P.: Empirical comparison of critical success factors for public-private partnerships in developing and developed countries: a case of Ghana and Hong Kong. Eng. Constr. Archit. Manag. **24**(6), 1222–1245 (2017)
15. Mukhtar, M.M., Amirudin, R.B., Sofield, T., Mohamad, I.B.: Critical success factors for public housing projects in developing countries: a case study of Nigeria. Environ. Dev. Sustain. **19**(5), 2039–2067 (2017)
16. Berssaneti, F.T., Carvalho, M.M.: Identification of variables that impact project success in Brazilian companies. Int. J. Project Manage. **33**(3), 638–649 (2015)
17. Ministry of Energy, Green Technology, and Water (KeTTHA): Green Technology Master Plan Malaysia 2017–2030 (2017)
18. Lee, Z.P., Rahman, R.A., Doh, S.I.: Success factors of design-build public sector projects in Malaysia. In: IOP Conference Series: Materials Science and Engineering, vol. 712, No. 1, p. 012045 (2020)
19. Rahman, R.A., Radzi, A.R., Saad, M.S.H., Doh, S.I.: Factors affecting the success of highway construction projects: the case of Malaysia. In: IOP Conference Series: Materials Science and Engineering, Vol. 712, No. 1, p. 012030 (2020)
20. Braun, V., Clarke, V.: Using thematic analysis in psychology. Qual. Res. Psychol. **3**(2), 77–101 (2006)
21. Rahman, R.A., Ayer, S.K.: Prevalent issues in BIM-based construction projects. Proc. Joint Conf. Comput. Constr. **1**, 645–652 (2017)
22. Radzi, A.R., Bokhari, H.R., Rahman, R.A., Ayer, S.K.: Key attributes of change agents for successful technology adoptions in construction companies: a thematic analysis. In: Computing in civil engineering 2019: Data, Sensing, and Analytics, pp. 430–437 (2019)
23. Radzi, A.R., Rahman, R.A., Doh, S.I., Esa, M.: Construction readiness parameters for highway projects. In: IOP Conference Series: Materials Science and Engineering, vol. 712, No. 1, p. 012029 (2020)

Success Factors for Construction Waste Recycling in Developing Countries: A Project Management Perspective

Rahimi A. Rahman, Abdulmalek K. Badraddin, Muzamir Hasan, and Nor'Aini Yusof

Abstract Construction industries around the world are generating a large number of wastes that end at landfills every year, and recycling is one of the approaches for minimizing that amount. While various strategies have been adopted in practice, recycling rates of construction projects are still at a low level in numerous countries. Therefore, identifying factors that influence the successful recycling of construction waste is crucial. This study identifies the success factors for recycling construction waste from industry practitioners' perspectives. To achieve this objective, interview data with project managers are analyzed using the thematic analysis. The major findings from the analysis are: (1) the success factors relate to both people or process; (2) people-related factors involve having individuals that are highly competent, aware on construction waste recycling, and knowledgeable; (3) process-related factors include having a detailed project planning, adequate education and training programs, clear project scope and design, effective procurement system, and consistent monitoring of the construction waste recycling system; and (4) the criticality of the success factors differs between developing countries. This research adds to the body of knowledge by providing a set of success factors for recycling construction waste, which can assist researchers and practitioners in developing strategies to increase recycling rates of construction projects.

Keywords Sustainable construction · Construction waste management · Developing countries · Project managers · Success factors

R. A. Rahman (✉) · A. K. Badraddin · M. Hasan
Faculty of Civil Engineering Technology, Universiti Malaysia Pahang, Lebuhraya Tun Razak, 26300 Kuantan, Pahang, Malaysia
e-mail: arahimirahman@ump.edu.my

R. A. Rahman · M. Hasan
Earth Resources and Sustainability Centre, Universiti Malaysia Pahang, Lebuhraya Tun Razak, 26300 Kuantan, Pahang, Malaysia

N. Yusof
School of Housing, Building, and Planning, Universiti Sains Malaysia, USM, 11800 Pulau, Pinang, Malaysia

V. M. C. F. Cunha et al. (eds.), *Proceedings of the 3rd RILEM Spring Convention and Conference (RSCC 2020)*, RILEM Bookseries 35,
https://doi.org/10.1007/978-3-030-76543-9_18

1 Introduction

Solid waste is a global issue affecting our society's ability to addressing environmental sustainability. Nowadays, organized systems have been developed to manage waste, including recycling. Recycling is defined as a process of repurposing waste materials into value applications than the material was used previously to it becoming waste [1]. The benefits of recycling include saving resources, saving energy, helping to protect the environment, and reducing incineration. Therefore, various sectors are identifying approaches to implement it, including recycling construction and demolition waste [2–4]. Therefore, recycling construction waste is crucial in sustaining the environment.

Construction waste recycling has been adopted and implemented at a high level in several countries such as in Australia and Japan at 90% and 99.5%, respectively [5, 6]. However, in general, recycling waste, including construction waste in developing countries is at a low level of around 5% [7, 8]. The major barrier of construction waste recycling includes lack of acceptance of recycling from the increase in management cost and documentation workload, and refusal to use the recycled material [9]. Furthermore, reluctance among individuals is suggested to inhibit the success of adopting innovations in construction projects by deciding not to implement them [10, 11]. Various factors can influence the success of construction waste recycling, including the past experiences and beliefs of project team members [12, 13]. Also, political, economic, social, technological, legal, and environmental factors are also suggested to influence the success of construction waste recycling [14, 15]. In other words, while studies can illustrate the benefits of recycling construction waste, the innovation might not be adopted in practice because of various factors. Therefore, understanding the success factors is crucial to understand the right context for recycling waste in construction projects.

This study aims to identify the success factors for recycling construction waste from industry practitioners' perspectives. To achieve that objective, this study analyzes individual interview data with project managers using the thematic analysis. This study contributes to the existing body of knowledge by providing an alternative list of success factors for recycling construction waste. Researchers and practitioners can use the findings to strategize in increasing waste recycling in construction projects. Any increment in construction waste recycling will reduce the volume of waste ending up at landfills. Lowering waste volume can benefit the environmental management atmosphere and sustain the environment.

2 Background

2.1 Construction Waste Recycling

Construction waste recycling is the process of separating and recycling of recoverable waste materials generated during construction and demolition works. Examples of products that are produced from recycling construction wastes include base course materials for building driveways and footpaths from concrete waste, construction materials from large untreated timber, and new asphalts from old asphalt paving. As this process recycling reduces the construction industry's impact on the environment, topics related to construction waste recycling have been investigated by both construction material and project management researchers and practitioners. Studies suggest that the technical performance and environmental impact of recycled aggregate concrete are promising when compared to standard concrete [1, 16–18]. Additionally, the potential benefits of construction waste recycling have been determined, including mitigating greenhouse gas and land-use change [2]. Studies have also identified the current regulations and practice of construction waste recycling in developed countries [3, 4, 9, 19]. Finally, the barriers of implementing construction waste recycling developed countries such as the United States, Japan, Australia, and Hong Kong have also been identified including lack of acceptance on recycled materials, increase in management cost, and increase in documentation workload [9, 20–22]. In summary, various studies have illustrated the potential benefits, current processes, and barriers to adopting construction waste recycling because the implementation is crucial to sustaining the environment.

2.2 Construction Waste Recycling in Malaysia

Similar to most countries, researchers and practitioners are also investigating topics related to construction waste management in Malaysia. Past studies have analyzed the cost–benefit of construction waste minimization to identify its economic feasibility [23], implementation of a waste management system to minimize construction waste [24], and attitude and behavioral factors in waste management [25]. On the other hand, recent studies are providing insights on the generation of construction waste for high-rise buildings [26], strategies to minimize construction waste [27], and industry practices for controlling the generation of construction waste [28]. In other words, while these studies have provided meaningful understandings towards construction waste management, the success factors for recycling construction waste that is specific to the local context have not been identified.

The government has also developed several policies to increase recycling, including the Solid Waste and Public Cleansing Management (Act 672), National Policy on the Environment (DASN), and The Green Technology Master Plan (GTMP). Act 672 provides for and regulates the management of controlled solid waste and public cleansing for the purpose of maintaining proper sanitation and for matters incidental thereto, including recycling. DASN established the continuous economic, social and cultural progress and enhancement of the quality of life of Malaysians through environmentally sound and sustainable development, including sustainable lifestyles and patterns of consumption and production. GTMP is an outcome of the Eleventh Malaysia Plan (11 MP), which has earmarked green growth as one of six game-changers altering the trajectory of the nation's growth, including sustainable construction practices. While the government has initiated several policies to increase construction waste recycling, the recycling rate is still at 2–3%. Therefore, understanding the success factors for recycling construction waste is crucial to provide context on the barriers of recycling construction waste.

2.3 Positioning This Study

While studies can provide worthy insights on implementing construction waste recycling in both theory and practice, its implementation might be hindered because of various factors. To address that, researchers and industry practitioners are already providing insights on construction waste management. However, the existing body of knowledge lacks the understanding of the success factors for construction waste recycling in local conditions. In other words, prior findings are unable to provide the fundamental knowledge of the factors that ensure the success of waste recycling in construction projects. Therefore, this study targets to identify an alternative list of success factors for recycling construction waste to fill this gap by analyzing individual interview data with project managers.

3 Methodology

3.1 Data Collection

In this study, success factors are considered as the necessary elements for an organization in achieving a specific or set of objectives (in this case, recycling waste of construction projects). Data on the success factors for recycling construction waste in construction projects are collected by interviewing construction project managers because this approach provides an understanding of individuals' perspectives and experiences. Also, this approach has been used to identify success factors in other construction management topics, including design-build in public sector projects [29] and construction of highway projects [30]. The investigators purposefully select project managers because these individuals have experience and knowledge related to construction projects to maintain a level of quality of the interviewees. Open-ended questions are provided to respondents because the questions can encourage participants to contribute as much detailed information as desired while allowing the investigators to ask probing questions as mean to follow-up. The main question is: What are the success factors for recycling waste at construction sites? The investigators took notes and provide follow-up questions, while participants are addressing this question. The interviews are not recorded to reduce hesitancy to respond among participants. After the interview, the investigators summarized the notes and sent the summary to the participants for validation to avoid misinterpreting or misquoting the interviewee. This process ends when based on the principle of data saturation in interviews.

3.2 Data Analysis

The interview data with project managers are analyzed to identify the success factors for recycling construction waste. The factors are identified by analyzing the interview data using the thematic analysis approach [31]. This approach is selected for this research because the approach allows the creation of new themes systematically. Examples of other construction management topics that use this method to analyze qualitative data include identifying problems in construction projects [32], attributes of change agents in construction companies [33], and parameters for highway construction projects [34]. The themes are formulated by coding the interview data, organizing the codes into concepts, forming categories of related concepts, elaborating patterns and linkages between categories, and developing themes that explain the data. The grouping process is followed by mapping the success factors to each respondent to identify the number of hits for each factor.

4 Results and Discussion

Table 1 summarizes the success factors for construction waste recycling from the analysis of individual interviews with project managers. The factors are themed into two main categories, people and process. People relate to the workforce that is necessary to ensure the success of construction waste recycling in projects. Conversely, process-related success factors can be associated with a series of activities to ensure that success. The success factors are further discussed in the subsequent subsections. Since qualitative studies have limitations in its generalizability, this study uses the total number of hits for each success factor in Table 1 to arrange the discussion rather than claiming the factors' significant importance.

Table 1 The total number of hits for success factors of construction waste recycling (CWR)

	Contractor 1	Contractor 2	Contractor 3	Contractor 4	Contractor 5	Contractor 6	Contractor 7	Contractor 8	Contractor 9	Contractor 10	Contractor 11	Contractor 12	Contractor 13	Total no of hits
People														
Highly competent			√	√	√	√		√		√	√	√		8
Aware of CWR		√	√	√										3
Knowledgeable	√	√										√		3
Process														
Detailed project planning	√		√			√	√	√	√	√	√	√	√	8
Adequate educational and training programs		√						√	√	√	√		√	6
Clear project scope and design					√	√								2
Effective procurement process					√		√							2
Consistent monitoring of CWR system	√			√										2

4.1 People-Related Success Factors

Highly competent. To ensure the success of construction waste recycling, construction projects should select highly competent project team members to complete the project. This selection of individuals should include having adequate experience, satisfying the minimum qualifications, and possessing the skills, knowledge, and abilities to work effectively. Selecting a highly competent workforce is suggested to be necessary because having skilled individuals can minimize mistakes during the process of separating waste materials. In other words, having a highly competent workforce can increase the recycling rate of construction waste. Example of the responses from the interview summaries that illustrate this factor include:

> Having highly skilled workers in construction also helps in recycling some material waste. For example, highly skilled individuals can take advantage of the small pieces of leftover irons, and recycle the waste into another product for the same project.
>
> Selecting subcontractors that are highly experienced in addition to satisfying the minimum qualifications is one of the most important factors for the success of construction waste recycling.
>
> Recycling of construction waste depends on the upper management's selection of workers and subcontractors that are highly competent in recycling construction waste as new material for the same project or another.
>
> Selecting skilled workers will help to reduce mistakes.
>
> The selection of skilled manpower has a positive impact on the reuse of some wastes.

Knowledgeable. In addition to having a highly competent workforce, project team members should be knowledgeable in differentiating materials that can be and cannot be recycled. This knowledge also includes having an understanding of the types of wastes that can be reused for other purposes at construction sites or recycled into other products at factories. This knowledge is necessary for the workforce to make appropriate and quick decisions when separating the waste. On the other hand, a lack of knowledge in individuals can hinder the workforce in making efficient decisions. Samples of responses from the interview summaries include:

> Workers should know about separating the waste from each other and knowing which material can be recycled at the project site and which material needs to be sent to the factory for recycling.
>
> It is important to employ individuals with that are highly knowledgeable in recycling construction waste materials such as bricks, tiles, timbers, and other waste materials.

Awareness of construction waste recycling. Besides being highly competent and knowledgeable, having a workforce that is aware of construction waste recycling is necessary to ensure successful waste recycling programs at construction sites. Specifically, having environmental awareness relates to understanding the impact of construction projects toward our fragile environment and the importance of protecting the environment from that impact. Therefore, awareness is the foundation to ensure that individuals in the project participate actively in ensuring the success of construction waste recycling programs. In other words, these findings suggest that preparing individuals with the required competency, necessary knowledge, and environmental awareness is crucial for the success of construction waste recycling.

4.2 *Process Related Success Factors*

In addition to people related success factors, process-related success factors include detailed project planning, clear project scope and design, consistent monitoring of the construction waste recycling system, effective procurement process, and clear project scope and design. These factors are discussed in the following subsections.

Detailed project planning involves preparing a detailed construction waste management plan, including identifying construction waste that is recyclable, controlling the waste from displacement, and ensuring the waste reaches the recycling factory. On the other hand, without adequate project planning, project team members might mistakenly send recyclable construction waste to landfills. Therefore, preparing detailed plans is crucial to avoid the reduction of construction waste recycling. Samples of responses from the interview summaries that illustrate this success factor include:

> A good waste management system is important to separate construction waste from each other. For example, projects with different waste containers for that targets different types of waste material can help the process of separating the recyclable and non-recyclable waste.
>
> Proper planning before executing the project, including identifying waste materials that can be recycled or and reused, is necessary. For example, planning to use the small leftover tiles generated from cutting tiles can be recycled or reused as wall skirting.
>
> During demolishing, proper planning is necessary for dismantling indiscriminately the waste generated so the waste can be recycled or reused in other projects.

Adequate educational and training programs. As suggested in previous subsections, people related success factors involve having project team members that are knowledgeable, highly competent, and aware of the importance of recycling construction in protecting the environment. While having team members that satisfy those requirements is ideal, that rarely happens in reality. Therefore, adequate education and training programs are necessary to prepare individuals that satisfy those requirements. For example, from the interview summaries, one respondent state:

> It is important to conduct training and awareness workshops for preparing contractors with the approaches in dealing with construction waste. For example, the process of reinforcing produces large pieces of iron waste that is reusable in other suitable places in the project. Also, irons from damaged scaffolds can be repaired directly at construction sites for other usages in the project.

In other words, a good program can provide individuals with knowledge of the approaches for identifying and separating construction waste from each other. Also, effective programs can increase project team members' environmental awareness that leads to intrinsic motivation to recycle construction waste. Therefore, providing educational and training programs on construction waste recycling is crucial.

Clear project scope and design. Besides preparing a detailed project plan, developing a clear project scope and design is also vital in ensuring the success of construction waste recycling. An example of a response from the interview summaries that relate to this success factor includes:

> We need to minimize construction mistakes from improper designs that change throughout the projects, and these changes generate construction waste. For example, there are cases where some mistakes in the design are identified during construction that requires demolishing a completed work that results in a large amount of that waste.

In other words, having a clear project scope and design is vital in ensuring the success of construction waste recycling and avoiding unnecessary generation of construction waste that may end in landfills.

Effective procurement processes involve having accurate estimations of materials necessary to complete the projects, verifying quantities and conditions of materials through regular stock-taking at sites, and buying materials at the right amount and time. While these activities are hardly associated with construction waste recycling in general, this factor increases the recycling rate of construction projects by reducing waste generation at sites. Buying materials at the right amount reduces the unnecessary generation of construction waste by avoiding negligence among individuals. On the other hand, having materials just in time avoids missing and damaged materials from thieves, misplacements of materials, and rain. Therefore, having effective procurement processes contributes to the success of construction waste recycling through the reduction of unnecessary generation of construction waste. Examples of responses from the interview summaries for this factor are:

> We should carefully purchase materials according to the necessary time and need, and ensuring good storage conditions during construction can minimize waste from excess raw materials.
>
> Performing regular inventory is important to know which materials are available and which materials to buy is important to avoid over-purchasing.
>
> Buying materials at the required quantities and sizes without over purchasing greatly helps to minimize the waste in construction.

Consistent monitoring of the construction waste recycling system is the last process-related success factors identified from this study's analysis of individual interview data with project managers. This factor includes monitoring, supervising, and managing project team members in obliging to the construction waste management plan, providing proper guidance on the process of recycling to individuals at sites, and ensuring proper execution of the construction waste management plan. In other words, while other process-related success factors can be associated with the planning phase of a construction project, having a good monitoring system during the execution phase is vital to ensure the success of construction waste recycling in construction projects. Responses from the interview summaries that illustrate this factor include:

> Construction workers should be monitored, supervised, and managed on their obligation to the established construction waste recycling plan for the project.

Workers are throwing damaged materials that can be repaired and reused. So, controlling workers and subcontractors help in recycling and repairing much unnecessary waste.

4.3 Comparison with Other Developing Countries

This section compares the success factors for construction waste recycling in Malaysia to those of other developing countries—China, Iraq, and Sudan. While a comparison with developed countries is feasible, comparing with countries that have a more similar social and economic background allows the identification of the potential critical success factors for construction waste recycling in Malaysia. To do that, the top five success factors for construction waste recycling in those countries are extracted and compared to this study's findings. Table 2 shows the result of that comparison.

The results show that only four of this study's eight success factors are identified as the top five success factors for construction waste recycling in other developing countries. Awareness of construction waste recycling among project team members is suggested to be crucial because managers and constructors with little awareness in saving resources and protecting the environment refuse construction waste recycling because it adds project costs and reduces company profits [35]. Conversely, consistent monitoring of construction waste recycling systems is necessary to avoid weak control of the system's implementation during the project [37]. Therefore, providing training and education that increases competencies in sustainable development and construction waste management to construction practitioners is necessary for improving both supervision and awareness of construction waste recycling [35, 36]. Lastly, detailed project planning is vital because changes on site can result in more construction waste and, therefore, a lower rate of construction waste recycling [35]. In a nutshell, while the findings can be found in other separate studies, these results illustrate that Malaysia, as a nation requires the same type and level of effort as other countries in enhancing waste recycling in its construction industry.

Table 2 Comparison with other developing countries' top five success factors for construction waste recycling (CWR)

Malaysia	China [35]	Sudan [36]	Iraq [37]
Aware of CWR	√	–	–
Detailed project planning	√	–	–
Adequate educational and training programs	–	√	–
Consistent monitoring of CWR system	–	√	√

4.4 *Limitations*

Although this study achieves its aim, there are some limitations worth stating. First, this study has a limited number of respondents. Therefore, the available data is not robust enough to represent the whole construction industry of developing countries, including Malaysia itself. However, rather than generalizing the success factors towards the local construction industry, this study aims to identify an alternative set of success factors for recycling construction waste in a developing country that might not be captured in the existing body of literature. Future studies can use this paper's results to develop questionnaire surveys, which usually have a higher sample size for interpretation and generalization. Also, subsequent studies can use evidence-based case studies to verify the identified success factors in this study.

5 Conclusion

This study identifies the success factors for recycling construction waste in developing countries by analyzing individual interview data with thirteen project managers using the thematic analysis. The major findings include:

- The success factors for recycling construction waste is related to either people or process.
- The people-related success factors involve having project team members that are highly competent, aware of construction waste recycling, and knowledgeable.
- The process-related success factors include having detailed project planning, adequate education and training programs, clear project scope and design, effective procurement system, and consistent monitoring of the construction waste recycling system.

These findings highlight the need to prepare appropriate project team members and processes in implementing waste recycling in construction projects. Research and industry practitioners can use these findings to develop strategies to ensure the success of construction waste recycling programs at sites. The key theoretical contribution of this research is by providing a set of alternative success factors that are affecting construction waste recycling in a developing country.

Acknowledgements The authors would like to thank the Ministry of Higher Education, Malaysia for supporting this study through the Fundamental Research Grant Scheme (FRGS/1/2019/TK06/UMP/02/1) as well as the project managers that agreed to participate in this work.

References

1. Lockrey, S., Verghese, K., Crossin, E., Nguyen, H.: Concrete recycling life cycle flows and performance from construction and demolition waste in Hanoi. J. Clean. Prod. **179**, 593–604 (2018)
2. Zhang, C., Hu, M., Dong, L., Xiang, P., Zhang, Q., Wu, J., Shi, S.: Co-benefits of urban concrete recycling on the mitigation of greenhouse gas emissions and land use change: A case in Chongqing metropolis, China. J. Clean. Prod. **201**, 481–498 (2018)
3. de Larrard, F., Colina, H.: Concrete Recycling: Research and Practice (2019)
4. Jin, R., Chen, Q.: Overview of concrete recycling legislation and practice in the United States. J. Constr. Eng. Manag. **145**(4), 05019004 (2019)
5. Pickin, J., Randell, P.: Australian National Waste Report 2016. Department of the Environment and Energy (2017)
6. Amemiya, T.: Current state and trend of waste and recycling in Japan. Int. J. Earth Environ. Sci. (2018)
7. Mah, C.M., Fujiwara, T., Ho, C.S.: Life cycle assessment and life cycle costing toward eco-efficiency concrete waste management in Malaysia. J. Clean. Prod. **172**, 3415–3427 (2018)
8. Mah, C.M., Fujiwara, T., Ho, C.S.: Concrete waste management decision analysis based on life cycle assessment. Chem. Eng. Trans. **56**, 25–30 (2017)
9. Jin, R., Chen, Q.: Investigation of concrete recycling in the US construction industry. Procedia Eng. **118**, 894–901 (2015)
10. Chan, A.P., Darko, A., Ameyaw, E.E., Owusu-Manu, D.G.: Barriers affecting the adoption of green building technologies. J. Manag. Eng. **33**(3), 04016057 (2016)
11. Darko, A., Chan, A.P.C., Ameyaw, E.E., He, B.J., Olanipekun, A.O.: Examining issues influencing green building technologies adoption: the United States green building experts' perspectives. Energy Build. **144**, 320–332 (2017)
12. Silvius, A.G., Kampinga, M., Paniagua, S., Mooi, H.: Considering sustainability in project management decision making; An investigation using Q-methodology. Int. J. Project Manage. **35**(6), 1133–1150 (2017)
13. Zhao, X., Hwang, B.G., Lee, H.N.: Identifying critical leadership styles of project managers for green building projects. Int. J. Constr. Manag. **16**(2), 150–160 (2016)
14. Govindan, K., Muduli, K., Devika, K., Barve, A.: Investigation of the influential strength of factors on adoption of green supply chain management practices: an Indian mining scenario. Resour. Conserv. Recycl. **107**, 185–194 (2016)
15. Gutierrez, A., Boukrami, E., Lumsden, R.: Technological, organisational and environmental factors influencing managers' decision to adopt cloud computing in the UK. J. Enterp. Inf. Manag. **28**(6), 788–807 (2015)
16. Lotfi, S., Eggimann, M., Wagner, E., Mróz, R., Deja, J.: Performance of recycled aggregate concrete based on a new concrete recycling technology. Constr. Build. Mater. **95**, 243–256 (2015)
17. Lotfi, S., Rem, P., Deja, J., Mróz, R.: An experimental study on the relation between input variables and output quality of a new concrete recycling process. Constr. Build. Mater. **137**, 128–140 (2017)
18. Serres, N., Braymand, S., Feugeas, F.: Environmental evaluation of concrete made from recycled concrete aggregate implementing life cycle assessment. J. Build. Eng. **5**, 24–33 (2016)
19. Everaert, M., Stein, R., Michaux, S., Goovaerts, V., Groffils, C., Delvoie, S., … Broos, K.: Microwave Radiation as a Pre-treatment for Standard and Innovative Fragmentation Techniques in Concrete Recycling. Materials **12**(3), 488 (2019)
20. Kasai, Y.: Barriers to the reuse of construction by-products and the use of recycled aggregate in concrete in Japan. In: Sustainable Construction: Use of Recycled Concrete Aggregate: Proceedings of the International Symposium, pp. 433–444 (1998)
21. Tam, V.W., Tam, L., Le, K.N.: Cross-cultural comparison of concrete recycling decision-making and implementation in construction industry. Waste Manage. **30**(2), 291–297 (2010)

22. Thompson, J.D., Bashford, H.H.: Concrete recycling and utilization of recycled concrete: an investigation of the barriers and drivers within the Phoenix Metropolitan Area. In: Construction Research Congress 2012: Construction Challenges in a Flat World, pp. 1682–1688 (2012)
23. Begum, R.A., Siwar, C., Pereira, J.J., Jaafar, A.H.: A benefit–cost analysis on the economic feasibility of construction waste minimisation: the case of Malaysia. Resour. Conserv. Recycl. **48**(1), 86–98 (2006)
24. Begum, R.A., Siwar, C., Pereira, J.J., Jaafar, A.H.: Implementation of waste management and minimisation in the construction industry of Malaysia. Resour. Conserv. Recycl. **51**(1), 190–202 (2007)
25. Begum, R.A., Siwar, C., Pereira, J.J., Jaafar, A.H.: Attitude and behavioral factors in waste management in the construction industry of Malaysia. Resour. Conserv. Recycl. **53**(6), 321–328 (2009)
26. Mah, C.M., Fujiwara, T., Ho, C.S.: Construction and demolition waste generation rates for high-rise buildings in Malaysia. Waste Manage. Res. **34**(12), 1224–1230 (2016)
27. Esa, M.R., Halog, A., Rigamonti, L.: Strategies for minimizing construction and demolition wastes in Malaysia. Resour. Conserv. Recycl. **120**, 219–229 (2017)
28. Wahi, N., Joseph, C., Tawie, R., Ikau, R.: Critical review on construction waste control practices: Legislative and waste management perspective. Procedia Soc. Behav. Sci. **224**, 276–283 (2016)
29. Lee, Z.P., Rahman, R.A., Doh, S.I.: Success factors of design-build public sector projects in Malaysia. In: IOP Conference Series: Materials Science and Engineering, Vol. 712, No. 1, p. 012045 (2020)
30. Rahman, R.A., Radzi, A.R., Saad, M.S.H., Doh, S.I.: Factors affecting the success of highway construction projects: the case of Malaysia. In: IOP Conference Series: Materials Science and Engineering, Vol. 712, No. 1, p. 012030 (2020)
31. Braun, V., Clarke, V.: Using thematic analysis in psychology. Qual. Res. Psychol. **3**(2), 77–101 (2006)
32. Rahman, R.A., Ayer, S.K.: Prevalent issues in BIM-based construction projects. Proc. Joint Conf. Comput. Constr. **1**, 645–652 (2017)
33. Radzi, A.R., Bokhari, H.R., Rahman, R.A., Ayer, S.K.: Key attributes of change agents for successful technology adoptions in construction companies: a thematic analysis. In: Computing in Civil Engineering 2019: Data, Sensing, and Analytics, pp. 430–437 (2019)
34. Radzi, A.R., Rahman, R.A., Doh, S.I., Esa, M.: Construction readiness parameters for highway projects. In: IOP Conference Series: Materials Science and Engineering, Vol. 712, No. 1, p. 012029 (2020)
35. Lu, W., Yuan, H.: Exploring critical success factors for waste management in construction projects of China. Resour. Conserv. Recycl. **55**(2), 201–208 (2010)
36. Omran, A., Eltayed, M.: Determining the critical success factors for waste management in construction projects in Khartoum City, Sudan. Acta Technica Corviniensis-Bull. Eng. **9**(3), 123 (2016)
37. Al-Agele, H.K.B., Al-Kaabi, S.A.: Identification of key factors affecting waste management in life cycle of the construction project by using Delphi technique. J. Eng. **22**(7), 19–34 (2016)

A Proposed Methodology of Life Cycle Assessment for Hot Water Building Systems

Arthur B. Silva, Mohammad K. Najjar, Ahmed W. A. Hammad, Assed N. Haddad, and Elaine G. Vazquez

Abstract Population growth and technological development in recent decades have made human activities largely responsible for structural changes in the built environment at regional and global levels. Civil construction, as an integral part of the chain of industrial activities, is also one of the segments responsible for energy consumption and potential greenhouse gas emissions throughout its life cycle. The building materials and their systems have a direct influence on energy consumption and impact assessment, both in the pre-operational, use and end-of-life and disposal phases. In this context, Hot Water Building Systems (HWBS) are included. The variability of possibilities available with regard to the choice of energy sources, water reserve and distribution systems and the selection of materials used in these building systems allows empowering the decision-making in the designing phase. The definition of the type of installation to be used in a building is defined by technical and/or economic requirements. However, the spectrum of possibilities should consider resource consumption and generation of environmental impacts throughout the life cycle. This research proposes a novel application of an environmental management method to empower the decision-making process and encourage the selection course of HWBS. This work insights a Life Cycle Assessment (LCA) methodology to compare a specific the environmental performance of two distinct HWBS (i.e. Natural Gas Heating System and Solar Heating System) for multi-family residential developments.

Keywords Life cycle assessment methodology · Hot water building systems · Environmental performance

A. B. Silva · M. K. Najjar · A. N. Haddad · E. G. Vazquez (✉)
Escola Politécnica, Department of Civil Construction, UFRJ, Rio de Janeiro, Brazil
e-mail: elaine@poli.ufrj.br

A. W. A. Hammad
Faculty of Built Environment, New South Wales University, Sydney, Australia

V. M. C. F. Cunha et al. (eds.), *Proceedings of the 3rd RILEM Spring Convention and Conference (RSCC 2020)*, RILEM Bookseries 35,
https://doi.org/10.1007/978-3-030-76543-9_19

1 Introduction

Population growth and technological development in recent decades have made human activities largely responsible for structural changes in the environmental landscape at regional and global levels [1, 2]. Regarding the aspects of natural resources consumption and the passive impact of human activities, the energy sector is responsible for a major part of greenhouse gas (GHG) emissions [3]. For instance, residential and commercial buildings account for around 41% of total energy consumption in the United States [4]. In these terms, building components have a direct influence on energy consumption and environmental impacts over the entire Life Cycle Assessment (LCA), basically, during the pre-operational phase (i.e. material manufacturing, transportation and construction), as well as at the end-of-life and disposal phase [5].

Buildings are major consumers of energy throughout their life cycle. Generation of energy primarily depends on conventional sources, which is the basic cause of environmental pollution [6]. The materials and their systems have a direct influence on energy consumption and impact generation, in the pre-operational phase (materials manufacturing, transportation and construction), also in the final and discarded life.

Hot Water Building Systems (HWBS) are directly related to energy consumption in residential buildings; performing the second largest energy consumer in buildings and, thus, representing an integral part of the water-energy nexus [3].

The conventional selection of a water heating system in residential buildings focuses on the financial evaluation rather than the sustainability pillars and life cycle consequences (i.e. economic x environmental pillars) [7]. At this level of the analysis, the application of LCA methodology at an early designing phase of residential buildings could empower the decision-making process and sustainability [8], as well as facilitating the selection criteria of HWBS [9], where professional and experts could evaluate the environmental performance of the installed water heating system [10, 11].

A novel application of an environmental management method is presented herein to empower the decision-making process and encourage the selection course of HWBS, taking into consideration the technical and economic aspects at an early designing phase of buildings. The aim of this work is to present a proposal for a method derived from the general LCA methodology in order to compare the environmental performance of two distinct HWBS for multi-family residential developments, through thermal heaters installed on the final roof of buildings, with supplementation of electrical supply, so that accurate information on the environmental performance of the systems can be obtained. However, the installed HWBS considered in this work are NGHS and SHS. In this work, a literature review of the LCA methodology is presented in Sect. 2. The proposed methodology to evaluate the LCA for HWBS is presented in Sect. 3. However, results are discussed in Sect. 4, while the conclusions and final recommendations are presented in the last section.

2 A Literature Review of LCA Methodology

LCA is described by as a scientifically based analysis and assessment of the environmental impacts of product systems [12]. Regarding to ISO standards, this methodology was revised in 2006 and started to be condensed into ISO 14,040 and ISO 14,044 standards [13]. In Brazil, the Brazilian Association of Technical Standards, published equivalent versions initially translated in 2001 (NBR ISO 14,040, NBR ISO 14,041, NBR ISO 14,042 and NBR ISO 14,043) and later in 2009 and 2014 (NBR ISO 14,040 and NBR ISO 14,044 replacing the previous ones) in a way to support the descriptive text and definitions, as well as facilitate the understanding of the theme [14].

At the level of the energy consumption and impact assessment of products, LCA is characterized as a management methodology that help computing inputs and outputs of a production system to evaluate the environmental performance over their entire lifespan [15]. The application of LCA methodology in the construction sector focuses mainly on the characteristics of the building typology and components [16]. However, such application is facing several challenges that are giving a wide spectrum of related variables, and making it necessary and interesting to define a standardized analysis structure in order to increase its accuracy [17]. In this context, HWBS should also be assessed over their entire lifespan, so that the energy incorporated into the biogenic emissions are considered and give greater dimension to the impacts of the systems [18]. The methodology adopted for this study is related to the LCA of HWBS in multifamily residential buildings, taking into consideration comparing the different types of systems in relation to their environmental performance at an early designing step. The general scope of such application will be conducted in four phases based on the LCA methodology: Goal and Scope definition; Life Cycle Inventory analysis (LCI); Life Cycle Impact Assessment (LCIA); and Interpretation of data and results obtained by the partial and final methodological processes [19, 20].

Defining the Goal and Scope of the study means determining the intended application for the analysis and the reason for carrying out the study, the target audience to whom the results are intended to be communicated and, therefore, the means for their dissemination [19]. In these terms, the scope of the study should include the definition of the product system to be studied; the functions of this system or compared systems; the determination of the functional unit/functional equivalent; the system boundary; allocation procedures; selected impact categories and methodology for impact assessment as well as subsequent interpretation to be used; data requirements; assumption; limitations; initial requirements for data analysis; type of critical analysis, if applicable; type and format of report required for study.

The next step is to build up the LCI based on ISO 14,044, which demonstrates a definition of the inventory analysis phase that involves the cradle-to-grave character of the method; "life cycle assessment phase involving the compilation and quantification of inputs and outputs of a product system over its lifetime. life cycle". Table 1 illustrates the output results of the LCI step [21], including a list of data of the environmental impacts to be evaluated at the next step of the LCA methodology,

Table 1 Characterization of impact categories commonly demonstrated in studies

Impact Categories	Geographic Scale		
	Global	Regional	Local
Global Warming	■		
Stratospheric ozone depletion	■		
Phoyochemical oxidant formation		■	■
Acidification		■	■
Nutrient enrichment		■	■
Ecotoxicity		■	■
Human toxicty		■	■
Working environment			■
Odour			■
Noise			■
Radiation	■	■	■
Resource consumption	■	■	■
Land use			■
Waste			■

Adapted from Stranddorf and Hoffmann [21]

which is LCIA. At this level of the analysis, LCIA aims to give an overview of the significance of the potential impacts of the examined product [22].

There are several methods to evaluate the extracted impacts from the LCI, hence, it is highly important to choose the most appropriate method for each case study [18]. LCIA can be distinguished within two levels: midpoints and endpoints [23]. At the midpoint impact assessment level, indicators are given along the environmental mechanism, while at the endpoint impact assessment level, "*Characterization considers the entire mechanism to its end point, ie. it refers to a specific damage related to the broader area of protection, which may be human health, natural environment or natural resources*" [24]. Finally, the interpretation level refers to permeate the entire analysis process, where the findings of LCI and LCIA are to be consistently combined with the defined Goal and Scope in order to draw conclusions and recommendations [19].

3 Proposed LCA Methodology for HWBS

The objective of this work is to Present a methodological work flowchart for the comparative application of LCA for HWBS in multi-family residential buildings, as a way to obtain data for analysis regarding the environmental impacts of such systems, taking into consideration the energy consumption during of the operation phase of HWBS and its impact on the living standards, combined with the LCA database of the of the applied materials. The developed method herein is developed based on nine main phases, which guide the elaboration and evaluation of the projects and their respective analysis and are the organization of the general phases of methodological application of the LCA recommended by ISO Standards. At this level of the analysis, the interpretation phase is divided into four distinct stages; Interpretation (A), Interpretation (B), Interpretation (C) and Interpretation (D). Each stage has been oriented and modified to verify the output results collected from the previous steps (LCI and LCIA), as well as evaluating their quality, coherence and importance to the study. Figure 1 describes in detail the steps contained in phases I, II, III, and IV of the study, which are Goal and Scope Definition (A), Interpretation (A), Goal and Scope Definition (B), and Interpretation (B), respectively.

The starting point of the analysis, as presented in Fig. 1 is to determine the environmental profile of a certain HWBS. Phase I, Goal and Scope Definition (A), means to conduct general definitions, function, and functional unit. At this level of the analysis, Item (I.1), presented in Fig. 1, means determining the general purpose of the analysis to be performed. Such an objective must be clear and consistent with the reality of the place of application so that it can be valid and have real importance in the context in which it will be applied. The results of the analysis of the performance of the HWBS during the operation phase can be used to make a decision about the use of a particular type of system still in the design phase, with the general objective being traced, for example, as the definition of the type of system, system to be used for the distribution of water in a specific building type, or if a copper or PVC pipe is defined in the project, given the observed impacts.

Item (I.2), presented in Fig. 1, means determining the target audience. In the case of the building installation project, the target audience can be defined as the end user, who will actually use the system and wants to know which one is most advantageous in this respect, or the builder who will do the work and needs the best cost- benefit, or it may be the own design team that needs the determination of the system that consumes the least environmental resources or generates the least impact in order to have a sustainable building profile that seeks environmental certification. Next, item (I.3) determines the scope of the analysis. At this point, a phase of refinement of objectives is to be included, defining the stages of application, scope, work team involved and other important aspects to the elaboration. Moreover, the evaluated function in the employed method can be traced in item (I.4). The main functions of cold water, hot water and piped gas systems can be analyzed, considering that the functions are closely linked to the uses of systems. The functional unit of the study, presented in item (I.5) in Fig. 1, deals with the quantification of the determined function or product performance characteristic. In these terms, the functional unit must, in the case of building systems, be worked as a performance unit and not a mass or metric unit.

Phase II, Interpretation (A), refers to identify stakeholders, focusing on a more managerial profile of the process. Construction projects involves many stakeholders (i.e. facility owners and users, project managers, project team members, facility managers, designers and architects, allotment companies, shareholders of

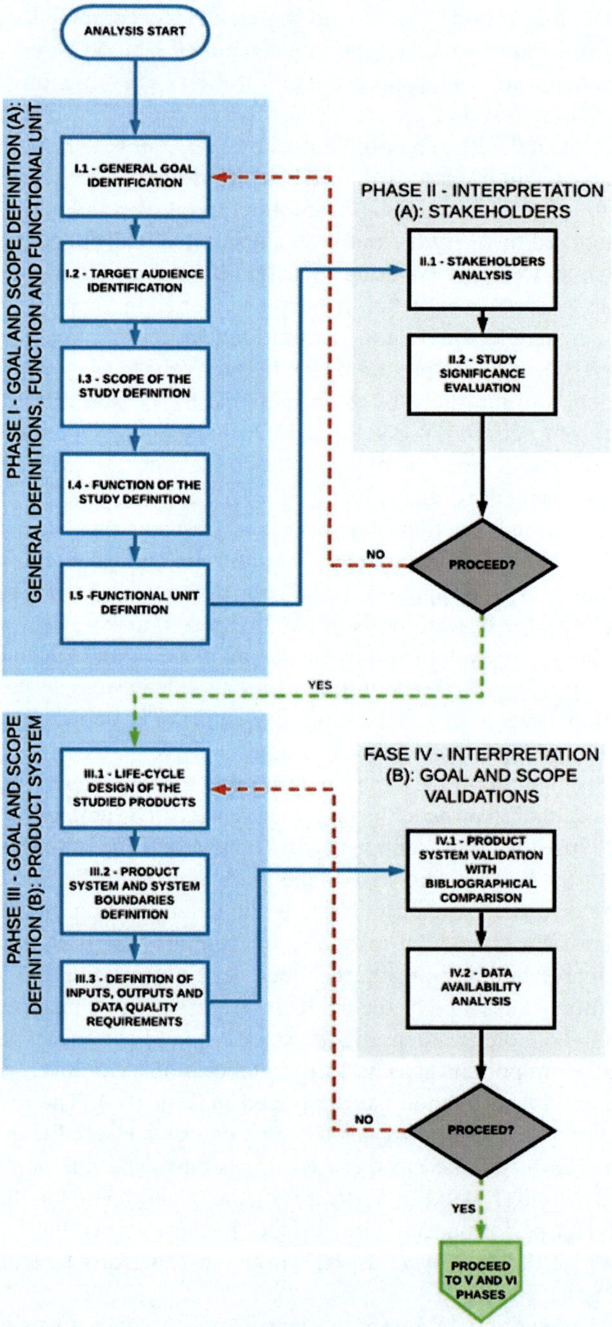

Fig. 1 Detailed flowchart of the proposed methodology of this work—Phases I to IV

the company that develops the enterprise, the public administration, construction workers, subcontractors and outsourced service providers, competitors, banks, insurance companies, representatives of the surrounding community, the general public, and others) [25].

Phase III, Goal and Scope Definition (B) begins by defining the product system, inputs and outputs, which determines the life cycle design of the studied facility systems, as well as the system boundaries. At this level of the analysis, it is necessary to trace the processes and phases involved in the life cycle of the hydraulic installations from the production of the component materials to the final disposal of the system after the operation phase. Conducting the life cycle design of the studied products leads to item (III.2), where the product system and system boundary are to be determined. Item (III.3), presented in Fig. 1, necessitates defining the inputs and outputs of the product system and data requirements. In this work, the most important issue when building up the inventory of database is the quality, relevance, accuracy completeness, and representativeness of the data due to technological and profile changes of the products used to HWBS, taking into account the type and location of installation, consistency and reproducibility of the products.

Phase IV, Interpretation (B), refers to validate the product system. For this, a comparison should be made between the determined life cycle and product system for analysis with other similar related studies and make adjustments that the professional deems necessary to make the system as objective as possible with respect to the results. This factor is extremely important to the legitimacy of the study, since the geographical location and the local social and environmental profile have a direct influence on the quantification and qualification of the impacts of a product system represented by a building installation.

The second sequence of the detailed flowchart of the proposed methodology if this work is illustrated in Fig. 2a, represented by Preliminary System Design Development (Phase V), Inventory Analysis (Phase VI), and Interpretation (C) (Phase VII).

Phases V and VI should begin concurrently or sequentially, as the first step in the inventory phase is to define and organize data sources to enable collection, organized according to requirements. In order to obtain these, it is oriented to use an internationally consolidated database, since the exact obtaining of the production processes involved denotes time and resources. However, such a process may lead to variability of results with respect to the actual life cycle of the examined system and compromise the reliability and requirements of the data. Data collection at this level of the analysis may involve the need for bibliographic and market research that fosters the assembly of product system processes and guides the volume of materials consumed for them.

Phase V consists of the preliminary design of building systems for analysis objects for the primary purpose of obtaining the data necessary to foster life cycle inventory analysis, as presented in Fig. 2a. Hence, four main sequential activities related to good engineering practices are to be defined, as follows:

(a) An architectural assessment of the needs and demands of the building should be carried out.

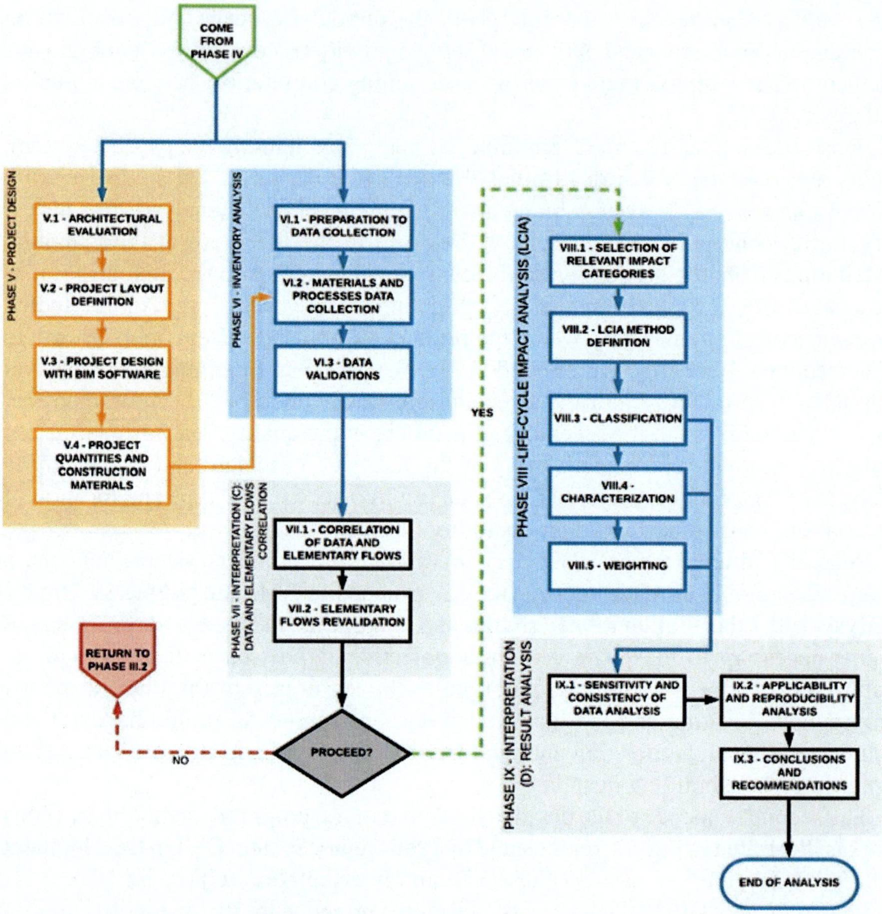

Fig. 2 Detailed flowchart of the proposed methodology of this work—Phases V to IX

(b) The project layout should be defined, taking into consideration elements such as technical reserve location, water or power inlet leasing, internal distribution pipe division, better leasing of equipment needed for system operation.

(c) The effective elaboration of the projects should be considered based on the current norms and available database. Hence, one should seek to understand the level of details that the project necessitates to meet the requirements of the LCA methodology such as the specifications of buildings components, taking into consideration that the LCA methodology involves processes such as "cradle to grave", that is, from the extraction and manufacturing phase until the end of life; "cradle to gate", that is, from the extraction until the end of manufacturing phase; "cradle to cradle", that is, from the extraction and manufacturing until the end of life and recycling to be reused again [13].

(d) The project materials and construction components should be quantified, considering the use of the products that will actually be installed to increase the reliability and completeness of the data obtained.

Phase VI, inventory analysis, is the elaboration phase of the projects that ends up with obtaining the list of materials that fosters the inventory data. The data collection requires obtaining data related to the life cycle of the components of the HWBS; a factor that is intended to be performed with the help of a database, such as Ecoinvent. This phase necessitates guiding the modeling of the product system with the aid of LCA software, such as OpenLCA. System modeling often involves the combination of basic processes and raw materials in the database to obtain the desired products [26], a factor that can create uncertainties in the process, given the insufficient knowledge obtained about the process. production or misuse of processes. At this level of the analysis, the data collection stages consider the database collections, materials and processes data collection, and culminate in the data validation. The data collected in the inventory should be evaluated against the data requirements defined at the beginning of the study in a way to determine the relevance or discard of collected data by screening the material.

Phase VII is Interpretation (C), where the correlation of data and elementary flows is ranked in the classification of the collected data according to the defined flows for the studied functional unit [27]. Thus, it is verified whether all flows considered have consistent and sufficient data for the elaboration of the LCIA, if other data are needed or if there is not enough data available for the definition of all flows. At the end of this analysis, a decision should be taken whether to continue the process or to redefine and redo the completed phases to ensure concise results. The carried inventory and the refined product system facilitate proceeding to the LCIA and final interpretations, which are detailed in the figure, Fig. 2b, which demonstrates the final phases of the proposed methodology of this work.

Phase VIII, LCIA, means selecting the relevant impact categories to the study, taking into consideration the history of application of LCA for analysis of hydraulic systems. The major impact categories, as previously presented in Table 1, can be exposed to global warming, human toxicity (carcinogenic and non-carcinogenic), shortage of fossil resources and mineral resources and waste, impacts considered directly related to the building systems of employment of the study. The next step after selecting the impact categories, the LCIA step is to be conducted according to the item (VIII.2), illustrated in Fig. 2b. Hence, it is important to consider the midpoint impact assessment, which has less data uncertainties [28], using an impact assessment method such as ReCiPe, which combines the Eco-indicator 99 and CML methods, giving them an update regarding the content, deriving characterization factors according to a midpoint approach (with 17 indicators) or endpoint (with 3 indicators) [29]. There is a fundamental relationship between midpoint indicators, direct impacts, and endpoint indicators. The structure overview of ReCiPe impact assessment method is presented in Fig. 3 [30]. Phase IX, Interpretation (D), which necessitates evaluating the completeness, sensitivity and consistency of the data.

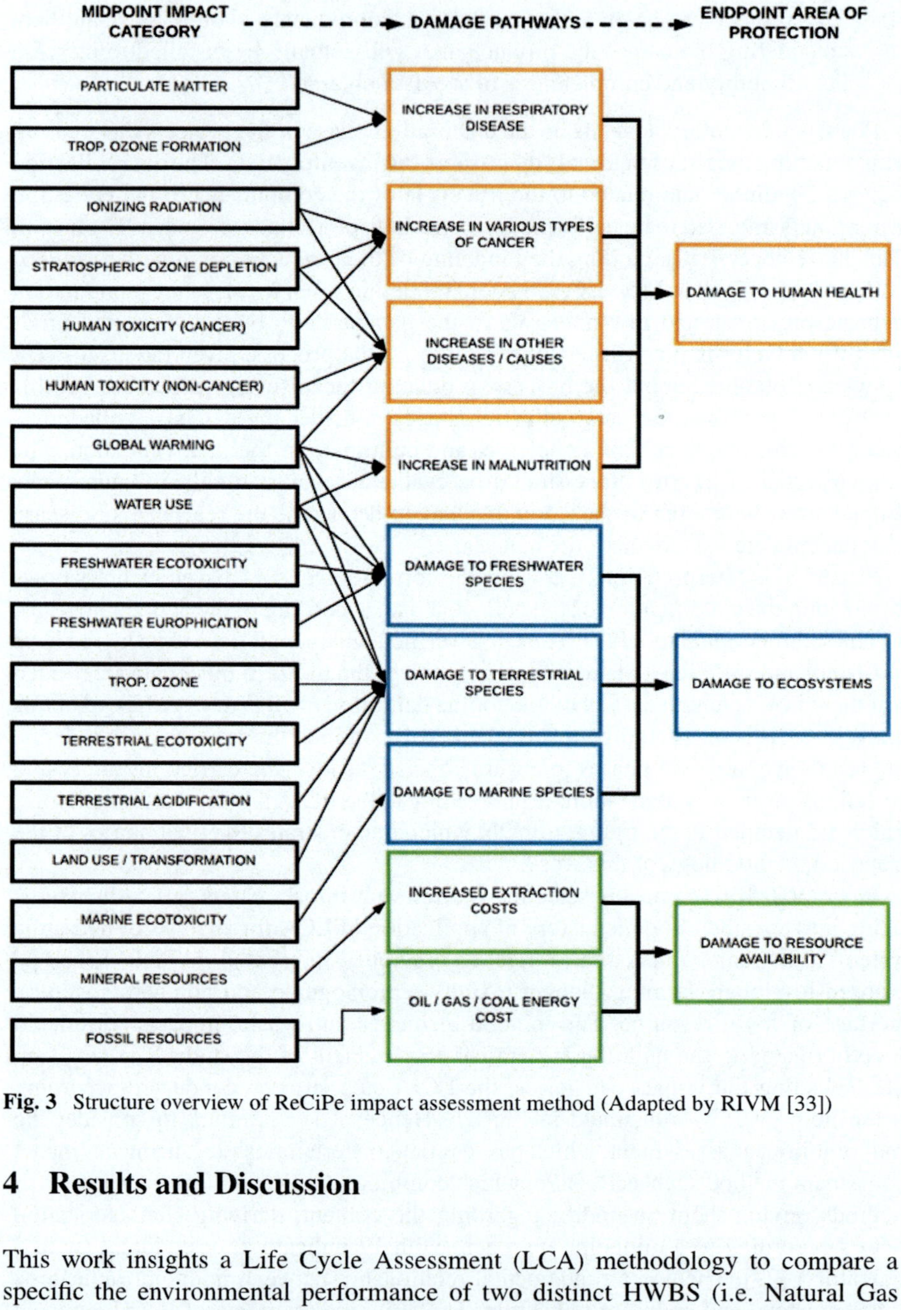

Fig. 3 Structure overview of ReCiPe impact assessment method (Adapted by RIVM [33])

4 Results and Discussion

This work insights a Life Cycle Assessment (LCA) methodology to compare a specific the environmental performance of two distinct HWBS (i.e. Natural Gas Heating System and Solar Heating System) for multi-family residential developments.

The proposed methodology presented in this work intends to facilitates analyzing the life cycle analysis of Hot Water Building Systems at early stages of building design, taking into consideration from a cradle-to-gate perspective; including primary extraction of component materials, beneficiation and production and transportation to the place of execution of the projected installation. Hence, the energy consumption incorporated to them and their biogenic emissions are considered and given larger dimensions of impact assessment.

Based on the expected results, an analysis of the impacts obtained can be performed and the study is concluded by checking the quality of the conclusive data and its reproducibility, defining improvements to the method and observed limitations, and culminating in guidelines for the methodological application.

5 Conclusions

Population growth and technological development have made human activities largely responsible for structural changes in the built environment at regional and global levels. Building materials and their systems in the construction sector have a direct influence on energy consumption and impact assessment over the entire lifespan of Hot Water Building Systems. Hence, the novelty of this work is in proposing an application of an environmental management methodology to empower the decision-making process and encourage the selection course of Hot Water Building Systems, taking into consideration the technical and economic aspects at an early designing phase of buildings. The main objective of this work is to apply the Life Cycle Assessment LCA methodology to compare the environmental performance of two distinct Hot Water Building Systems; Natural Gas Heating System and Solar Heating System, in a multi-family residential development.

The construction sector has a vast and direct influence on the environmental impacts caused by human activities in the built environment at various scales. The application of Life Cycle Assessment methodology at early stages of designing buildings could provide more information related to the environmental performance of the building materials, building systems and building installations. Such useful information could culminate in the reduction of passive impacts of construction projects over their entire lifespan. At this level of the analysis, the life cycle impacts are highly interdependent, so that one phase can influence others. For example, selecting building materials can reduce the need for space heating, but can also increase built-in energy and transportation-related impacts or change the lifespan of the building as a whole.

The proposed methodology presented herein has some limitations as to its implementation. The results of the application will provide a complex set of numerical values for environmental impact indicators and a report with all related assumptions made during the analysis, which makes the interpretation of the results by non-specialists in Life Cycle Assessment are difficult to be conducted. Particularly, if there is insufficient comparative standard, which is considered as a determining factor in the results of the analysis performed for buildings and building systems. Another limitation of this work refers to the lack of an inventory database for South

America and for local studies. The application of Life Cycle Assessment methodology depends on using existing data in global bases. Hence, it is highly important to highlight the important role of the regionality of the case studies, taking into consideration the heterogeneity of the natural, cultural and economic profiles among the localities of the region. On the other hand, the determination of a methodological standardization regarding the application of Life Cycle Assessment methodology for building systems allows more professionals to be involved in this process, and data can be obtained from the results of building applications, which can be a parameter for other studies.

References

1. Meyer, W.B., Turner, B.L.: Human population growth and global land-use/cover change. Annu. Rev. Ecol. Syst. **23**(1), 39–61 (1992)
2. Harte, J.P.: Human population as a dynamic factor in environmental degradation. Popul. Environ. **28**(4–5), 223–226 (2007)
3. Riahi, K., Rao, S., Krey, V., Cho, C., Chirkov, V., Fischer, G., Kindermann, G., Nakicenovic, N., Rafaj, P.: RCP 8.5—A Scenario of Comparatively High Greenhouse Gas Emissions. Clim. Change Clim. Change **109**, 33 (2011)
4. Fumo, N., Mago, P., Luck, R.: Methodology to estimate building energy consumption using EnergyPlus Becnhmark Models. Energy Build. **42**(1), 2331–2337 (2010)
5. Constantinos, A.B., Droutsa, K., Dascalaki, E., Kontoyiannidis, S.: Heating energy consumption and resulting environmental impact of European apartment buildings. Energy Build. **37**(5), 429–442 (2005). https://doi.org/10.1016/j.enbuild.2004.08.003
6. Valdehi, A.D., Ralegaonkar, R.V., Mandavgane, S.: Improving environmental performance of building through increased energy efficiency: a review. Sustain. Urban Areas **1**(4), 211–218 (2011). https://doi.org/10.1016/j.scs.2011.07.007
7. Martinopoulos, G., Papakostasa, K.T., Papadopoulos, M.: A comparative review of heating systems in EU countries, based on efficiency and fuel cost. Renew. Sustain. **90**(1), 687–699 (2018). https://doi.org/10.1016/j.rser.2018.03.060
8. Najjar, M., Figueiredo, K., Hammad, A.W.A., Haddad, A.: Integrated optimization with building information modeling and life cycle assessment for generating energy efficient buildings. Appl. Energy **250**, 1366–1382 (2019)
9. Glick, S., Guggemos, A.A.: Life-cycle assessment and life-cycle cost as collaborative tools in residential heating system selection. J. Green Build. **5**(3), 107–115 (2010)
10. Randi, H.B., Marc, A.E.: A review of the sustainability of residential hot water infrastructure: public health, environmental impacts, and consumer drivers. J. Green Build. Fall **6**(4), 77–95 (2011)
11. Moore, C.C.S., Rego, E.E., Kulay, L.: The Brazilian electricity supply for 2030: a projection based on economic, environmental and technical criteria. Environ. Nat. Resour. Res. **7**(4), 17–29 (2017). https://doi.org/10.5539/enrr.v7n4p17
12. Klöpffer, W.: Introducing life cycle assessment and its presentation in 'LCA Compendium'. In: Klöpffer, W. (ed.) Background and Future Prospects in Life Cycle Assessment. LCA Compendium – The Complete World of Life Cycle Assessment. Springer, Dordrecht, pp. 39–84 (2014)
13. UNEP: Avaliação de Políticas Públicas para Redução da Emissão de Gases de Efeito Estufa em Edificações. São Paulo (2012). Disponível em: http://www.cbcs.org.br/userfiles/comitestematicos/outrosemsustentabilidade/UNEP_capa-miolo-rev.pdf

14. De Souza, C.G., Barbastefano, R.G., Teixeira, R.C.: Life cycle assessment research in Brazil: characteristics, interdisciplinarity, and applications. Int. J. Life Cycle Assess. **22**(1), 266–276 (2017). ISSN: 09483349
15. Coelho Filho, O., Saccaro Jr, N.L., Luedemann, G.: A avaliação de ciclo d vida cmo ferramenta para a formulação de políticas públicas no Brasil. Brasília: Ipea (2016)
16. Meex, E., Hollberg, A., Knapen, E., Hildebrand, L., Verbeeck, G.: Requirements for applying LCA-based environmental impact assessment tools in the early stages of building design. Build. Environ. **133**(1), 228–236 (2018)
17. Ochsendorf, J., et al.: Methods, Impacts, and Opportunities in the Concrete Building Life Cycle. Cambridge (2011). Disponível em: https://www.greenconcrete.info/downloads/MITBuildingsLCAreport.pdf
18. Da Silva, G.A. et al.: Avaliação do ciclo de vida: ontologia terminológica. Instituto Brasileiro de Informação em Ciência e Tecnologia – Ibict, 72p (2014)
19. ISO—International Organization for Standardization: ISO 14040. Environmental Management—Life Cycle Assessment—Principles and Framework. Geneva: ISO, 20p (2006a)
20. ISO—International Organization for Standardization: ISO 14044. Environmental Management—Life Cycle Assessment—Requirements and Guidelines. Geneva: ISO, 46p (2006b)
21. Stranddorf, H.K., Hoffmann, L.: Update on impact categories, normalization and weighting in LCA-Selected EDIP97-data. Danish Environmental Protection Agency, pp. 995, 290 (2005a)
22. Najjar, M. et al.: Integration of BIM and LCA: evaluating the environmental impacts of building materials at an early stage of designing a typical office building. J. Build. Eng. **14**(1), 115–126 (2017)
23. Najjar, M.K., Figueiredo, K., Evangelista, A.C.J., Hammad, A.W.A., Tam, V.W.Y., Haddad, A.: Life cycle assessment methodology integrated with BIM as a decision-making tool at early-stages of building design. Int. J. Construct. Manage. 1–15 (2019b)
24. Crespo, N.M., Bueno, C., Ometto, A.R.: Avaliação de Impacto do Ciclo de Vida: revisão dos principais métodos Palavras-chave. Production, no x (2013)
25. Machado, G.: Aprenda como funciona a gestão de Stakeholders na Construção Civil. Halo Notoriedade Empresarial (2017). Disponível em: http://halonotoriedade.com.br/aprenda-como-funciona-a-gestao-de-stakeholders-na-construcao-civil/. Acesso em: 07/abr./19
26. Oyarzo, J., Peuportier, B.: Life cycle assessment model applied to housing in Chile. J. Clean. Prod. **69**(1), 109–116 (2014). ISSN: 0959-6526
27. Inaba, A. et al.: Chapter 4: Data documentation, review, and management. In: Global Guidance Principles for Life Cycle Assessment Databases: A Basis for Greener Processes and Products, pp. 85–95 (2011). ISBN: 978-92-807-3174-3
28. Bare, J.C., Hofstetter, P., Pennington, D.W., Udo de Haes, H.: Int. J. LCA **5**(319) (2000)
29. Acero, A.P., Rodríguez, C., Ciroth, A.: LCIA Methods: Impact assessment methods in Life Cycle Assessment and Their Impact Categories. Greendelta, 23p. (2015)
30. RIVM. LCIA: The ReCiPe Model (2018) Disponível em: https://www.rivm.nl/en/life-cycle-assessment-lca/recip

Screening Regionally Available Natural Resources and Waste Streams as Potential Supplementary Cementitious Material

Matea Flegar, Marijana Serdar, Diana Londono-Zuluaga, and Karen Scrivener

Abstract The usage of industrial by-products and natural resources has already been acknowledged as a viable option for lowering the CO_2 emissions in cement production. The utilization of locally available materials as supplementary cementitious material (SCM) presents an opportunity to produce more sustainable cements that could satisfy the growing concrete demand. However, the composition of these materials is origin related and can differ from region to region. For that reason, this study correlates the physical and chemical properties of local materials with their mechanical performance when used as SCM. Twelve different samples were collected: three slags, two fly ashes, one silica fume, and six clays. The R^3 reactivity test was performed together with compressive strength tests on mortars after 2, 7 and 28 days. Finally, the paper presents the most suitable materials prioritising the results according to the obtained data. With the screening approach, this research brings attention to the key parameters that are vital for preliminary SCM identification.

Keywords Supplementary cementitious materials (SCMs) · Screening of raw materials · Reactivity · R^3 test · Compressive strength

1 Introduction

Currently, the concrete consumption has grown such that it has become the second most used material in the world after water [1]. This is due to the increase in demand for housing caused by the increase in population. The developing countries contribute to the increase in population and as a result these countries consume almost 70% of all world cement [2]. With a "business as usual" strategy, it is estimated that cement industry will solely be responsible for 24% of total global CO_2 emissions by year 2050

M. Flegar · M. Serdar (✉)
Faculty of Civil Engineering, Department of Materials, University of Zagreb, Zagreb, Croatia
e-mail: mserdar@grad.hr

D. Londono-Zuluaga · K. Scrivener
Laboratory of Construction Materials, École Polytechnique Fédérale de Lausanne, Lausanne, Switzerland

V. M. C. F. Cunha et al. (eds.), *Proceedings of the 3rd RILEM Spring Convention and Conference (RSCC 2020)*, RILEM Bookseries 35,
https://doi.org/10.1007/978-3-030-76543-9_20

[3]. Due to this, there is a strong need to unlock the potential of accessible materials that can supply the growing demand for cement and lower the CO_2 emissions without compromising the performance and durability of concrete.

CO_2 is emitted into two ways during the production of Portland cement clinker, namely, (a) from the energy supply needed to heat the cement kiln, and (b) from decarbonation of calcium carbonate ($CaCO_3$) [4]. The first case accounts for 40–50% of emissions in the production, while the chemical reaction of calcium carbonate releases the rest. Considering the fact that the raw material in Ordinary Portland cement (OPC) consists mostly of $CaCO_3$ (75–79%) [4], high volume replacement of clinker with supplementary cementitious materials (SCMs) can significantly lower the mentioned emissions.

Nowadays, different types of cements are being used in the construction industry. A certain amount of these cements contain a relatively small portion of SCMs which are added primarily to reduce the cost or to improve the long-term mechanical and durability properties of concrete [5]. Higher cement replacements and utilisation of locally available materials could potentially serve as a way to reduce the carbon footprint and produce more ecological solutions.

The availability of well-known and widely used materials, such as Fly ash and ground-granulated blast-furnace slag (GGBFS), is expected to decrease in near future and therefore, is opening the quest for potential alternative SCMs that meet the growing demands [6, 7]. Industrial waste and natural resources are promising options which can be found locally, and which possess physical and chemical properties that could be beneficial. Some of these materials are discussed in this paper as a part of a Swiss-Croatian collaborative project ACT, that is promoting the potential usage of locally available materials as SCMs.

The potential beneficial effect of a certain raw material, when used in concrete, cannot be recognised by a single property. Numerous studies have suggested different chemical and physical tests that can be used for preliminary characterisation [8–10] and for assessment of pozzolanic reactivity [11–13]. This study is focusing on the chemical characterisation of raw state materials and the correlation of one reactivity test (R^3 test) with the compressive strength of blends with a 30% cement substitution.

2 Materials and Methodology

In this study, twelve potential SCMs were evaluated: two fly ashes (FA_T and FA_K), two steel slags—basic oxygen furnace and ladle furnace slag (SL_BOF, SL_LF), one densified silica fume (SF_D), one sample of iron silica fines (SL_Aru) and six clay samples (C_Cu1, C_Cu2, C_Ora, C_Ilo, C_Mar, C_Top1). For the sake of comparison, a sample of quartz powder (Q) was used to distinguish the pozzolanic properties from the inert filler effect.

All materials originate from the South East Europe region and were collected as a part of the ACT project. Before characterisation and testing, all materials, except silica fume and fly ash (Fa_K), were dried in an oven for 24 ± 2 h at 60 ± 5 °C and then grinded in a disc mill for 30 s.

2.1 Material Characterisation

Chemical composition is considered as one of the starting points for characterisation of any raw materials. Most widely used SCMs generally have a lower calcium content than the OPC which results in a different hydrates formation and can have an effect on the strength and durability [14].

Figure 1 shows the percentage of each chemical compound that is represented in the material sample. The results were obtained by X-ray Fluorescence (XRF), in accordance with the norm ISO/TS 16,996:2015 [15].

The difference between the types of materials can be seen from Fig. 1. All SCMs show a higher amount of silica (SiO_2) and a much smaller amount of calcium oxide (CaO) when compared to OPC. Silica fume, as expected, holds the highest amount of silica (>90%), while the slags have the least silica content. Fly ashes (Fa_T and Fa_K) have a similar amount of silica and alumina, both having some portion of iron oxide and other impurities. The CaO amount in these two samples is 11.5 and 13.5%, respectively. The clays show similar chemical composition even though they originate from different locations. The BOF and LF slags have a significant amount of CaO around 30% which is much greater than in other samples. They also hold the highest amount of magnesium oxide (MgO). Additionally, the BOF slag consists of a large portion of iron oxide (Fe_2O_3). Iron silica (SL_Aru), on the other hand, is produced from iron slag so the content of Fe_2O_3 is expectedly high, while the second-largest component is SiO_2.

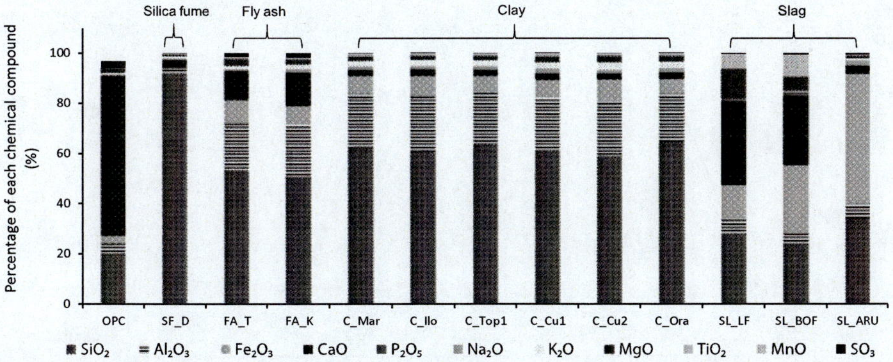

Fig. 1 Percentage of specific chemical compounds in each material

2.2 The R^3 Test

The R^3 test, developed for RILEM TC-267 committee, was used to determine the pozzolanic reactivity of the chosen SCMs. This was carried out at 40 °C by isothermal calorimetry, obtaining the total heat release of hydration for 7 days of pastes composed of the SCMs.

Prior to the test, all the SCMs and reagents were weighted, mixed and held at 40 ± 2 °C for 24 h. The formulation of the solid blends relies on the ratio of $Ca(OH)_2$/SCM and $CaCO_3$/SCM of 3 and ½ respectively. In addition, an alkali solution 3 M of K were prepared with KOH and K_2SO_4.

The pastes were prepared in a high shear mixer at 1600 ± 50 rpm for 2 min until a homogenous paste was obtained. Immediately, they were cast in glass vial and placed into the isothermal calorimeter. The cumulative heat release acquired were reported per g of SCM, as can be observed in Fig. 2.

Figure 2 indicates the results of reactivity of each material by the R^3 test method. We can assume the pozzolanic property of each SCM, when they are compared with the inert quartz. According to the given data, silica fume shows the highest amount of heat release after 7 days but surprisingly has lower initial reactivity. Both fly ashes have a similar performance of heat release, even though Fa_T shows the biggest initial reaction (first 10 h), and second largest 7-day reaction overall. The results for Fa_K

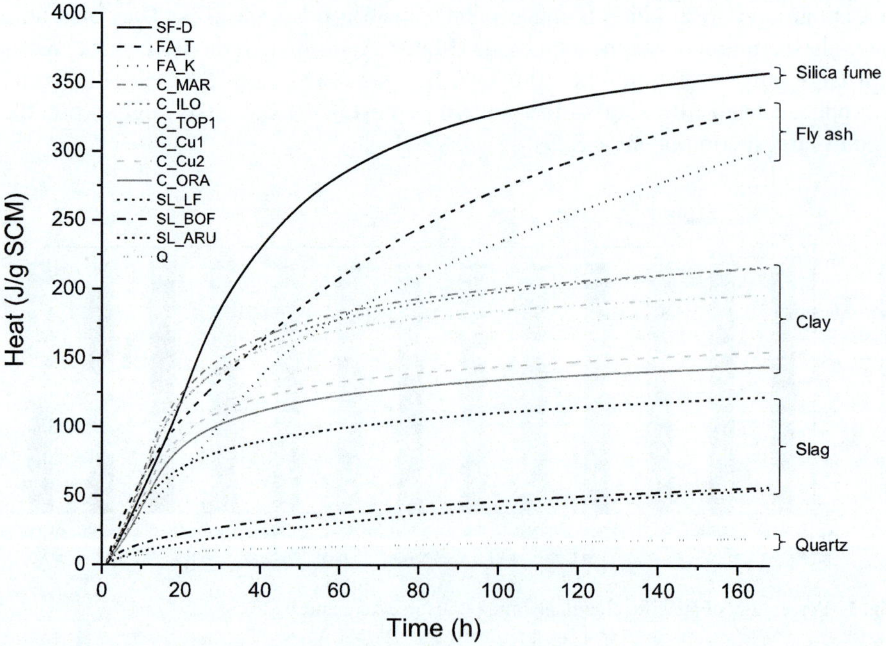

Fig. 2 Results obtained by the heat release measurement of the R^3 test using isothermal calorimetry

indicate a slower initial reaction and a lower long-term reactivity (7 days). All of the clays exhibit similar heat release performance, slightly faster heat release in the first 15 h of reaction followed by a slower reaction after 24 h. C_Mar and C_Ilo showed the most promising results. According to the R^3 test, C_Ora shows the lowest results among clay samples. All of the slags showed very low reactivity. SL_LF shows some amount of heat release (still lower than the least reactive clays), while SL_BOF and SL_Aru are performing slightly better than the inert quartz powder. The SCMs were labelled in the upper left corner of Fig. 2 starting with the material holding the highest values of the test.

2.3 Mortar Strength Test

For mortar preparation mixing proportions of the samples were kept constant, with a 30% cement mass replacement for each SCM. The water to binder ratio was also constant, 0.5. Standardized sand was used as aggregate and there was no need for water adjustment. For each mixture, a volume of 1.5 L was mixed following the norm HRN EN 196-1 (2016) [16], a sufficient amount to cast molds of 40 × 40 × 160 mm. After casting the samples were covered with a plastic wrap and held 24 h in laboratory conditions. When demolded they were cured under water until testing time. Compressive strength test was carried out on 2 prisms after 2, 7 and 28 days of curing age, according to the norm mentioned above. OPC mortars of CEM 1 45.2 R were also prepared for comparison.

The compressive strength results are shown in Fig. 3. All samples of blended cement exhibited lower early age compressive strength (2-day) than OPC mortars. Even though showing a slightly lower initial strength, Silica Fume outperforms the

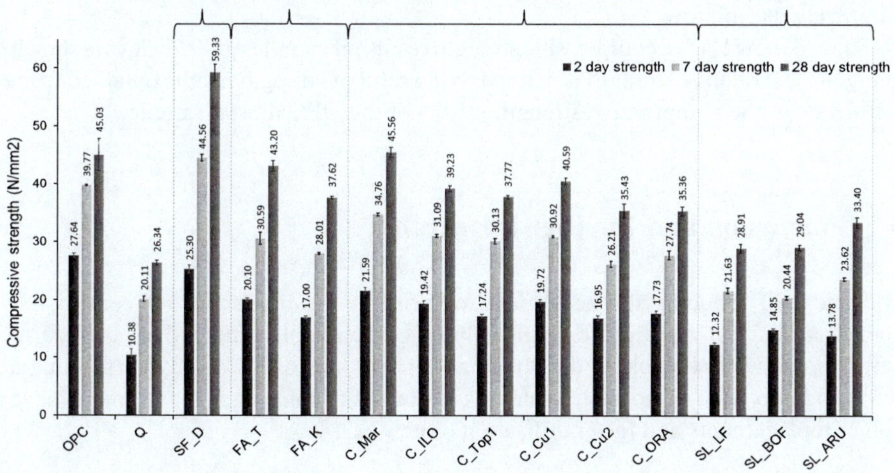

Fig. 3 Compressive strength results for mortar samples after 2, 7 and 28 days

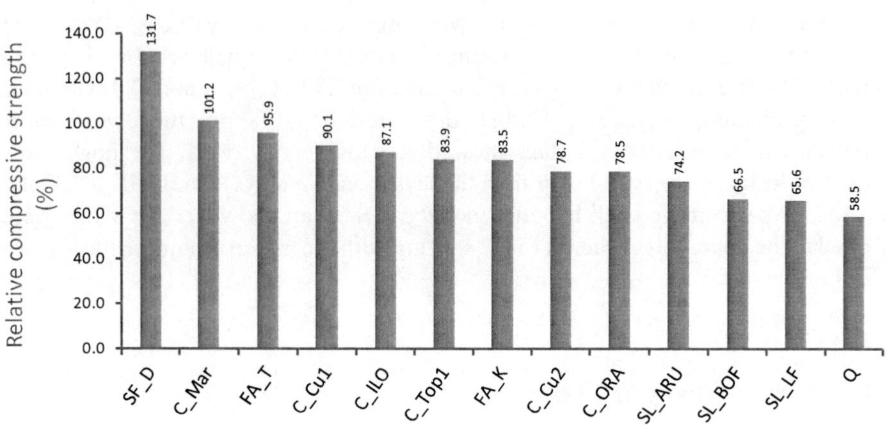

Fig. 4 Relative strength results after 28 days aligned from the highest to the lowest values

reference sample of OPC after 7 and 28 days. Among the fly ashes, Fa_T shows more promising results after 7 days of hydration. C_Mar shows the second-best compressive strength results after 7 days of hydration, and the highest compressive strength amongst other clays. The 2-days and 7-days compressive strength of C_Mar is almost identical to the OPC (34.76 and 54.56 MPa). With the 28-days strength close around 40 MPa, C_Cu1 and C_Ilo are ranked at the 3rd and 4th place, respectively, with lower initial strength then the C_Mar clay sample. Sample C_Top1 shows lower compressive strength results at all ages, than the before mentioned clays. Clay samples C_Cu2, and C_Ora performed poorly when compared to other clays. The slags, as the R^3 test showed, presented the lowest values. The mean compressive strength after 28 days is 28.91, 29.04 and 33.40 MPa for SL_LF, SL_BOF, and SL_Aru, respectively. Again, the slags show slightly better results than the inert quartz powder mixture.

The order of relative compressive strength results of samples after 28 days is shown in Fig. 4. The relative strength is defined as the ratio of strength results obtained from the blends to the compressive strength results of the OPC mortar sample.

3 Conclusion and Further Research

The intent of reducing the ecological footprint in the production of cement has opened a quest in research for high volume clinker replacement. This has led to new approaches in unlocking the potential of new types of SCMs. By performing a screening approach of possible materials for cement replacements, we can achieve bigger replacements and faster utilisation of new SCMs.

This study aimed at combining the 3 testing methods, chemical characterisation, the R^3 reactivity test and compressive strength, with the goal of evaluating the potential of different materials as SCMs. It was observed that the materials holding a higher amount of silica, such as silica fume, fly ash and clays, performed better in both the calorimetry measure and the compressive strength test. With the highest amount of SiO_2 and the best results of both the tests, silica fume demonstrates the highest reactivity. When considering other material samples, FA_T had proven to be more reactive than the FA_K sample of fly ash, while the C_Mar sample is the most reactive amongst clays. The slag samples showed the lowest strength and cumulative heat results, performing just slightly better than the inert quartz referent mixture.

It has been proven that the outputs of the R^3 test can relate to the strength development of mortar mixes and that the most promising SCMs for future research are silica fume SF_D, fly ash Fa_T and clay C_Mar. These samples will be used in the next steps of the ACT project where they will be evaluated on a concrete level.

Acknowledgements The presented research is a part of project "*Advanced low CO_2 cementitious materials,* ACT", financed within the Croatian-Swiss Research Program of the Croatian Science Foundation and the Swiss National Science Foundation with funds obtained from the Swiss-Croatian Cooperation Program and project "*Alternative binders for concrete: understanding microstructure to predict durability*, ABC", financed by Croatian Science Foundation.

References

1. World Business Council for Sustainable Development: Cement Industry Energy and CO_2 Performance—Getting the Numbers Right (2009)
2. Selim, T.H., Salem, A.S.: Global Cement Industry: Competitive and Institutional Frameworks (2010)
3. Provis, J.L.: Green concrete or red herring?—future of alkali-activated materials. Adv. Appl. Ceram. **113**(8), 472–477 (2014)
4. Scrivener, K.L., Nonat, A.: Hydration of cementitious materials, present and future. Cem. Concr. Res. **41**(7), 651–665 (2011)
5. Juenger, M.C.G., Siddique, R.: Recent advances in understanding the role of supplementary cementitious materials in concrete. Cem. Concr. Res. **78**, 71–80 (2015)
6. Enriquez-Flores, J.J., Gervacio-Arciniega, C.I., Flores-Ruiz, F.J., Cardona, D., Camps, E., Muñoz-Saldaña, J., Espinoza-Beltrán, F.J.: Supplementary cementitious materials for concrete: characterization needs. Mater. Res. Soc. Symp. Proc. **1477**, 61–66 (2012)
7. Serdar, M., Bjegovic, D., Stirmer, N., Banjad Pecur, I.: Alternative binders for concrete: opportunities and challenges. In: Scientific Symposium Future Trends in Civil Engineering, pp. 199–218. Zagreb (October 2019)
8. Snellings, R.: Assessing, understanding and unlocking supplementary cementitious materials. RILEM Tech. Lett. **1**, 50–55 (2016)
9. Measham, F.: Lea's chemistry of cement and concrete, 4th edn. Butterworth-Heinemann, Oxford (2003)
10. Carević, I., Serdar, M., Štirmer, N., Ukrainczyk, N.: Preliminary screening of wood biomass ashes for partial resources replacements in cementitious materials. J. Clean. Prod. **229**, 1045–1064 (2019)
11. Snellings, R., Scrivener, K.L.: Rapid screening tests for supplementary cementitious materials: past and future. Mater. Struct. **49**(8), 3265–3279 (2016)

12. Donatello, S., Tyrer, M., Cheeseman, C.R.: Comparison of test methods to assess pozzolanic activity. Cem. Concr. Compos. **32**(2), 121–127 (2010)
13. Avet, F., Snellings, R., Alujas Diaz, A., Ben Haha, M., Scrivener, K.: Development of a new rapid, relevant and reliable (R3) test method to evaluate the pozzolanic reactivity of calcined kaolinitic clays. Cem. Concr. Res. **85**, 1–11 (2016)
14. Lothenbach, B., Scrivener, K., Hooton, R.D.: Supplementary cementitious materials. Cem. Concr. Res. **41**(12), 1244–1256 (2011)
15. International Organization for Standardization (ISO): ISO/TS 16996:2015 Solid Biofuels: Determination of Elemental Composition by X-ray Fluorescence
16. European Committee for Standardization (CEN): EN 196-1:2016 Methods of Testing Cement—Part 1: Determination of Strength

Thermal Performance of Compressed Blocks Made from Construction and Polyurethane Foam Waste

A. Briga-Sá, V. Neiva, D. Leitão, T. Miranda, and N. Cristelo

Abstract It is intended with the research work here presented to contribute to the scientific knowledge regarding the incorporation of C&D and polyurethane foam wastes in compressed blocks. Two mixtures were defined considering the percentages of 2.5 and 5% of polyurethane foam waste. Fly ash was used as a precursor and the two mixtures were activated with a solution based on sodium hydroxide and sodium silicate. Experimental work was carried out to assess the blocks thermal performance. Experimental measurements allowed determining indoor and outdoor temperatures, heat flux and inner surface temperatures. The data acquired allowed to estimate the values of the thermal transmission coefficient. It was concluded that an increase in the percentage of polyurethane foam waste in the composition leads to an improvement of the thermal behaviour. Values of 2.93 and 2.57 W/m^2 °C were estimated for the percentages of 2.5 and 5%, respectively, presenting higher values of thermal resistance when compared with a common solution of ceramic solid block with the same dimensions.

Keywords C&D waste · Polyurethane foam waste · Compressed waste blocks · Alkaline activation · Thermal performance · Sustainability

A. Briga-Sá (✉) · N. Cristelo
CQVR, Department of Engineering, University of Trás-Os-Montes E Alto Douro, Vila Real, Portugal
e-mail: anas@utad.pt

N. Cristelo
e-mail: ncristel@utad.pt

V. Neiva
ECT—School of Science and Technology, University of Trás-Os-Montes E Alto Douro UTAD, Quinta de Prados, Vila Real, Portugal

D. Leitão
Department of Civil Engineering, C-TAC, University of Minho, Guimarães, Portugal
e-mail: dleitao@civil.uminho.pt

T. Miranda
Department of Civil Engineering, ISISE, University of Minho, Guimarães, Portugal
e-mail: tmiranda@civil.uminho.pt

V. M. C. F. Cunha et al. (eds.), *Proceedings of the 3rd RILEM Spring Convention and Conference (RSCC 2020)*, RILEM Bookseries 35,
https://doi.org/10.1007/978-3-030-76543-9_21

1 Introduction

The rapid depletion of natural resources in modern society and the high amount of generated waste have motivated research in this field towards more sustainable and alternative building materials.

Construction and demolition (C&D) waste represents approximately 25–30% of the amount of waste produced in the European Union (EU). In this context, C&D waste has been defined as a priority by the EU with the main goal that by 2020 a minimum of 70% of non-hazardous C&D wastes shall be reused, recycled or recovered, including as a substitute of other materials [1].

Furthermore, the demand for polyurethane is continually growing in EU. More than 3.5 million tonnes of this insulation material are used each year and approximately 70% is presented in the form of foam. It is also estimated that about 20% of it ends up as waste and a vast majority is taken to landfill, corresponding to about 460 000 tonnes per year.

Several studies have been carried out on the use of C&D and polyurethane foam wastes in the construction industry, showing a high potential of these materials.

Research work has been conducted in order to analyse the possibilities of using waste materials to produce bricks [2]. Different authors analysed the production of fired bricks by using class F fly ash to replace clay and concluded that it increased the compressive strength and decreased the water absorption [3, 4]. Freidin [5] analysed the production of bricks using polymerization and showed that they can be produced from mixtures of fly ash, using sodium silicate solution as the alkali activator. Robayo-Salazar et al. [6] also developed research on the use of alkali-activated C&D wastes and concluded that building materials, such as masonry mortars and prefabricated elements, can be obtained. Sabai et al. [7] investigated the possibility of recycling C&D waste to produce blocks. It was observed that the blocks using 100% of recycled aggregates were weaker than those with natural aggregates. However, given that a compressive strength of 7 N/mm^2 was achieved, C&D waste could be a potential resource for building material production rather than discarding it. Fewer studies were found regarding thermal performance analysis of blocks. Bravo et al. [8] concluded that using recycled aggregates from C&D wastes improves the thermal performance of concrete mixes. Zhu et al. [9] also investigated the thermal properties of recycled aggregate concrete and recycled concrete blocks. A value of 0.93 W/m^2 K was obtained for the heat transfer coefficient of recycled concrete block masonry.

The behaviour of different polyurethane foams waste was analysed to assess their potential use in construction materials. The study showed that the waste generation process determines the foam's microstructure and the physical and chemical properties, affecting the building materials performance and, consequently, offering different possibilities of application [10]. The option of reusing polyurethane foam waste, combined with pitch binders, and PU foam waste as a dry aggregate in different cement or gypsum matrices has also been researched [11]. Other studies show the possibility to use this polymer to reduce the amount of sand in cement mortars [12–14]. A study on the properties and thermal behaviour of plaster with polyurethane foam wastes showed that increasing the polyurethane quantity decreases the density and mechanical properties of the plaster [15].

Nevertheless, there is still research that should be carried out in order to analyse the thermal properties of building materials using these types of wastes.

In this context, this study aims to analyse the thermal performance of compressed waste blocks (CDW) using C&D and polyurethane foam (PU) wastes and fly ash (FA). Two mixtures were defined using different percentages of PU waste (2.5 and 5%), activated with a solution based on sodium hydroxide and sodium silicate. The influence of different percentages of PU was discussed.

2 Materials and Methods

2.1 Materials Characterization

Construction and Demolition (C&D) Waste. The C&D waste (Fig. 1a) used in this study was provided by a Portuguese recycling company. According to the technical sheet provided, the material comes mainly from the demolition of small buildings, recovered from illegal deposition [16]. It is a mixture of concrete, mortar, bricks, glass, ceramics and wood, among others, and in accordance with the EU Council Decision 2003/31/EC [17] it has the code EWC 1701 07 referring to a "Selected C&D waste". In order to prepare the C&D waste for the blocks production, it was subjected to a drying oven at the temperature of 85 °C for 24 h. Then, it went through a process of breaking of the clods (Fig. 1b).

C&D waste was characterized in what concerns to the particle size distribution (LNEC E196 1966) [18], Fig. 2, maximum dry unit weight, optimum water content (LNEC E197 1966) [19], Table 1, specific gravity (NP 83 1965) [20], Table 1, and distribution of the constituents (EN 933-11 2009) [21], Table 2.

Polyurethane Foam Waste. PU waste was provided by a polyurethane producing company with the particle size that can be observed in Fig. 3a. Given that it was

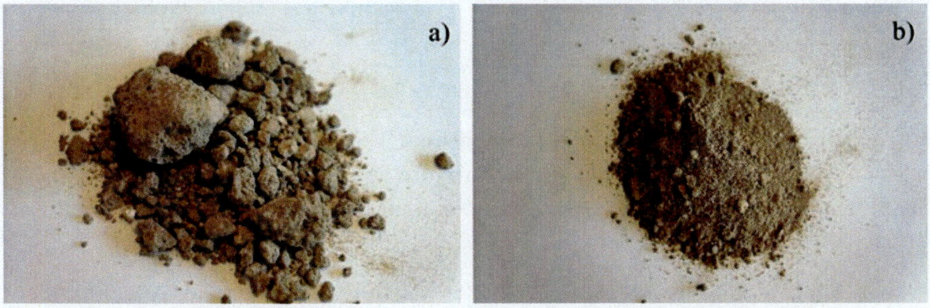

Fig. 1 General view of C&D waste: **a** before breaking of the clods; **b** after breaking of the clods

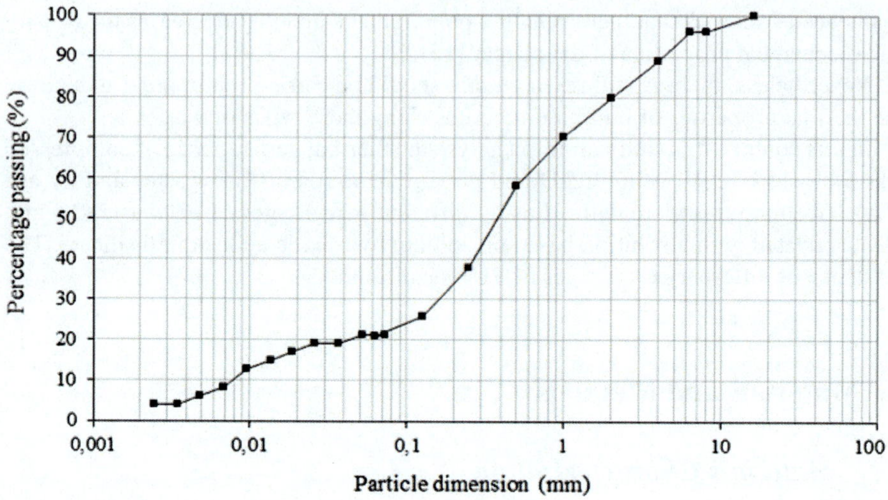

Fig. 2 C&D waste particle size distribution

Table 1 C&D waste geotechnical properties

Property	Value
Maximum dry unit weight (kN/m^3)	20.2
Optimum water content (%)	8.33
Specific gravity (kN/m^3)	25.8

Table 2 Distribution of the constituents of C&D waste

Constituents	Value
Concrete, concrete products, mortar, concrete mansory units, Rc (%)	25
Unbound aggregate, natural stone, hydraulically bound aggregate, Ru (%)	32
Clay and calcium silicate masonry units, aerated non-floating concrete, Rb (%)	12
Bituminous materials, Ra (%)	3
Glass, Rg (%)	2
Other materials, X (%)	26
Floating particles, FL(cm^3/kg)	28.5

composed by particles with different dimensions, it was necessary to homogenize it for its subsequent incorporation into the mixture and for that, it was sieved using a 1.60 mm square mesh sieve, Fig. 3b. In Fig. 4, it can be observed the microstructural characterization of PU waste.

Fly ash. The fly ash, type F (Fig. 5), used in this work as a precursor, was provided by a Portuguese thermo-electric plant and it has already been considered in other research works [22] where its characterization was carried out. As regards to the

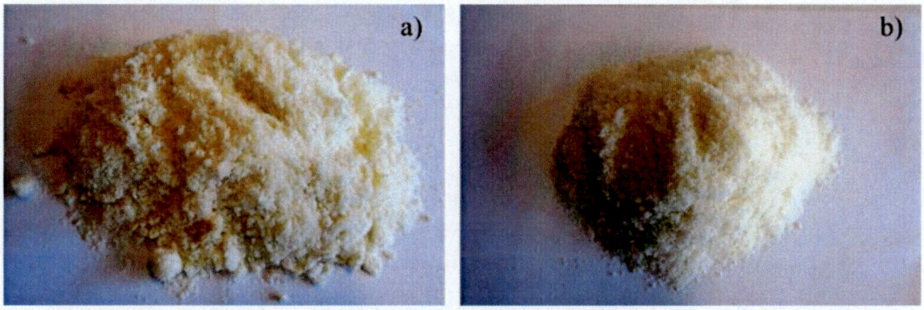

Fig. 3 General view of PU waste: **a** before and **b** after using a 1.60 mm square mesh sieve

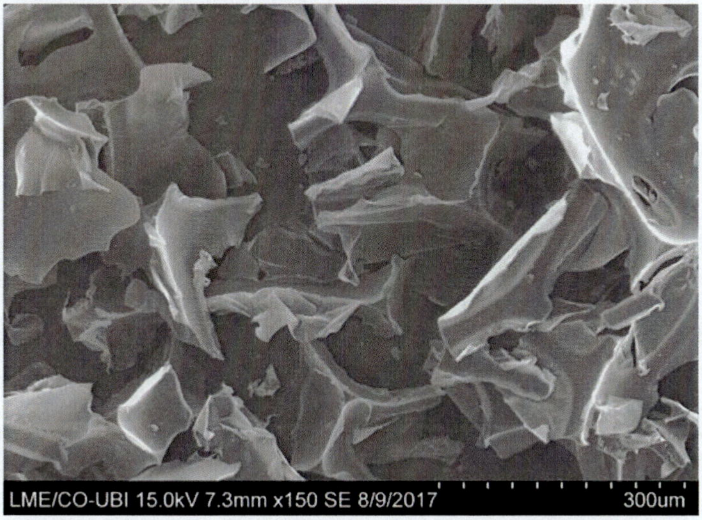

Fig. 4 Microstructural image of PU waste

Fig. 5 FA of type F

Table 3 Chemical characterization of FA

Element	Si	Al	Na	Mg	P	S	K	Ca	Fe	Ti
%	48.81	21.77	1.31	1.56	0.58	1.17	4.42	3.85	14.74	1.79

particle size, it was verified that this fly ash presents a diameter of less than 45 μm, which confirms its fineness. A chemical characterization was also done, whose results are shown in Table 3, leading to the conclusion that fly ash has an activation potential of approximately 75%.

Alkaline Activators. The alkaline activator solution was a combination of sodium hydroxide and sodium silicate. Sodium hydroxide, originally in flake form, with a specific gravity of 2.13 at 20 °C and 95–99% purity, was dissolved in distilled water up to the desired concentration of 10 molal.

2.2 Compressed Waste Blocks Production

The production of compressed waste blocks (CWB) was carried out according to the methodology that is normally used for the manufacture of compressed earth blocks (CEB) [23]. Some recommendations presented in international standards ((UNE 41,410 2008) [24], (HB 195 2009) [25], (NZS 4298 1998) [26]) and research work developed by some authors in this field [27–29] were taken into account in order to verify the suitability of some of the materials used as regards, for example, with the particle size distribution. As referred above, two different mixtures were considered for the CWB's production (Fig. 6), varying the percentage of PU wastes: 2.5% (PU2.5) and 5% (PU5), Table 4, in order to analyse its influence on the blocks thermal performance. FA was used as precursor and the resulting mixtures were activated with a solution based on sodium hydroxide and sodium silicate.

Microstructural images of PU2.5 and PU5 mixtures are presented in Fig. 7. EDX analysis were carried out at several points, near the particles of polyurethane and away from them, and it was concluded that polyurethane is randomly distributed in the mixture and there is no connection between it and the matrix. Thus, no preferred chemical elements are identified near or far from the polyurethane particles.

Table 4 Composition of the mixtures used in the CWB's production

Mixture	Solids (%)			Liquids (%)		Solid/Liquid ratio (%)
	C&D waste	FA	PU waste	NaOH	Na_2SiO_3	
PU2.5	67.5	30	2.5	66.67	33.33	6.30
PU5	65	30	5	66.67	33.33	6.30

Fig. 6 CWBs with different percentage of PU waste: **a** PU2.5, **b** PU5

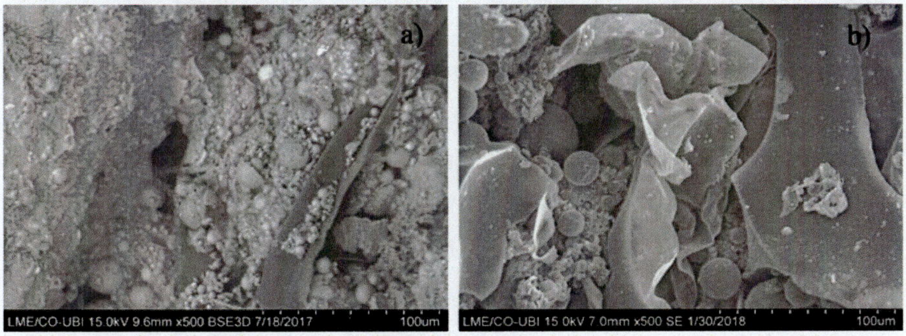

Fig. 7 SEM images of the mixtures: **a** PU2.5; **b** PU5

2.3 Thermal Performance Assessment

Experimental Setup. The experimental work to analyse the CWB's thermal performance was carried out in a test room with the dimensions of 4.00 m × 3.00 m × 2.54 m, whose indoor conditions were controlled. The experimental procedure was performed according to ISO 9869 [30] and Pereira [31] and has been successfully applied in similar research works. A uniform high thermal gradient between indoor and outdoor environments is desirable in order to guarantee a significant heat flow, always occurring in the same direction across the analysed sample. In this case, these conditions were obtained heating the test room, guarantying an interior temperature with lower oscillation patterns. The measurement period occurred between the 3rd and the 16th of June. Two panels made of CWB's were built in the north oriented external wall, Fig. 8, carefully fixed to the wall, surrounded by polyurethane foam to avoid thermal bridges, non-insulated headers and other faults. Two heat flux sensors, four inner surface temperature sensors (thermocouples) fixed in each panel and two temperature sensors, one inside the test room and the other outside to measure the exterior environment conditions, compose the experimental device. The heat flux and the surface temperature sensors were fixed in the panel's inner surface, in the hollow zones of the blocks, as it can be seen in Figs. 8 and 9. The temperatures of the

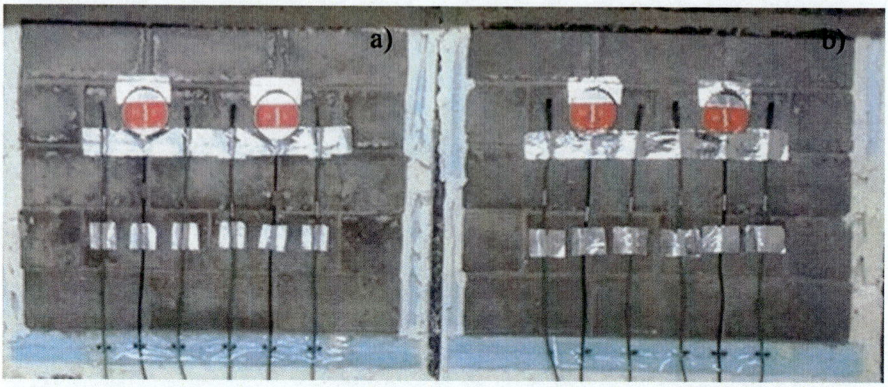

Fig. 8 Experimental setup for the CWB's panels thermal performance assessment: **a** PU2.5; **b** PU5

Fig. 9 Location of the sensors in the CWS blocks: 1—inner surface temperature sensors (Tsi_{11} and Tsi_{21}); 2—heat flux sensor ($q1$ or $q2$)

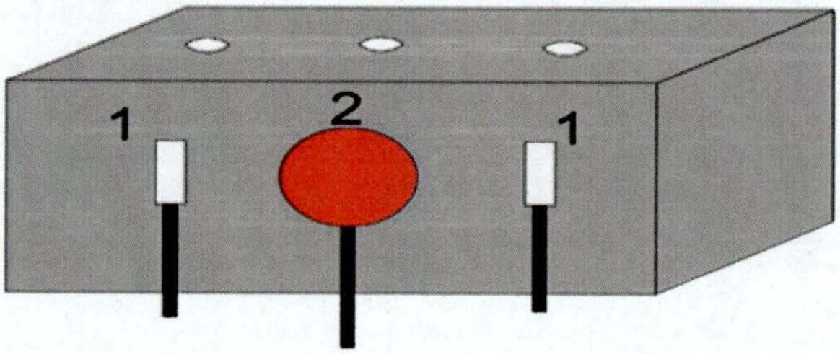

internal and external environment (*Ti*(*n*) and *Te*(*n*)), the heat flux (*q1*(*n*) and *q2*(*n*)) and the inner surface temperatures (*Tsi*(*n*)) were measured continuously with 10 min intervals (*n*).

Methodology. According to ISO 9869 [30] the thermal transmission coefficient (U) of a material or a building system can be quantified given the heat flow that occurs through the element when subjected to a temperature differential, as defined in Eq. (1):

$$U(ntotal) = \frac{\sum_{n=1}^{ntotal} q(n)}{\sum_{n=1}^{ntotal} (Ti(n) - Te(n))} \tag{1}$$

where *q*(*n*) is the heat flow across the wall sample for the instant *n*; *Ti*(*n*) and *Te*(*n*) are the interior and the exterior temperatures in the instant *n*, respectively; *ntotal* is the total number of instants acquired during the measurement period. In ISO 9869 [30], it is recommended the use of interior and exterior temperatures instead of the surface temperatures for the U value calculation. The use of two flux meter sensors in each panel allows to *obtain q1*(*n*) and *q2*(*n*). Considering the values of the temperature differential between the interior, *Ti*(*n*), and the exterior environments, *Te*(*n*), it is possible to estimate the thermal transmission coefficients, $U1(ntotal)$ and

*U*2(*ntotal*), respectively, by applying Eq. (1). The thermal transmission coefficient of the panel, *U*′(*ntotal*), is the average value of *U*1(*ntotal*) and *U*2(*ntotal*), as presented in Eq. (2).

$$U'(ntotal) = \frac{U1(ntotal) + U2(ntotal)}{2} \quad (2)$$

This calculation methodology was applied for the two CWBs panels and allowed to estimate the values of the thermal transmission coefficient resulting from experimental measurements, which were compared with the values already known for current construction solutions.

3 Results and Discussion

Data acquired during the experimental measurements allowed to obtain the variation of the indoor and outdoor conditions in what concerns to the temperatures *Ti* and *Te*. In Fig. 10, it is possible to observe that the stabilization of the interior temperature was guaranteed, as much as possible, as required by the experimental procedure, whose values varied between a minimum of 32 °C and a maximum of 37 °C, given the temperature range that characterize the climatic zone where the experimental test took place. The obtained values for the exterior temperature show this variation, where a minimum of 10 °C and a maximum of 29 °C were verified. Furthermore, *Ti* values were always higher than the ones obtained for *Te*, leading to obtain the necessary conditions for the occurrence of heat flow through the panels and, therefore, the reliability of the experimental test. A minimum and a maximum differential between *Ti* and *Te* of 5 °C during the diurnal periods and 24 °C during the night, respectively,

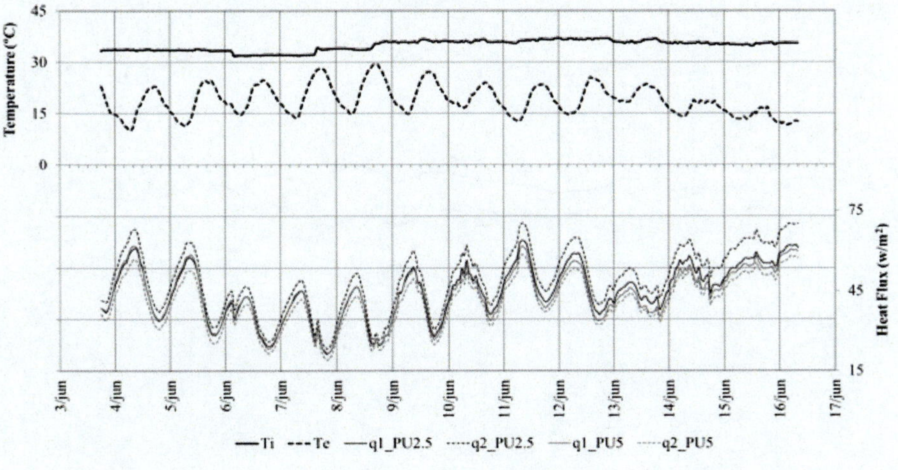

Fig. 10 Variation of interior temperature (*Ti*), exterior temperature (*Te*) and heat flux (*qi*) values

were verified. In what concerns to the heat flux variation, it is observed that higher values are obtained for the CWB panel with 2.5% of PU and an average decrease in q values of approximately 11% characterizes the panel with 5% of PU, Fig. 10. This difference in the thermal behaviour of the two panels is supported by the values obtained for the surface temperatures (see Fig. 11). As expected, higher values are presented for the PU5 mixture. An average of 1.2% is verified for Tsi differential between the two panels.

In order to obtained the thermal transmission coefficient of the CWB's panels, Eqs. (1) and (2) were applied. The variation of U values during the measurement period is presented in Fig. 12. The panel with 5% of PU is characterized by lower

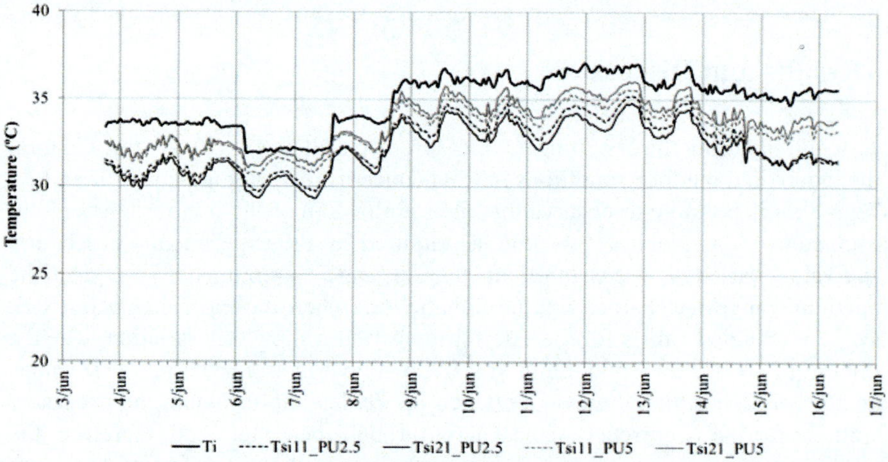

Fig. 11 Variation of interior (Ti) and inner surface temperatures (Tsi) values

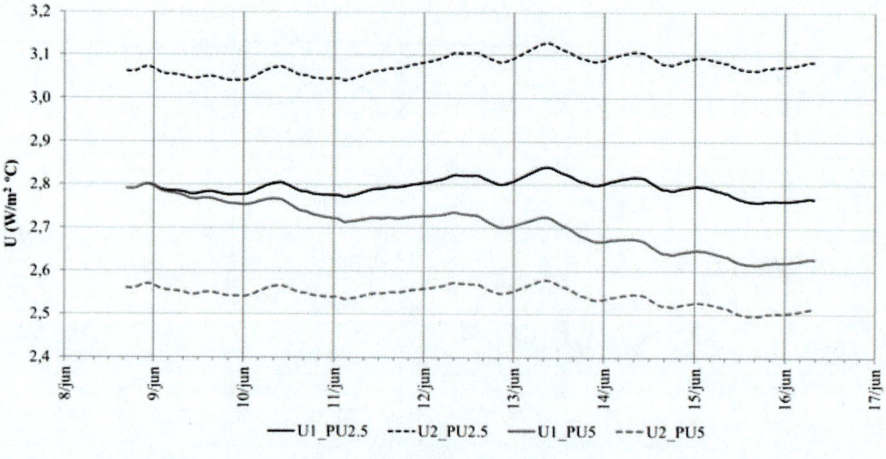

Fig. 12 Variation of the thermal transmission coefficient (U) values

values of U when compared with CWB with 2.5% of PU. According to Eq. (2), U' values of 2.57 W/m^2 °C and 2.93 W/m^2 °C are estimated for PU5 and PU2.5, respectively, leading to the conclusion that higher percentage of PU improves the thermal performance of the CWBs. So, thermal resistance (R) values of 0.34 m^2 °C/W and 0.39 m^2 °C/W can be achieved for the blocks with PU2.5 and PU5, respectively. When compared with a current solution of a ceramic massive block with the same thickness ($R = 0.13$ m^2 °C/W), CWBs present an improvement of the thermal behaviour.

4 Conclusions

The aim of this research was to analyze the thermal performance of CWBs, whose mixtures containing C&D waste, PU waste and FA were alkali activated. PU waste was used in the composition with the percentages of 2.5% and 5%. Two panels made of these blocks were subjected to experimental measurements and the results obtained for the different percentages of PU were compared. The results showed that panels PU5 presented lower values of heat flux and a lower oscillation pattern in the inner surface temperature curves, achieving higher values, comparing with the panel PU2.5. Given these results, lower values of the thermal transmission coefficient were estimated for CWBs with higher percentage of PU. It leads to the conclusion that increasing the percentage of PU will contribute to an improvement of 15% in the CWBs thermal resistance.

References

1. EU: Directive 2008/98/EC of the European Parliament and of the Council of 19 November 2008 on waste, Official Journal of the European Union (2008)
2. Zhang, L.: Production of bricks from waste materials—a review. Constr. Build. Mater. **47**, 643–655 (2013)
3. Lingling, X., Wei, G., Tao, W., Nanru, Y.: Study on fired bricks with replacing clay by fly ash in high volume ratio. Constr. Build. Mater. **9**, 243–247 (2005)
4. Kute, S., Deodhar, S.V.: Effect of fly ash and temperature on properties of burnt clay bricks. J. Civ. Eng. **84**, 82–85 (2003)
5. Freidin, C.: Cementless pressed blocks from waste products of coal-firing power station. Constr. Build. Mater. **21**, 12–18 (2007)
6. Robayo-Salazar, R.A., Rivera, J.F., Mejía de Gutiérrez, R.: Alkali-activated building materials made with recycled construction and demolition wastes. Construct. Build. Mater. **149**, 130–138 (2017)
7. Sabaia, M.M., Coxa, M.G.D.M., Matob, R.R., Egmonda, E.L.C., Lichtenberga, J.J.N.: Concrete block production from construction and demolition waste in Tanzania. Resour. Conserv. Recycl. **72**, 9–19 (2013)
8. Miguel Bravo, M., Britio, J., Evangelista, L.: Thermal performance of concrete with recycled aggregates from CDW plants. Appl. Sci. **7**(740), 2–21 (2017)
9. Zhu, L., Dai, J., Bai, G., Zhang, F.: Study on thermal properties of recycled aggregate concrete and recycled concrete blocks. Constr. Build. Mater. **94**, 620–628 (2015)

10. Gómez-Rojo, R., Alameda, L., Rodríguez, Á., Calderón, V., Gutiérrez-González, S.: Characterization of polyurethane foam waste for reuse in eco-efficient building materials. Polymers **11**(359), 1–16 (2019)
11. Junco, C., Rodríguez, A., Calderón, V., Muñoz-Rupérez, C., Gutiérrez-González, S.: Fatigue durability test of mortars incorporating polyurethane foam wastes. Constr. Build. Mater. **190**, 373–381 (2018)
12. Mounanga, P., Gbongbon, W., Poullain, P., Turcry, P.: Proportioning and characterization of lightweight concrete mixtures made with rigid polyurethane foam wastes. Cement Concr. Compos. **30**, 806–814 (2008)
13. Kismi, M., Mounanga, P.: Comparison of short and long-term performances of lightweight aggregate mortars made with polyurethane foam waste and expanded polystyrene beads. In: Proceedings on 2nd International Seminar on Innovation and Valorization in Civil Engineering and Construction Materials, Vol. 2, pp. 2–99 Paris, France (2012)
14. Gadea, J., Rodríguez, A., Campos, P.L., Garabito, J., Calderón, V.: Lightweight mortar made with recycled polyurethane foam. Cement Concr. Compos. **32**, 672–677 (2010)
15. Gutiérrez-González, S., Gadea, J., Rodríguez, A., Junco, C., Calderón, V.: Lightweight plaster materials with enhanced thermal properties made with polyurethane foam wastes. Constr. Build. Mater. **28**, 653–658 (2012)
16. Cristelo, N., Fernández-Jiménez, A., Vieira, C., Miranda, T., Miranda, P.Á.: Stabilisation of construction and demolition waste with a high fines content using alkali activated fly ash. Construct. Build. Mater. **170**, 26–39 (2018)
17. EU: Council Decision 2003/33/EC of 19 December 2002, Official Journal of the European Union (2003)
18. LNEC E196: Laboratório Nacional de Engenharia Civil, 1966. Solos. Análise Granulométrica. LNEC Espec. Lisboa (1966)
19. LNEC E197: Solos. Ensaio de Compactação. 10 (1966)
20. NP 83: Solos: Peso Específico Dos Grãos. Instituto Português Da Qualidade. Caparica, Portugal (1965)
21. EN 933-11: Methods of test for masonry. Determination of Compressive Strength (2009)
22. Soares, E.: Melhoria do solo residual granítico com recurso a activação alcalina de cinzas volantes (In Portuguese), Universidade de Trás-os-Montes e Alto Douro
23. Leitão, D., Barbosa, J., Soares, E., Miranda, T., Cristelo, N., Briga-Sá, A.: Thermal performance assessment of masonry made of ICEB'sstabilised with alkali-activated fly ash. Energy Build. **139**, 44–52 (2017)
24. UNE 41410: Bloques de Tierra Comprimida Para Muros Y Tabiques - Definiciones, Especificaciones Y Métodos de Ensayo. AENOR. Madrid, Espanhã (2008)
25. HB 195: The Australian Earth Building Handbook. Sydney, NSW 2001, Australia: Standards Australia International Ltd. (2009)
26. NZS 4298: NZS 4298 (1998): Materials and Workmanship for Earth Buildings. New Zealand Technical Committe 91 (1998)
27. United Nations Educational Scientific and Cultural Organization France: Compressed Stabilised Earth Block Manufacture in Sudan. p. 101
28. Waitakere City Council: Earth building. In: Waitakere City Council's Sustainable Home Guidelines. Henderson, Waitakere City 0650, New Zealand (2008)
29. Rigassi, V.: Compressed Earth Blocks : Manual of Production (1985)
30. ISO 9869: Thermal Insulation—Building Elements—In-situ Measurement of Thermal Resistance and Thermal Transmittance, International Organization for Standardization (ISO) (1994)
31. Cunha, S.: In situ evaluation and certification of the thermal quality of residential buildings (in Portuguese). In: PhD Thesis, UTAD, Vila Real, Portugal (2011)

A Short Review of Researches on Mechanical Properties of Traditional Chinese Timber Joints: From Experimental Aspect

Lipeng Zhang and Qifang Xie

Abstract Mortise-tenon joints play important roles in Chinese traditional timber structures. In fact, all the researchers in the field of timber building heritage protection have ever conducted investigations to acquire the mechanical properties. However, the seemingly overwhelming FE simulation and theoretical methods are ideal calculations, depending highly on many assumptions and the reliability of the adopted material model, which make them hard to consider the real configurations and stress state of the joints in tests and practical timber engineering. The other hand, most tests were carried out by different researchers according to various standards, with many uncertain factors, which makes the results loss their mutual comparability. Thus related testing and analytical results are merely limited in the research stage, and cannot be smoothly put into application. This dilemma faced in this field must be broken to favor further researches and applications. Some resent voice comes towards a more comprehensive and accurate testing protocols of timber joints, which would definitely be helpful to reach a mutually obeyed standard. In this paper, the authors attempt from experimental aspect to discuss part of the existing investigations conducted on traditional Chinese mortise-tenon joints. Useful information such as adopted wood materials (species, density, moisture content, etc.), model dimensions (size effects), test measuring setups, loading methods and related operations, testing results analysis methods, etc. are summarized and compared. Some future research efforts are proposed. Related content will provide favorable reference for the future researching and testing activities of both traditional and modern type timber joints.

Keywords Chinese traditional timber joints · Experiment researches review · Uncertainties · Future research efforts

---

L. Zhang (✉) · Q. Xie
School of Civil Engineering, Xi'an University of Architecture and Technology, Xi'an 710055, China
e-mail: zhanglipeng@live.xauat.edu.cn

Key Lab of Structure Engineering and Earthquake Resistance, Ministry of Education (XAUAT), Xi'an 710055, Shaanxi, China

Q. Xie
e-mail: xieqfgh@xauat.edu.cn

V. M. C. F. Cunha et al. (eds.), *Proceedings of the 3rd RILEM Spring Convention and Conference (RSCC 2020)*, RILEM Bookseries 35,
https://doi.org/10.1007/978-3-030-76543-9_22

1 Introduction

Traditional Chinese timber structures (TCTSs) have a long history and are one of the most precious culture heritages of the world, enriching the diversity of building heritages, for example, the earliest existing TCTS South Temple Hall of Wutai Mountain in China (Shanxi Province, Fig. 1a), the earliest existing Pavilion style TCTS Guanyin Pavilion of dule Temple (Tianjin Fig. 1b), the highest traditional pagoda Yingxian wood pagoda (In Shanxi Province, China, Fig. 1c), and the heaviest traditional timber building complex Beijing Imperial Palace (Beijing, Fig. 1d), etc. The construction of such TCTSs were primarily formed the designing specifications at that time called Ying-Zao-Fa-Shi by Li [1] in 1100 using traditional Chinese, but not the simplified one. The only knowledge of TCTSs were recorded in this difficultly understood book.

Although there is a long history of these typology of TCTSs, however, the protection activities of which are just things in recent several decades. Thus, Liang in 1984 initially started the earliest protective activities to translate it in modern Chinese to make a much easier understanding by modern specialists. In the early 90s, Ma carried out investigations from the construction aspects of this architecture. In 1992, further summarized the vertical load transfer mechanisms through key components (beams,

(a) South Temple Hall of Wutai Mountain (rebuilt in 782)

(b) Guanyin Pavilion of dule Temple (partly rebuilt in 984)

(c) the highest traditional pagoda Yingxian wood (built in 1056)

(d) traditional timber building complex Beijing Imperial Palace (1420)

Fig. 1 The outstanding representatives of TCTSs in China

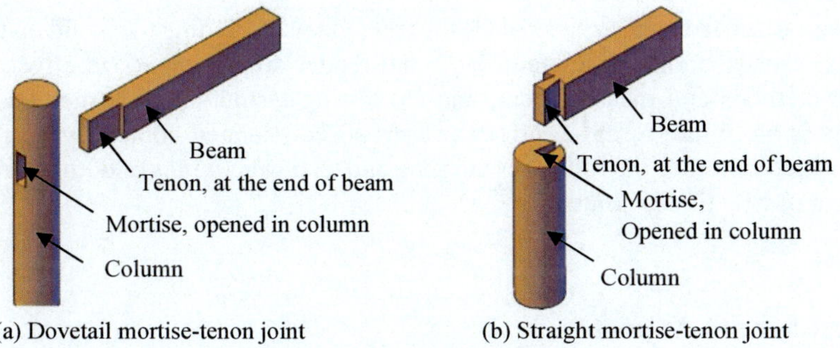

Fig. 2 Commonly applied mortise-tenon joints in TCTSs

columns, bracket sets, and so forth) using concepts of static mechanics. Such early groundbreaking works gave subsequent researchers accurate knowledge of the integral packing process of traditional Chinese timber buildings. Subsequent researches were conducted by King in 1996 to reveal the significance of the mortise-tenon joints, which profoundly affected directed the following focuses. So it can be said that the real understanding of TCTSs by means of modern mechanical tools are things in recent 25 years.

Mortise-tenon joints are the weakest link in TCTSs, connecting vertical columns and horizontal beams together as lateral force bearing system, as displayed in Fig. 2. They consume large amounts of energy in case of earthquakes through the compression and friction interactions between the mortises and tenons. Thus the investigation of such joints are greatly concerned by researchers.

In fact, basically all the researchers in the field of timber building heritage protection have ever conducted investigations to investigate the mechanical properties from test, theoretical and finite element aspects. However, the seemingly overwhelming FE simulation and theoretical methods are ideal calculations, depending highly on many assumptions and the reliability of the adopted material model, which make them hard to consider the real configurations and stress state of the joints in tests and practical timber engineering. Thus related analytical results are merely limited in the research stage, and cannot be smoothly put into application, which renders the experiments to be the available methods under some complex circumstance. However, we face another key issue that the wood species (including the physical and mechanical properties) of those typical structures differ in different places and that the loading schemes were different. Thus all the different factors were not controlled to be similar, making the analysis results cannot be easily compared with each other. Such situation is unfavourable to the achievement of an homogeneous results and must be broken to favor further researches and applications.

A more comprehensive and accurate testing protocols of timber joints would definitely be helpful to reach a mutually obeyed test agreement. Thus in this paper, the authors attempt from experimental aspect to discuss part of the existing investigations conducted on traditional Chinese mortise-tenon joints. Some detailed information was summarized from the aspects of (1) the test models, including the wood materials (species, density, moisture content, etc.) and model dimensions (size effects), (2) loading methods and measurement, and (3) testing results analysis methods, etc. Some possible future research efforts are proposed. Related content will provide favorable reference for the future researching and testing activities of both traditional and modern type timber joints.

2 Test Model Aspect

Test studies on mortise-tenon joints are dedicated to grasp the mechanical and energy-consumption mechanisms. Therefore different TCTSs prototypes are usually selected from three sources, namely *Ying zao Fa Shi* (YZFS), *Engineering practice of Ministry of works in Qing Dynasty* (ZFZL), and recently some practical timber structures (PTS). A simple summary of related investigations were given in Table 1.

2.1 Wood Material

Since the TCTSs prototypes were widely selected in the national range, meaning that different wood materials may be used to fabricate the models. However, they could be divided into two parts according to the location, in northern China or Southern China. The former case decided the use of pine species in TCTSs while the latter spruce, which can be seen from Table 1. This is good for conducting results comparison based on the available test data up to now, however, which was weaken by the fact that the physical mechanics if these species had a big scatter.

Observed from Table 1, less attention was payed to the physical condition of the wood material, especially the early tests. Thus in future researches, more attention shall be payed to the conditioning of the timber before the manufacture of the joint and also to the conditioning of the joints after their fabrication.

2.2 Model Dimensions

Related researches usually use different scaled models from different prototypes of TCTSs since the restrictions of the laboratory conditions, the commonly adopted schemes are listed in Table 1. It can be seen that the different prototypes and scale ratios made the comparison between existing data a complicated work. Also, the size effects on the rotational stiffness and hysteresis properties needed to be determined.

Lessons from Table 1 can be summarized as that more attention shall be payed to the conditioning of the joints after their fabrication, e.g. the joint dimensions, and fabrication details, the elapsed time between the fabrication and test, the detailed

Table 1 Summary of related investigations on the mortise-tenon joints

References	Mortise-tenon joints				Materials		
	Prototypes	Joints type	Numbers	Scale ratio	Species	Density /g/cm^3	MC /%
[1]	YZFS	B	4	3.52	Red pine	–	–
[2]	YZFS	B	10	3.52	Red pine	–	Air-dried
[3]	YZFS	B	4	3.52	Red pine	–	–
[4]	YZFS	B C	3 3	3.52 3.52	–	–	–
[5]	ZFZL	B	8	2.65	Spruce	–	–
[6]	YZFS	C	6	3.52	–	–	–
[7]	ZFZL	B	12	8	Red pine	–	–
[8]	PTS	C	6	1	Spruce	–	–
[9]	PTS	A	6	2	Spruce	0.409–	–
[10]	ZFZL	C D	4 4	1.76	Spruce	–	–
[11]	ZFZL	D	18	6	White pine	–	–
[12]	YZFS	A	7	4.8	Dahurian larch	–	Air-dried
[13]	ZFZL	A	6	2	Spruce	–	–
[14]	YZFS	B	3	3.2	Dahurian larch	–	–
[15]	YZFS	B	1 5 1	4.8 3.2 2.4	Dahurian larch	–	14.3
[16]	YZFS	A A A A A A C D A C D	1 1 1 1 1 2 2 2 1 1 1	1.6 2.4 3.2 4.8 4.8 4.8 4.8 4.8 4.8 4.8 4.8	Dahurian larch Dahurian larch Dahurian larch Dahurian larch Dahurian larch Dahurian larch Dahurian larch Dahurian larch Spruce Spruce Spruce	–	Air-dried

(continued)

Table 1 (continued)

References	Mortise-tenon joints				Materials		
	Prototypes	Joints type	Numbers	Scale ratio	Species	Density /g/cm^3	MC /%
[17]	ZFZL	B	6	3.2	Pine	–	–
[18]	YZFS	B C D	4 4 4	3.52 3.52 3.52	Pine	–	–
[19]	PTS	D	10	1.33	Spruce	–	–
[20]	PTS	D	3	1.76	Spruce	–	–
[21]	PTS	B B	6 6	–	Pine Spruce	0.573 0.451	11.66 12.97
[22]	ZFZL	C	4	1	South pine	0.619	21.85

A Straight mortise-tenon joints, *B* Dovetail joints, *C* Thorough mortise-tenon joint, *D* Semi-thorough mortise-tenon joints

closure between the mortise and tenon. These detailed information decides the accuracy and reasonability of the mutual comparison between tests conducted by different investigators.

On the other hand, the number of the replicates was likely not enough to support the repeatibility of test results; thus later investigations shall be improve replicates.

3 Loading and Measuring System Aspect

The aims of joint tests are to understand the semi-rigid mechanical mechanisms, grasp the rotational performances and obtain the rotational moment-rotation relationship, providing the required rotational stiffness data for the structural analysis of the timber frames and even the whole structures. Therefore, related tests should arrange some digital meters for the measurement of the rotation.

According to the literature survey, existing loading schemes can be primarily classified into two kinds, namely the scheme of directly loading the mortise-tenon joints and the scheme of first loading a frame and then indirectly calculating the moment, as shown in Fig. 3.

The measurement system can usually record the forces along the loading direction and the rotation between the beam and column. It can be clearly observed the loading methods in Fig. 3a are more efficient than those shown in Fig. 3b because the conversion methods of the latter is more complicated than the former. Both these caused great uncertainties on the calculation of the joint models.

Regarding the loading schemes, the several commonly used can be given as the following pictures, Fig. 4. In fact, the proper selection of the loading schemes help to assess the stiffness degradation of the joints. At present, most tests used Fig. 1a

(a) Directly loading the joints.

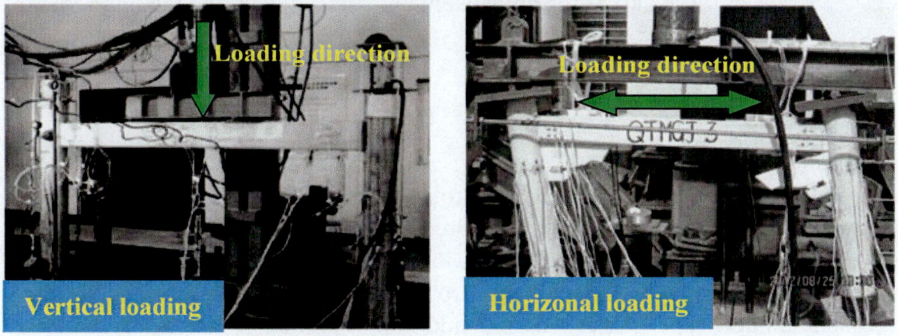

(b) Loading single-span timber frame.

Fig. 3 Loading schemes of the joints involved in existing literature

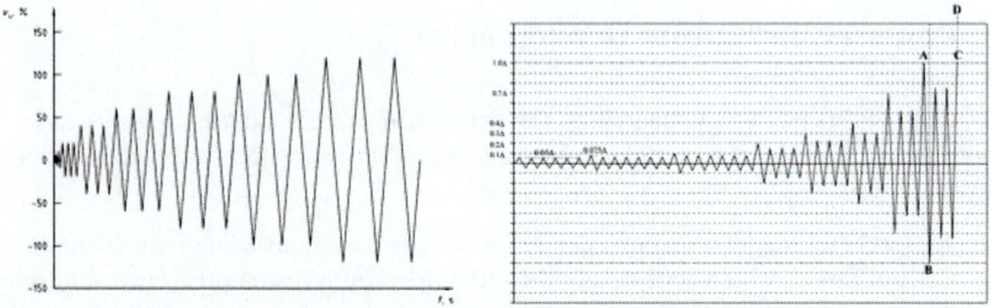

Fig. 4 Loading schemes of the joints involved in existing literature

scheme to loading the joints, proposed by ISO 16,670-2003; while another series of protocols were also described by Krawinkler et al. [23]. The cyclic loading protocols are intended for the simulation of to some extent the ground motion; thus the present protocols may not really reflect the realistic seismic loading history.

4 Results Analysis Aspect

4.1 Documentation of Test Results

Regarding the test results, basically all the investigations will analyze the failure observation, including the whole degradation process and the associated phenomena. The documentation of such test data as detailed as possible can raise the reasonability of the comparison works between different researches. On the other hand, the skeleton curves and hysteresis loops were automatically recorded by the measurement system. Thus the result of these curves to some extent are controlled by the loading scheme.

4.2 Result Analysis

Regarding the mortise-tenon joints, the rotational moment-rotation envelope and hysteresis relationships are usually required and if possible shall be fitted using some models for the sake of using them in the simulations of timber frames and more complex structures. Therefore, if proper loading scheme are selected, more reasonable models can be determined. In addition, if unified protocols standards are adopted, test results may become easier to be compared, which for sure will boost the research process.

5 Proposals for Future Development

The aims of this paper is at searching the deficiencies of the existing investigations about the TCTSs to present some favourable suggestions for the future conduction of related tests.

1. Different test results can be compared so as to reach a homogeneous researching circumstance, which will be good for the upgrading of the efficiency through the concerted efforts of relevant scholars in this field. It is easier to be observed that more important unification works should be put into the test preparation of the replicates and the recording the detailed information of these specimens.
2. More detailed information about the prototype, material properties, and if possible the initial contact state of the mortise and tenon should be given, which will post favourable advantage for the comparison between various tests.
3. In future investigation, more attention shall be payed to the conditioning of the joints after their fabrication, e.g. the joint dimensions, and fabrication details, the elapsed time between the fabrication and test, the detailed closure between the mortise and tenon. These detailed information decides the accuracy and

reasonability of the mutual comparison between tests conducted by different investigators.

References

1. Gao, D.: Aseismic characteristics of bucket arch and mortise-tenon joint of ancient Chinese timber building: experimental research. J. Nat. Disasters **17**(2), 58–64 (2008)
2. Xie, .: An experimental study on the strengthening of mortise-tenon joints in ancient Chinese wooden buildings. Chin. Civil Eng. J. **41**(1), 28–34 (2008)
3. u, .: Experimental study on seismic behavior of mortise-tenon joints of Chinese ancient wooden buildings strengthened with CFRP sheets and flat steel. World Earthq. Eng. **24**(3), 112–117 (2008)
4. Sui, .: Experimental study on mortise-tenon joints in historic timber buildings. World Earthq. Eng. **26**(2), 88–92 (2010)
5. Xu, M.: Experimental study on seismic behavior of mortise-tenon joints in Chinese ancient timber buildings. J. Build. Struct. **s2**, 345–349 (2010)
6. Zhao, H.: The experimental study on characteristics of mortise-tenon joint historic timber building. J. Xian Univ. Archit. Technol. (Nat. Ed.) **42**(3), 315–318 (2010)
7. an, W.: Aseismic strengthening experiments on Chinese ancient tenon-mortise joints. Earthq. Resist. Eng. Retrofit. **33**(2), 89–95 (2011)
8. Lu, W.: Seismic performance of the typical timber structure and mortise-tenon joints in villages. **34**(3), 82–85 (2012)
9. Lu, W.: Experimental research on seismic performance of wooden mortise-tenon joints before and after reinforcement. J. Earthq. Eng. Eng. Vib. **32**(3), 109–116 (2012)
10. Chen, C.: Experimental study on positive and reverse flexural behavior of asymmetric mortise-tenon joints. J. Southeast Univ. **44**(6), 1224–1229 (2014)
11. Gao, D.: Pseudostatic experimental study on mortise and tenon joints of timber structures of Chinese Ming and ing Dynasties. World Earthq. Eng. **30**(4), 8–16 (2014)
12. Xie, .: Experimental study on the seismic behavior of mortise-tenon joints of ancient timber buildings. Struct. Eng. Int. **35**(11), 143–150 (2014)
13. Sun, W.: Seismic performance of through - tenon joint strengthened with carbon fiber reinforced polymer in wood structure. J. Nanjing Tech. Univ. **37**(2), 91–96 (2015)
14. Xie, .: Tests for aseismic behaviors of damaged dovetail mortise-tenon joints of ancient timber buildings. J. Vib. Shock **34**(4), 165–210 (2015)
15. Xie, .: Experimental study on seismic behavior and size effects of dovetail mortise-tenon joints of ancient timber buildings. J. Build. Struct. **36**(3), 112–120 (2015)
16. Xie, .: Experimental study on seismic behavior of straight mortise-tenon joints of ancient timber buildings. Earthq. Eng. Eng. Dyn. **35**(3), 232–241 (2015)
17. Xue, J.: Experimental study on seismic performance of dovetail joints with different loose degrees in ancient buildings. J. Build. Struct. **37**(4), 73–79 (2016)
18. Ma, D.: Experimental study of aseismic behaviors of flexural tenon joint, through tenon joint and dovetail joint reinforced with flat steel devices. J. Beijing Univ. Technol. **45**(8), 763–771 (2019)
19. Xue, J.: Experimental study on seismic performance of cross-shaped ban-tenon joint reinforced by angle steel in traditional dwellings. J. Build. Struct. (published online) (2020)
20. Chun, .: Reearch on mechanical properties of ban mortise-tenon joint of the traditional timber buildings in the south yangtz river regions. J. Hunan Univ. **43**(1), 124–131 (2016)
21. Tao, Z.: Experimental study on the dovetail mortise-tenon joints with the viscoelastic dampers of traditional timber structure. J. Build. Struct. (published online) (2020)

22. Xu, .: Experimental research on strengthening methods for through—mortise and tenon joints of ing-style timber buildings. Sci. Conserv. Archiol. **30**(4), 70–79 (2018)
23. Krawinkler, H., Parisi, F., Ibarra, L., Ayoub, A., Medina, R.: Development of a testing protocol for woodframe structures, vol. 102. Richmond, CA: CUREe (2001)

The Effect of Recycled Fine Aggregate Sourced from Construction and Demolition Waste on the Properties of Epoxy Resin Coatings

Kamil Krzywiński and Łukasz Sadowski

Abstract Worldwide development is currently accompanied with a fast growing construction industry that is responsible for generating the highest amount of wastes in the European Union (EU). On the other hand, the extraction of natural aggregate is also high, with sand and gravel extraction coming out on top, generating a huge amount of pollutants. This extraction has a negative influence on the environment and does not provide a renewable material. In some countries, this resource is almost finished. By 2020 the majority of the EU countries want to achieve at least 50% recycling of construction and demolition waste due to its negative impact on the environment. Thus, the main aim of this article is to analyze the impact of sustainable utilization of the recycled fine aggregate (RFA), as a substitute for ordinary fine aggregate (OFA; river sand), which was used as an extender in epoxy resin coatings, on the properties of the epoxy resin coating. For that purpose, samples with different partial substitution of the OFA with RFA: 0, 20, 40, 60, 80 and 100% were prepared and macro tests for pull-off strength were performed. Based on the obtained results, it can be concluded that RFA has a beneficial impact on the pull-off strength of the epoxy resin coating. The summary evaluation of mechanical performance shows that substitution of RFA by OFA in epoxy resin coatings has visible impact on reduction of dispersion of strength results. The coating modified with RFA wastes become more homogenous material in comparison to the specimens with OFA.

Keywords Coating · Epoxy resin · Recycled fine aggregate · Sustainable technology · Reused wastes

1 Introduction

Recently the worldwide development is mainly based on fast growing construction industry [1–4]. This grow created countless job places, houses, block of flats etc. This tendency can still be observed. However, due to this tendency, the construction

K. Krzywiński (✉) · Ł. Sadowski
Department of Building Engineering, Wrocław University of Science and Technology, Wybrzeże Wyspiańskiego 27, 50-370 Wroclaw, Poland
e-mail: kamil.krzywinski@pwr.edu.pl

V. M. C. F. Cunha et al. (eds.), *Proceedings of the 3rd RILEM Spring Convention and Conference (RSCC 2020)*, RILEM Bookseries 35,
https://doi.org/10.1007/978-3-030-76543-9_23

industry is responsible for generating the highest amount of wastes in the European Union (924 million tons in 2016; Fig. 1a). Most of the constructions are made from concrete, which is in 75% made from aggregates [5]. Thus, the extraction of the aggregates is very high (Fig. 1b) and still is growing. Moreover, extraction of the aggregate (especially river sand) has very negative impact on the environment. Therefore, actions should be taken to reduce this negative process.

The utilization of the recycled fine aggregate (RFA) is problematic and leads to destroying the environment. There are well known approaches of possible utilization of wastes [8–11]. The RFA has similar properties like the old concrete [12–16]. It makes the application of this material considerably easy.

On the other side, the rapid growth of the industry is accompanied with growing numbers of large area storage halls. Mostly, their indoor floors could be covered with the cement mortar [17–19] or the epoxy resin coatings [20, 21]. The manufacturers allowed to add a river aggregate (1–2 mm) as a filler inside the epoxy resin coating to increase its volume. However, the extraction of river aggregate is mostly located far away from places where it is used.

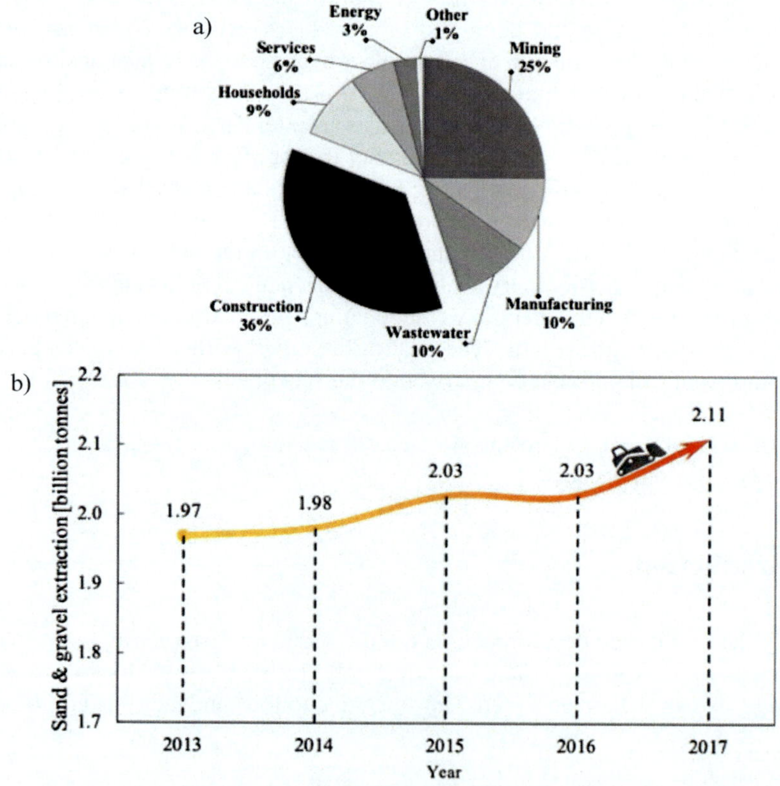

Fig. 1 Statistic for European Union countries: **a** waste generation in 2016 by economic activities and households [6]; **b** recent sand and gravel extraction [7]

Summarizing above, the main aim of this paper is to investigate the effect of recycled fine aggregate sourced from construction and demolition wastes on the properties and sustainable production of the epoxy resin coatings.

2 Materials and Methods

2.1 Concrete Substrate

One type of concrete substrate was prepared in wooden form with dimensions of 900 × 300 × 40 mm. This substrate was divided into twelve smaller parts. Each part had a dimensions of 150 × 150 × 40 mm. The substrate was prepared using a commercially available ready-mix concrete mixture class C16/20.

2.2 Epoxy Resin Coating

The coating was made of commercially available three component-based epoxy resin. The first (Component A, base) is a bisphenol-based epoxy resin. The second component (Component B) is an aliphatic polyamines-based hardener. The third component (Component C) is a fine aggregate. For 3 mm coating the weight ratio of A:B:C is usually 100:25:75. All components were mixed together for 3 min to obtain a uniform consistency. Curing of the concrete sample with epoxy resin was performed in a controlled laboratory conditions with relative humidity less than 60%, at the temperature of 21 ± 2 °C. The tests were performed after 7 days.

2.3 Aggregate

In order to reduce the volume of river aggregate in the epoxy resin the recycled fine aggregate (RFA) has been used to replace partially the ordinary fine aggregate (OFA). Therefore, six different samples were prepared (Fig. 2).

In this study OFA has been used. The properties of the OFA declared by the manufacturer are presented in Table 1.

The grain size of RFA was very similar to OFA, what allowed to compare the test results.

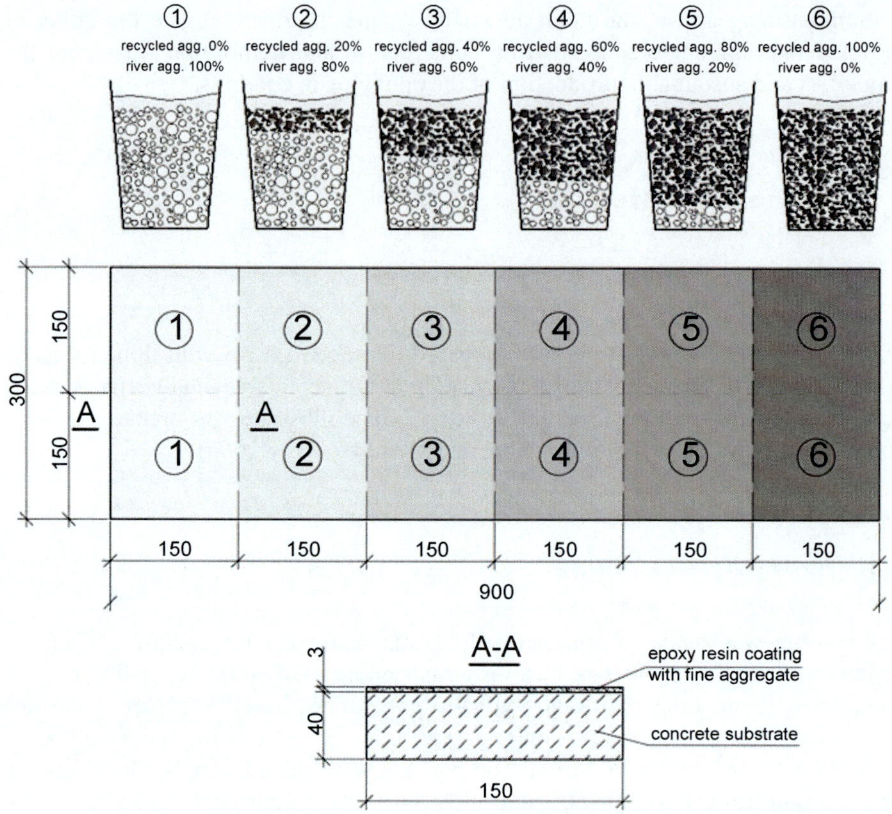

Fig. 2 Samples preparation process for each participation of aggregate added to epoxy resin

Table 1 River aggregate properties

Property	Description
SiO_2 (%)	87.78
Loose bulk density of material (g/cm^3)	1.50
Aggregate bulk density (g/cm^3)	2.60

2.4 *Pull-off Strength Test*

It is possible to evalutate the pull-off strength using ultrasonic methods [22, 23], however, in this study the destructive methotd was used. Impact of different ratio of the river aggregate and the recycled aggregate added to epoxy resin on pull-off strength of coatings was measured with pull-off strength test according to ASTM D4541 [24]. For this purposethe automatic adhesion tester has been used. Each sample was tested three times to obtain the average value of the pull-off strength, as it was done in [25]. The loading rate was 0.05 MPa/s.

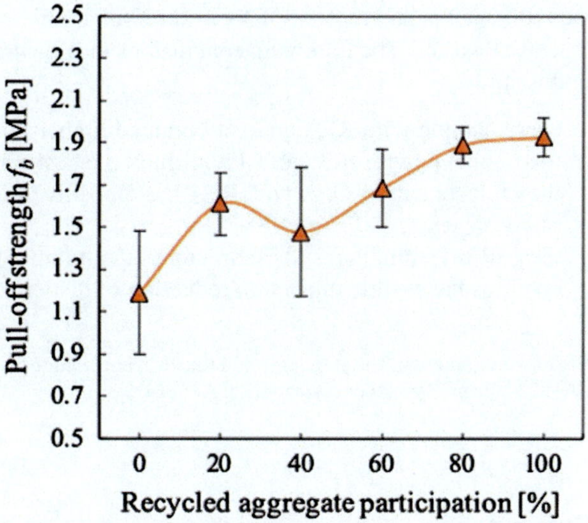

Fig. 3 Pull-off strength test results

3 Results

Figure 3 presents the obtained pull-off strength test results. It is visible from Fig. 3 that the lowest value of pull-off strength has been obtained for the sample no. 1 with 100% OFA added to epoxy resin mixture (1.19 MPa ± 0.29). Slightly better results have been obtained for sample with 40% RFA participation with (1.48 MPa ± 0.30). Samples No. 2, 5 and 6 have similar pull-off strength. The jump and drop of the strength for 20 and 40% of RFA is probably a reason of the two factors which affected the results. The first reason is the dispersion of the selected places for the test. The whole area cannot be verified because of the limits of the pull-off strength test. Moreover, the observed failure mode for each tested sample was cohesive [26]. Thus, the impact of the concrete substrate near-surface zone area properties, which is heterogeneous material, could also affects the results. Presented above factors had an influence on the obtained results. However, each sample with RFA obtained better result in comparison to reference sample with OFA. The concrete substrate detached thickness varied from 2 to 7 mm. In the previous study [27] the thickness of detached concrete substrate was much higher for textured surfaces, what had also impact on higher pull-off strength.

4 Conclusions

Environmental protection plays a key role for sustainable development. Processes that have damaging impact on environment, should be regulated or changed in order to reduce negative effects. In this work the negative effects were avoided by substituting the ordinary fine aggregate (OFA) with recycled fine aggregate (RFA) sourced from industrial wastes. The following conclusions may be drawn from the work presented in this study:

- every sample with RFA content obtained higher pull-off strength results, where the sample 6 (with 100% of RFA) exhibit the best result (1.93 MPa $\pm$ 0.15). Thus, the replacement of OFA with RFA has a positive impact on the pull-off strength of epoxy resin,
- the pull-off strength results shows that substitution of RFA by OFA in epoxy resin coatings has visible impact on reduction of dispersion of strength results.

Acknowledgements The authors would like to acknowledge the contribution of the COST Action CA18224. https://www.cost.eu/actions/CA18224.

References

1. Du, Q., Zhou, J., Pan, T., Sun, Q., Wu, M.: Relationship of carbon emissions and economic growth in China's construction industry. J. Clean. Prod. **220**, 99–109 (2019)
2. Govindan, K., Shankar, K.M., Kannan, D.: Sustainable material selection for construction industry—a hybrid multi criteria decision making approach. Renew. Sustain. Energy Rev. **55**, 1274–1288 (2016)
3. Hu, X., Liu, C.: Carbon productivity: a case study in the Australian construction industry. J. Clean. Prod. **112**, 2354–2362 (2016)
4. Martinico-Perez, M.F.G., Fishman, T., Okuoka, K., Tanikawa, H.: Material flow accounts and driving factors of economic growth in the Philippines. J. Ind. Ecol. **21**(5), 1226–1236 (2017)
5. De Brito, J., Saikia, N.: Recycled Aggregate in Concrete: Use of Industrial, Construction and Demolition Waste. Springer Science & Business Media (2012)
6. Waste Generation: https://ec.europa.eu/eurostat/statistics-explained/. Accessed 15 Nov 2019
7. Material Flow Accounts: http://appsso.eurostat.ec.europa.eu/. Accessed 15 Nov 2019
8. Galińska, A., Czarnecki, S.: The effect of mineral powders derived from industrial wastes on selected mechanical properties of concrete. IOP Conf. Ser. Mater. Sci. Eng. **245**(3), 032039 (2017)
9. Sadowska-Buraczewska, B., Barnat-Hunek, D., Szafraniec, M.: Influence of recycled high-performance aggregate on deformation and load-carrying capacity of reinforced concrete beams. Materials **13**(1), 186 (2020)
10. Andrzejuk, W., Barnat-Hunek, D., Fic, S., Styczeń, J.: Wettability and surface free energy of mineral-asphalt mixtures with dolomite and recycled aggregate. IOP Conf. Ser. Mater. Sci. Eng. **471**(3), 032011 (2019)
11. Chowaniec, A., Ostrowski, K.: Epoxy resin coatings modified with waste glass powder for sustainable construction. Czas. Tech. **8**, 99–109 (2018)
12. Rodrigues, F., Carvalho, M.T., Evangelista, L., De Brito, J.: Physical–chemical and mineralogical characterization of fine aggregates from construction and demolition waste recycling plants. J. Clean. Prod. **52**, 438–445 (2013)
13. Martínez, P.S., Cortina, M.G., Martínez, F.F., Sánchez, A.R.: Comparative study of three types of fine recycled aggregates from construction and demolition waste (CDW), and their use in masonry mortar fabrication. J. Clean. Prod. **118**, 162–169 (2016)
14. Evangelista, L., de Brito, J.: Mechanical behaviour of concrete made with fine recycled concrete aggregates. Cem. Concr. Compos. **29**(5), 397–401 (2007)

15. Khatib, J.M.: Properties of concrete incorporating fine recycled aggregate. Cem. Concr. Res. **35**(4), 763–769 (2005)
16. Evangelista, L., De Brito, J.: Durability performance of concrete made with fine recycled concrete aggregates. Cem. Concr. Compos. **32**(1), 9–14 (2010)
17. Tabsh, S.W., Abdelfatah, A.S.: Influence of recycled concrete aggregates on strength properties of concrete. Constr. Build. Mater. **23**(2), 1163–1167 (2009)
18. Sadowski, Ł., Krzywiński, K., Michoń, M.: The influence of texturing of the surface of concrete substrate on its adhesion to cement mortar overlay. J. Adhes. 1–14 (2019)
19. Jin, X., Zhang, X., Luo, Y.: A calculation method for the floor surface temperature in radiant floor system. Energy Build. **42**(10), 1753–1758 (2010)
20. Mikami, T., Kang, Y.H., Choi, S.K.: Judging indices for evaluating the exfoliation of synthetic resin floorings using the impact acoustics method. Constr. Build. Mater. **95**, 345–354 (2015)
21. Song, G.S.: Buttock responses to contact with finishing materials over the ONDOL floor heating system in Korea. Energy Build. **37**(1), 65–75 (2005)
22. Czarnecki, S.: Ultrasonic evaluation of the pull-off adhesion between added repair layer and a concrete substrate. IOP Conf. Ser. Mater. Sci. Eng. **245**(3), 032037 (2017)
23. Czarnecki, S.: Non-destructive evaluation of the bond between a concrete added repair layer with variable thickness and a substrate layer using ANN. Procedia Eng. **172**, 194–201 (2017)
24. ASTM D4541-95e1: Standard Test Method for Pull-off Strength of Coatings Using Portable Adhesion Testers. ASTM International, West Conshohocken, PA, USA (2002)
25. Krzywiński, K., Sadowski, Ł., Łaszczak, M.: The effect of the type of substrate and its surface treatment on the pull-off strength of gypsum plasters. J. Adhes. 1–14 (2019)
26. Szymanowski, J.: Evaluation of the adhesion between overlays and substrates in concrete floors: literature survey, recent non-destructive and semi-destructive testing methods, and research gaps. Buildings **9**(9), 203 (2019)
27. Krzywiński, K., Sadowski, Ł: The effect of texturing of the surface of concrete substrate on the pull-off strength of epoxy resin coating. Coatings **9**(2), 143 (2019)

Performance Evaluation of Warm Recycled Surface Mixtures with Steel Slag

P. Georgiou and A. Loizos

Abstract Development of sustainable asphalt mixtures for implementation during the life cycle stages of road pavements is of supreme importance for minimizing the environmental burdens. To this aim, recycling and reuse techniques offer significant benefits in terms of resource conservation and emission reduction. Also, reduction in production temperature of asphalt mixtures by means of warm mix asphalt (WMA) technology results in reduced emissions, fuel usage and binder oxidation, and hence may boost further the environmental but also economic benefits. This study investigates the performance of surface course mixtures designed with various contents of reclaimed asphalt (RA) and EAF steel slag aggregates, and produced at reduced temperatures by using organic additives. Performance evaluation of warm recycled surface mixtures focused on stiffness and durability. The test results are discussed to address whether the concurrent use of WMA and recycled/solid waste materials is feasible in developing asphalt mixtures with high level of performance.

Keywords Asphalt mixtures · Warm mix asphalt · Reclaimed asphalt · EAF steel slag · Sustainability

1 Introduction

The criticality of implementing sustainability policies has become more acute in light of growing evidence suggesting that human activities are jeopardizing the health of the planet. A shift to the circular economy development is a crucial step for decoupling economic growth from further unsustainable resource use and increased CO_2 emissions. With sustainability becoming a key part of the global agenda, the European Commission has set a goal of a total CO_2 saving of 55% by 2030. To this aim, it should be noted that the transportation sector has been identified to account for about a fifth of total CO_2 emissions in the European Union [1]. Also, road network, which is mostly comprised of asphalt pavements, serves over 80% of transportation

P. Georgiou (✉) · A. Loizos
Laboratory of Pavement Engineering, National Technical University of Athens (NTUA), Athens, Greece
e-mail: georgp@central.ntua.gr

V. M. C. F. Cunha et al. (eds.), *Proceedings of the 3rd RILEM Spring Convention and Conference (RSCC 2020)*, RILEM Bookseries 35,
https://doi.org/10.1007/978-3-030-76543-9_24

needs and road transport emissions have been generally increasing since 1990. With this in mind, development of sustainable asphalt mixtures for implementation during the life cycle stages of road pavements is of supreme importance for minimizing the environmental burdens.

Recycling and reuse techniques offer significant benefits in terms of resource conservation and energy reduction. Reclaimed asphalt (RA) milled off road asphalt layers with rotary drum cold milling machine or from a ripping/crushing operation, has long recognized as a valuable component of asphalt mixtures. It may be used as a useful surrogate of aggregates and bitumen required in the production of asphalt mixtures [2]. The utilization of RA material in road pavement construction has been of great strategic importance for several reasons, including reduction of (a) construction debris disposal into landfills, (b) quarries exploitation and increased use of nonrenewable natural resources such as virgin aggregate and asphalt binder and (c) cost of asphalt mix production. However, there are concerns over undesired heterogeneity and intrinsic properties of the RA, such as aged binder and poor aggregate quality. This has led road pavement engineering community to recommend characterization of RA as a constituent material of asphalt mixtures [3]. On the other side, the blending phenomena of aged RA bitumen with virgin bitumen and recycling agents are not fully understood, thus hindering the incorporation of high RA contents and design of highly recycled asphalt mixtures [4–6].

The use of secondary (recycled) solid waste materials in road construction projects is also considered a sustainable strategy with great potential in further easing landfill pressures and reducing the demand for quarrying minerals. Research has shown that the Electric arc furnace (EAF) slag, which is one of the major by-products of the steel industry, can be reused in asphalt mixes [7, 8]. The inherent properties of iron and steel slag make it an ideal aggregate for surfacing asphalt products. Steel slag aggregates has proven to appear good physical and mechanical properties, especially demonstrating good resistance to fragmentation, surface wear by abrasion as well in the polishing effect of vehicle tires. As such, the use of EAF slag in surface course asphalt mixes is typically reported in partial or total substitution of the coarse aggregates in order to enhance the frictional performance of these mixtures.

Over the last decades there is also a considerable striving in lowering mixing and compaction temperatures of asphalt mixtures. Warm mix asphalt (WMA) technology is an emerging technology, recently gaining more acceptance and popularity among asphalt industry compared to hot mix asphalt (HMA) as being an effective method to reducing mixing and compaction temperatures and subsequently the energy consumption and emissions associated with conventional HMA production [9, 10]. The WMA temperature reduction is the result of developed technologies that involve the use of organic additives, chemical additives, and water-based or foaming processes.

In light of the above, a challenge emerges to put emphasis in designing asphalt mixtures by appropriately combining recycled aggregates in high contents and WMA for most enhancing pavement sustainability without sacrificing mixtures' performance. In this respect, this study investigates the performance of surface course mixtures designed with various contents of reclaimed asphalt and EAF steel slag aggregates, and produced at reduced temperatures by using organic additives. Evaluation of warm recycled surface mixtures mainly focused on durability and stiffness performance, complemented with assessment of degree of bitumen blending. The test results are discussed to address whether the concurrent use of WMA and recycled/solid waste materials is feasible in developing asphalt mixtures with high level of performance.

2 Experimental Program

2.1 Description

To accomplish the study objectives, a set of asphalt mixtures for wearing course was designed with EAF steel slag and varying amounts of RA (25, 40 and 50% by mass), and incorporating WMA organic additive for lowering the production temperatures (designated hereafter as 'WMA_25R', 'WMA_40R' and 'WMA_50R'). Performance properties of these mixtures subjected to short-term ageing were compared to a reference mixture produced using HMA techniques (designated hereafter as 'HMA').

2.2 Materials

Steel slag aggregates, limestone fine aggregates, RA, polymer modified bitumen (PmB) and WMA organic additive were used to produce surface course asphalt mixtures. In all mixtures, EAF slag, produced during ferrous scrap melting in steelworks, was used as coarse aggregates and limestone as fine aggregate and filler. Reclaimed asphalt used in this study was produced from milling operations taken place on a motorway. In order to characterize the properties of the RA material a suite of tests was performed, including black and white particle size analyses and determination of RA mineral aggregates properties after bitumen extraction, as recommended in [3].

Table 1 lists the basic properties of aggregates.

Moreover, two SBS polymer modified bitumen were employed for preparing the HMA and WMA-RA mixtures. A 25–55/75 PmB was used for the reference HMA mixture, whereas for the WMA-RA mixtures a 'softer' PmB designated as 45–80/65 was selected to compensate for the aged bitumen of RA materials. For lowering the production temperatures with respect to the HMA, the latter bitumen was mixed with a synthetic Fischer–Tropsch (FT) wax, which is completely soluble in bitumen and significantly reduces viscosity of the resulting bitumen blend (i.e. WMA bitumen), thereby providing the potential for reduction in manufacturing temperatures of the

Table 1 Physical properties of aggregates

Property	Steel slag (coarse)	Limestone (fine)	RA aggregates
Particle density (kg/m^3)	3247	2603	2809
Water absorption (%)	1.8	1.6	1.4
Flakiness	5.7	–	9.6
LA abrasion (%)	13.8	–	–

Table 2 Characteristic properties of bitumens

Property	PmB (reference)	WMA bitumen	RA bitumen
Penetration (0.1 mm)	45	44	15
Softening point (°C)	75	68	70

asphalt mixtures up to 30 °C. Table 2 lists the main basic properties of virgin, WMA as well as the RA bitumen.

2.3 *Mix Design and Specimen Preparation*

The target mixture of this study is a semi-open graded Asphalt concrete (AC) with a nominal maximum aggregate size of 12.5 mm, which is commonly used in wearing courses of Greek motorways. Four mixtures were designed in laboratory. Specifically, the aggregate grading curves were designed to conform to the grading limits prescribed in the national technical specifications, and aimed to optimize both the structural as well surface texture characteristics of asphalt mixtures. Reference mix included steel slag as substitute for coarse aggregates and limestone sand, whereas the remaining mixtures further included fractionated RA of increasing percent (i.e. 25, 40 and 50%). As shown in Fig. 1, all mixtures have almost the same grading.

The optimum bitumen content of all mixtures was determined using the Marshall mix design methodology. With regard to the HMA, the optimum bitumen content was 4.8% by aggregate weight; the same was also for the WMA-RA mixtures.

For preparing the HMA, steel slag and limestone where mixed with the specific bitumen by means of laboratory planetary equipment. Aggregates were mixed in 'dry state' for 30 s, then bitumen was added and the mixing time lasted in total 2 min. Similar mixing time was followed for the recycled mixtures, although three cycles were adopted in the mixing sequence; initially virgin aggregates were mixed, then

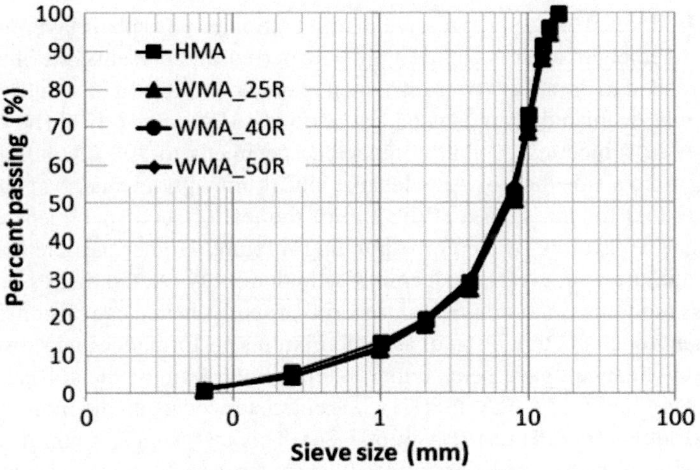

Fig. 1 Grading of HMA and WMA-RA mixtures

virgin with RA aggregates and finally all-in aggregates with WMA bitumen. It should be noted that RA was previously fractionated in five fractions in order to minimize potential variability related to RA materials, which can affect gradation and mixture's performance. Moreover, the WMA bitumen was pre-blended through a mechanical stirrer with the organic additive following the producer's recommendations and then added to the all-in aggregates in quantity assuming full reactivation of RA bitumen.

After mixing, the loose mixtures were transferred to flat shallow pans and placed in an air-drafted ventilated oven for short-term conditioning. The oven temperature was adjusted to the desired compaction temperatures of 160 °C for HMA and 130 °C for WMA-RA mixtures. Recycled WMA mixtures were conditioned at the planned compaction temperature for 2 h, whereas HMA for 4 h and then compacted using the impact compactor according to EN 12697-30 by applying 50 or 75 blows on each side of the cylindrical sample depending on the subsequent testing procedure.

2.4 Testing Procedures

For each studied mixture, the following performance properties were evaluated; compactability, raveling resistance and stiffness modulus.

Compactability properties of asphalt mixtures were evaluated through impact compaction data by calculating the specimens' air voids (according to EN 12697-8) after being compacted using 50 and 75 blows, respectively. To this purpose, the bulk density of test specimens was measured following the by dimensions procedure prescribed in EN 12697-6, whereas the maximum density of the mixtures was determined using the volumetric procedure defined in EN 12697-5.

Raveling, defined as the progressive disintegration of an asphalt layer as a result of the dislodgement of asphalt coated aggregate particles, represents one of the major distresses related to durability of (semi) open graded mixtures. In this study, raveling resistance was evaluated by means of Cantabro test (EN 12697-17). The Cantabro test consists in subjecting asphalt compacted specimens to 300 revolutions, at 30 revolutions/min, inside the Los Angeles machine drum without charge of steel balls. The particle loss (or mass loss) at the end of the test of such specimens indicates the resistance to raveling of the corresponding mixture. Higher particle loss values indicate a mixture with greater susceptibility to durability related distresses.

Stiffness modulus is a fundamental material property, hence can be considered as a key element for characterization of asphalt mixtures performance and subsequently the response of asphalt pavements under loading. In this study, the stiffness performance of HMA and WMA-RA mixtures was characterized using the Indirect Tensile Stiffness Modulus (ITSM) test (EN 12697-26). This test, which is non-destructive, consists of sinusoidal load pulses applied along the vertical diameter of specimen and characterized by the rise-time and transient horizontal deformation measured along the horizontal diameter using linear variable differential transformers (LVDT). Typically, for 100-mm diameter specimens, the target peak transient horizontal deformation is 5 μm and the target load pulse rise time is 124 ms. Five load pulses are applied to the specimen and the average modulus is determined. The specimen is then rotated 90° around its horizontal axis and further five load pulses are applied and the resulting mean stiffness modulus is obtained. The stiffness modulus is calculated as the average of these two mean values.

3 Results and Discussion

3.1 *Compactability*

Compactability properties were evaluated through impact compaction data by calculating the specimens' air voids of HMA and WMA-RA mixtures after being compacted using 50 and 75 blows per side. Data are shown in Fig. 2 as average of four replicate specimens along with the standard deviation (error bars).

The results indicate clear differences between the reference HMA compacted at 160 °C and the two WMA mixtures with the highest RA contents (i.e. WMA_40R, WMA_50R) compacted at 130 °C. The latter are characterized by air voids values significantly lower than the HMA. Regarding the WMA_25R mix, higher compaction level compared to the HMA was achieved for the higher compaction energy (i.e. 2 × 75 blows). Hence, it can be concluded that the organic additive incorporated in the recycled mixtures ensured better compactability with respect to the HMA mixture, despite reduced production temperatures. It is also interesting to note that no clear differences in compactability are generally detected due to varying compaction energy, especially for the highly recycled asphalt mixtures.

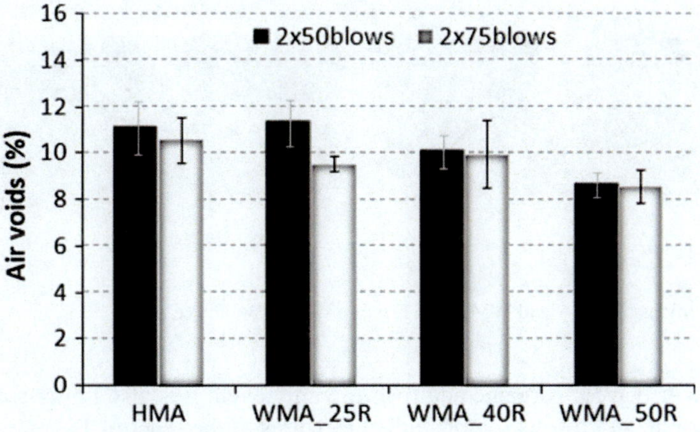

Fig. 2 Air voids of HMA and WMA-RA mixtures

3.2 *Raveling Resistance*

As mentioned previously, raveling resistance was evaluated by means of Cantabro tests. Prior to Cantabro testing, all specimens were conditioned for 4–5 h at 25 °C in a temperature controlled cabinet.

Figure 3 presents the results from the Cantabro tests.

In terms of particle loss results, similar raveling performance is depicted for the warm recycled mixtures compared to the reference HMA. This is also affirmed from Fig. 4, which illustrates the test specimens after being subjected to testing. This demonstrates that the incorporation of high recycling rates in asphalt mixtures produced at significantly lower temperatures than HMA does not affect adhesion

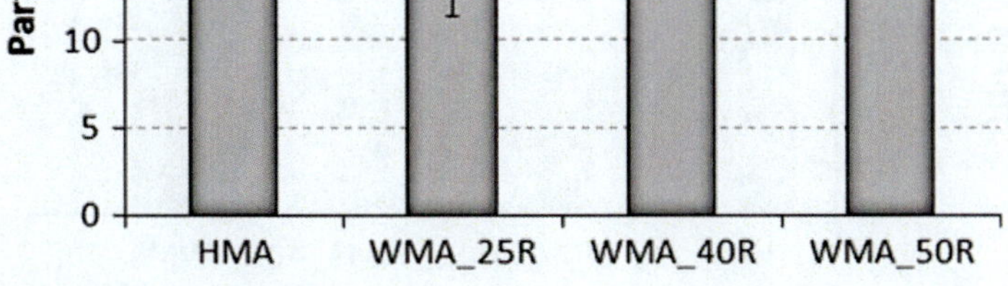

Fig. 3 Particle loss results of studied mixtures

Fig. 4 Specimens of HMA and WMA-RA mixtures after Cantabro tests

properties when WMA organic additives are employed. It is also important to note that the particle loss results for all studied mixtures of the order of 14% satisfies the recommended maximum permitted value being 20%, as prescribed in ASTM D7064.

3.3 *Stiffness Modulus*

The stiffness performance of HMA and WMA-RA mixtures was characterized using the ITSM test. Controlled deformation stiffness tests at a target deformation of 5 μm were performed at a test temperature of 20 °C. Figure 5 presents the stiffness modulus as a function of compaction energy applied to the test specimens.

As observed, the stiffness modulus of all studied mixtures increases with the compaction energy. This manifests more in the case of WMA_25R mix and can be attributed to the significant reduction of air voids achieved with increasing the compaction energy (i.e. 75 blows). For this mixture the magnitude increase is of

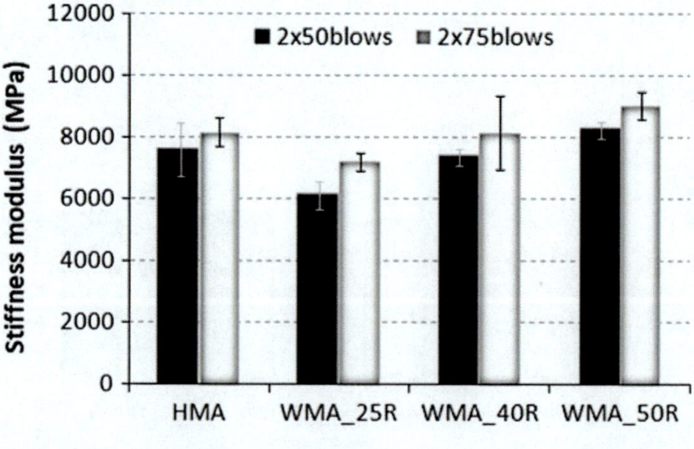

Fig. 5 Stiffness modulus results of studied mixtures

the order of 18%, whereas for the remaining mixtures range between 8 and 10%. Moreover, comparison among the WMA-RA mixtures indicates that the stiffness modulus increases with the increase of RA proportion, as expected. Specifically, the stiffness modulus of WMA_40R and WMA_50R mixtures exhibited an increase of 13% and 25% with respect to the 25% recycled WMA mixture, respectively. The improvement in stiffness modulus is attributed to the stiffening of blended bitumen characteristics. It should be also noted that these mixtures show similar or even higher stiffness modulus compared to the conventional HMA mixture, thus denoting that the lower production temperatures of recycled mixtures does not adversely affect their performance. In any case, it is highlighted that all recycled warm mixes exhibited adequate modulus values, which further suggests that can significantly contribute in the bearing capacity of asphalt pavements.

3.4 Assessment of Degree of Blending

Based on the ITSM results, a complementary analysis is elaborated to assess the degree of bitumen blending (DoB) particularly for the highly recycled mixtures, for which limitations for broader use still exist around the world, based on a practical approach suggested in a recent study [6].

The hypothesis behind is that the ITSM test is sensitive to bitumen stiffness. As such, the ratio of stiffness modulus of recycled WMA mixtures to that of reference HMA can be used as indicator to quantify the degree of interaction and blending between bitumen reactivated from the RA and virgin bitumen during the manufacturing process of asphalt mixtures. In this study, if the WMA-RA mixtures show similar stiffness modulus values to HMA, then the DoB is considered 100%. Lower stiffness ratio values indicate less degree of blending which can be attributed to less diffusion of virgin bitumen into the aged RA bitumen. Figure 6 presents the DoB results.

It can be seen that for a given mixing time and temperature, the degree of blending achieved for the WMA_40R and WMA_50R mixtures was estimated to be 100–110%. This finding indicates that a significant part of RA bitumen was reactivated during manufacturing process, which in turn suggests that the incorporation of high contents of RA on asphalt mixtures is feasible, even for those produced at reduced temperatures with respect to HMA.

4 Conclusions

This study investigates the performance of surface course mixtures designed with various contents of reclaimed asphalt and EAF steel slag aggregates, and produced at reduced temperatures by using organic additives, compared to a reference HMA. Based on the present study results, the following conclusions can be drawn.

Fig. 6 DoB results

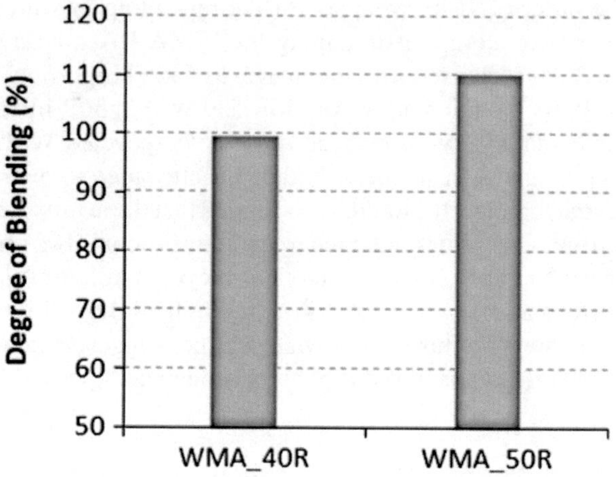

The compactability (in terms of air voids) of recycled surface mixtures with steel slag was better compared to HMA, despite reduced production temperatures. Therefore, by incorporating organic additives the compaction temperatures of recycled mixtures have the potential to be reduced by 30 °C.

The Cantabro test results suggest similar raveling performance for the warm recycled mixtures compared to HMA. Also, the particle loss results satisfy technical requirements for material acceptance.

The WMA-RA mixtures showed comparable stiffness modulus at 20 °C to the HMA mixture, with increasing trend in modulus values associated with incorporation of higher RA contents. Also, based on the stiffness ratio of recycled WMA mixtures to that of HMA, it can be considered that a significant part of RA bitumen was reactivated during the manufacturing process resulting in a high degree of blending between aged and virgin bitumen.

The above preliminary results indicate that coupling WMA and recycled/solid waste materials are feasible in developing asphalt mixtures for surface courses with a high level of performance. These results being complemented with evaluating resistance in moisture damage and fracture along with functional characteristics will enable us to fully characterize these mixtures performance.

Acknowledgements This study is part of a project which has received funding from the Hellenic Foundation for Research and Innovation (HFRI) and the General Secretariat for Research and Technology (GSRT), under grant agreement No 216.

References

1. EAPA: A European Green Deal: the asphalt industry's contributions to climate neutrality and preservation of Europe's natural environment. https://eapa.org/wp-content/uploads/2019/12/EAPA-manifesto.pdf
2. Copeland, A.: Reclaimed Asphalt Pavement in Asphalt Mixtures: State of the Practice. FHWA, Publication No. FHWA-HRT-11-021 (2011)
3. Tebaldi, G., Dave, E., Cannone Falchetto, A., et al.: Recommendation of RILEM TC237-SIB: protocol for characterization of recycled asphalt (RA) materials for pavement applications. Mater. Struct. **51**, 142 (2018)
4. Navaro, J., Bruneau, D., Drouadaine, I., Colin, J., Dony, A., Cournet, J.: Observation and evaluation of the degree of blending of reclaimed asphalt concretes using microscopy image analysis. Constr. Build. Mater. **37**, 135–143 (2012)
5. Rad, F., Sefidmazgi, N., Bahia, H.: Application of diffusion mechanism: degree of blending between fresh and recycled asphalt pavement binder in dynamic shear rheometer. Transp. Res. Rec. J. Transp. Res. Board **2444**(1), 71–77 (2014)
6. Abed, A., Thom, N., Lo Presti, D.: Design considerations of high RAP-content asphalt produced at reduced temperatures. Mater. Struct. **51**(4), 91 (2018)
7. Sorlini, S., Sanzeni, A., Rondi, L.: Reuse of steel slag in bituminous paving mixtures. J. Hazard. Mater. **209–210**, 84–91 (2012)
8. Skaf, M., Manso, J.M., Aragón, Á., Fuente-Alonso, J.A., Ortega-López, V.: EAF slag in asphalt mixes: a brief review of its possible re-use. Resour. Conserv. Recycl. **120**, 176–185 (2017)
9. Rubio, M.C., Martínez, G., Baena, L., Moreno, F.: Warm mix asphalt: an overview. J. Clean. Prod. **24**, 76–84 (2012)
10. Shiva Kumar, G., Suresha, S.N.: State of the art review on mix design and mechanical properties of warm mix asphalt. Road Mater. Pavement Des. **20**(7), 1501–1524 (2018)

Sustainable Polyurethane Plasterboard for Construction

Víctor Miguel, Carlos Junco, Sara Gutiérrez, Lourdes Alameda, and Alba Rodrigo

Abstract The introduction of polyurethane waste from the production of cars into gypsum plasterboards in its matrix is studied. This new plasterboard is compared to commercial gypsum plasterboard; therefore, the doses and the uses of both materials are the same. The prefabricated material is entirely characterized under the Standard EN 520:2005+A1 by the following tests: bulk density, maximum breaking load under flexion stress, total water absorption and surface hardness. The results indicate that the use of polyurethane waste makes the plasterboard lighter as the density of the polyurethane is lower than the gypsum one. The water absorption increased when the amount of residue increased. The lower density leads to a higher porosity, what permits a higher absorption of water and much better thermal isolation. It also reduces its mechanical performance while preventing the board from breaking since only small cracks appear. Besides, the elastic properties of the polyurethane make the surface hardness decrease. With respect to mechanical properties, plasterboard is susceptible to the mechanical impact damage. Although the flexural strength of the plaster specimens decrease as the amount of the waste increase, it remains within the minimum reference value required by standard. Non-combustibility test is determinate on the basis of experimental data obtained according to Standard EN 13501-1. Turning waste into a resource is one key to a circular economy, the employ of this polyurethane waste for the fabrication of plasterboards could contribute to maximize the reuse of this kind of waste.

Keywords Recycled polyurethane · Plasterboard · Density · Water absorption · Surface hardness · Fire behavior

V. Miguel (✉) · C. Junco · S. Gutiérrez · L. Alameda · A. Rodrigo
Department of Architectural Construction, Higher Polytechnic School, University of Burgos, 09001 Burgos, Spain
e-mail: vmb0015@alu.ubu.es

V. M. C. F. Cunha et al. (eds.), *Proceedings of the 3rd RILEM Spring Convention and Conference (RSCC 2020)*, RILEM Bookseries 35,
https://doi.org/10.1007/978-3-030-76543-9_25

1 Introduction

According to PlasticsEurope-The facts 2019 [1], the European's plastic demand in 2018 reached 51.2 MT, 2.5% higher than in 2016 [2] from which 7.9% is polyurethane. Polyurethane is a thermostable polymer that changes its chemical composition when heated. Due to this process, the polyurethane cannot be melted and reformed, what leads to a difficult and not profitable recycling process; therefore, this waste is sent to landfills or burnt in order to produce electricity sending out pollutants to the atmosphere. The European Union is promulgating directives as a way to control the environmental impact of burning wastes (Directive 2010/75/EC) [3] and sending them to landfills (Directive 2008/98/EC) [4], this control is through taxes, making both process more expensive and encouraging the search for new recycling and reusing processes.

Several researches have studied the improvement of one or several characteristics of gypsum plasterboards when added different types of polymeric wastes [5, 6]. The addition of these polyurethane wastes into the gypsum matrix has been proven to produce lighter products, what reduces the transportation costs and makes the assembly easier. It also enhances its thermal properties, making it possible to avoid the use of insulation layers or reducing their width [7–9]. There are no previous studies in which polyurethane is mixed with other waste products, what leads to the use of "dirty" wastes, increasing the reuse in polyurethane-producer industries.

The aim of this study is to produce gypsum plasterboards with polyurethane waste from the car industry, creating lighter products with enhanced thermal properties while reusing a residue. This prevents that the waste ends up in landfills or burnt.

This work is a continuation of the results obtained in the LIFE REPOLYUSE Project "Recovery of polyurethane for reuse in eco efficient materials" and it aims to improve the replicability of the project since this new polyurethane waste has different origin and composition in comparison with the polyurethane waste used in the project.

2 Experimental Process

2.1 Raw Materials

Gypsum. The gypsum used is provided by Placo, Saint-Gobain [10], classified as A1 (Gypsum conglomerate) by Standard EN 13279:2009 [11] and as E-35 according to traditional naming.

It is classified as Euroclass A1 (no fire contribution), with a purity value >90% Re (Fig. 1).

Polyurethane Waste PU (AT). The polyurethane waste (PU (AT)) comes from the excess of car ceiling production in the car industry Grupo Antolín, in Burgos.

Fig. 1 Gypsum used in the investigation

The waste appears in panels with the following composition Re (Fig. 2):

- Cardboard with fiberglass.
- Polyurethane foam.
- Cardboard with fiberglass.
- Polyurethane foam.
- Reinforcement plate.
- Unweaved fabric.

Characteristics of the grinded waste:

Bulk and true density: 94.9 kg/m^3 and 1761.6 kg/m^3 respectively.

Table 1 shows the elemental analysis of the waste material.

2a 2b

Fig. 2 Waste before (**a**) and after (**b**) the grinding

Table 1 Elemental analysis (CHNS)

Carbon	Hydrogen	Nitrogen	Sulfur
47.33%	4.20%	3.05%	0.00%

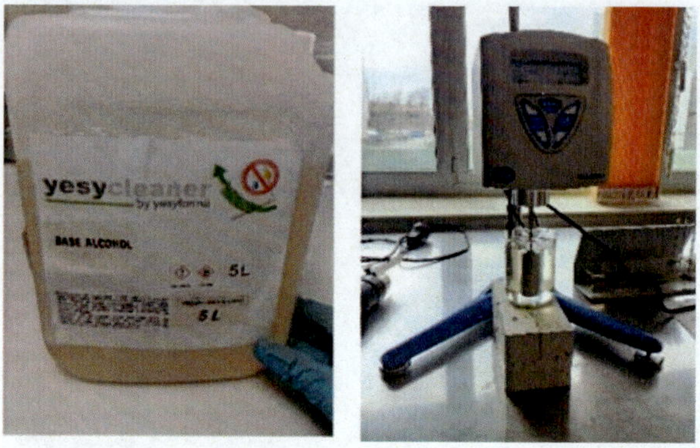

3a **3b**

Fig. 3 Additive used in the investigation (**a**), viscosity test (**b**)

It was not possible to get a powered sample in order to study its grading with the method of X ray diffraction.

Additive. A super-fluidifier, alcohol-based is used. Re (Fig. 3a) In order to study its viscosity, 48.5 mPa s, the Standard EN ISO 2555 [12] is followed Re (Fig. 3b).

2.2 *Justification of the Dosage Used*

The purpose is to use as much residue as possible while enhancing the properties of the final product, therefore, a volume replacement ratio of (1/1.5) of gypsum A1 with PU (AT) is used. Previous studies with different polyurethane wastes [8] have shown that with different dosages (1/3 and 1/4) its mechanical properties decrease and with the dosage (1/2), its fire performance is inadequate. On the contrary, with a lower replacement ratio, the thermal enhancement and the environmental impact are negligible.

The industrialized process requires a liquid mixture of the materials, with a water/gypsum ratio of 0.95 and a super-fluidifier (0.5% regarding the water weight). In order to study the effect on the properties of polyurethane plasterboard in relation to a reference gypsum plasterboard, in which the only variable is the addition of waste, the parameters will be the same to the ones mentioned above.

2.3 *Fabrication Process*

Two types of plasterboard are produced, one with 1 part gypsum and 1.5 parts PU (AT) waste, and super-fluidifier additive (PYL Y1.5AT), and another one as a reference with gypsum A1 and super-fluidifier additive (PYL Y0).

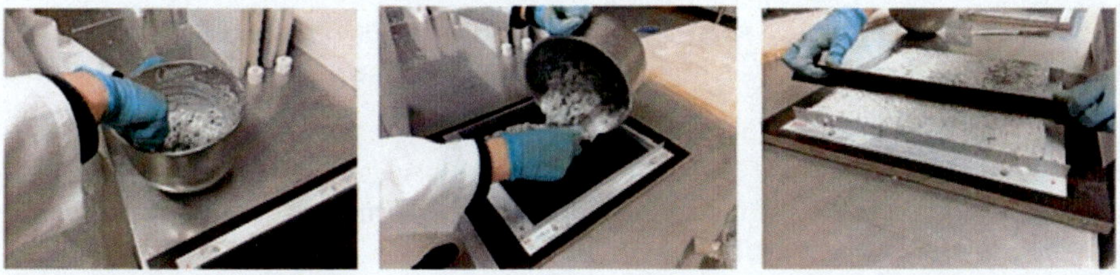

Fig. 4 Manual mix (**a**), mold filling (**b**) and leveled out (**c**)

Process. The same process is employed for both plasterboards. The gypsum is dried in the stove (40 ± 2) °C for 48 h, and then it is homogenized and stored in a moisture-tight container. Later, the gypsum and the PU (AT) residue are weight in the same container and homogenized. A release agent is applied only to the mold.

Afterwards, the water and the additive are weight and mixed, and then the gypsum and the polyurethane are added to the mixture. The mixture is manually mixed during 30 s Re (Fig. 4a) and mechanically mixed during 90 s. Then it is poured into the mold Re (Fig. 4b) and levelled out to get rid of the excess of material Re (Fig. 4c). It hardens for 24 h; it is removed from the mold and introduced into the stove (40 ± 2) °C until a constant mass is obtained.

2.4 *Gypsum Plasterboard Characterization*

The gypsum plasterboards are characterized under the Standard EN 520:2005+A1 [13] by the following tests: bulk density, maximum breaking load under flexion stress, total water absorption and surface hardness. This standard specifies that the tested product should be cut out from a commercial gypsum plasterboard with different sizes according to the test.

For this tests, plasterboards of (400 × 300 × 15) mm are used due to the difficulty of making commercial size ones in the laboratory, and each test is done in 3 different plasterboards.

The Standard EN 12667:2002 [14] is followed for the thermal performance test, in which plasterboards of (300 × 300 × 20) mm are used.

Non-combustibility test is determined under Standard EN 13501-1:2019 [15]. The test tubes must be cylindrical, with a volume of (76 ± 8) cm^3, a diameter of (45^{+0}_{-2}) mm and a height of (50 ± 3) mm.

Bulk Density. It is obtained by weighting the tested product. It is done first while the plasterboards are wet after removing them from the mold and then once they are dry in the stove (40 ± 2 °C) until a constant mass is obtained.

Maximum Breaking Load Under Flexion Stress. It determines the load that the prefabricated product can support before breaking. A force is applied in the center, with a speed of (250 ± 5) N/min. The plasterboard stands on two parallel cylinders with a radius of 10 mm and separated (350 ± 1) mm. The precision has to be 1 N.

Total Water Absorption. The mass increase percentage in relation to the initial one is determined after the plasterboard is immersed in water (23 ± 2) °C for 2 h ± 2 min. Once it is removed from the water, the excess of surface water is cleared, and weight with a precision of 0.1 g.

Surface Hardness. This test determines the superficial mark caused by a steel ball of 50 mm diameter (510 ± 10) g, projected from a free fall height of (500 ± 5) mm. This process is done 3 times per plasterboard. Once it is done, two perpendicular diameter of the mark are measured with a precision of 1 mm.

Thermal Performance. It determines the thermal resistance by means of guarded hot plate and heat flow meter methods, from which the thermal conductivity is obtained. The thermal resistance is calculated with the equation Eq. (1).

$$\text{TR} = \text{Width}\,(m)/\text{Thermal conductivity}\,(\lambda) \tag{1}$$

Non-combustibility. This test allows studying the performance to non-combustibility of the test tubes in specific conditions under the Standard EN ISO 1182.2011 [16]. This performance can be extrapolated to a real situation. The test tubes must be prepared according to the Standard EN 13238:2010 [17].

3 Results

3.1 Bulk Density

The bulk densities of the plasterboards, both wet and dry are shown in Table 2.

The bulk density of the gypsum plasterboards with PU (AT) is lower, this could be due to two factors. There is a substitution of the main material (gypsum) with a new material (PU (AT)) with a lower density, and the new material absorbs more water, which is evaporated during the hardening, leading to a higher porosity [8].

Table 2 Bulk density results

Naming	Results (kg/m^3)	
	Wet	Dry
PYL Y1.5AT	1407.20	884.42
PYL Y0	1508.33	963.86

3.2 Maximum Breaking Load Under Flexion Stress

Table 3 shows the average value of the results of the test.

The results show that the breaking load decreases when the residue is added to the mixture. This can be because of three aspects: PU (AT) is less dense (94.9 kg/m^3) than gypsum (879.6 kg/m^3); the residue absorbs more water that evaporates during the hardening, creating holes, and making the plasterboard lighter but less resistant; by replacing the resilient material, gypsum, by another one that is not resilient, polyurethane, the resistance capacity of the final product decreases.

Figure 5 shows the behavior of the plasterboard during the test.

The plasterboard with PU (AT) has better elastic properties, being able to maintain the fracture load over time even after the breakage and minimizing the risk of collapse. This is due to the fiberglass of the polyurethane waste. As we can see in PYL Y1.5AT plasterboards Re (Fig. 6a), cracks are opened after the breakage, whereas in PYL Y0 plasterboards Re (Fig. 6b), the plate collapses.

Table 3 Maximum breaking load under flexion stress results

Naming	Results (N)
PYL Y1.5AT	240
PYL Y0	387

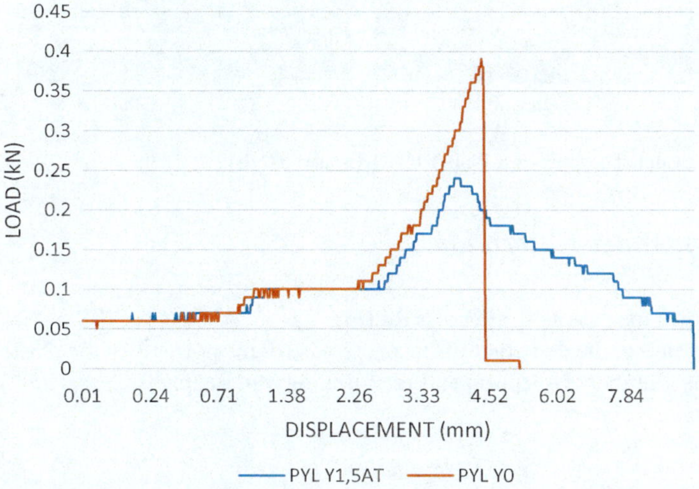

Fig. 5 Breaking load graphic

6a 6b

Fig. 6 Breaking load test PYL Y1.5AT (**a**), PYL Y0 (**b**)

Table 4 Total water absorption results

Naming	Results (%)
PYL Y1.5AT	9.03
Y0	5.25

7a 7b

Fig. 7 Total water absorption test PYL Y1.5AT (**a**) and Y0 (**b**)

3.3 Total Water Absorption

Table 4 shows the results for the test Re (Fig. 7).

The decrease of the density leads to an increase of the porosity of the plasterboards, what means a higher absorption and retention of water capacity.

3.4 Surface Hardness

Table 5 shows the test results.

The results show that by introducing PU (AT) the surface hardness decreases due to the elastic nature of the waste, making the size of the mark bigger. This elastic nature also allows that the plasterboard does not break Re (Fig. 8) contrary to the

Table 5 Surface hardness results

Naming	Results (mm)
PYL Y1.5AT	14.44
PYL Y0	14.28

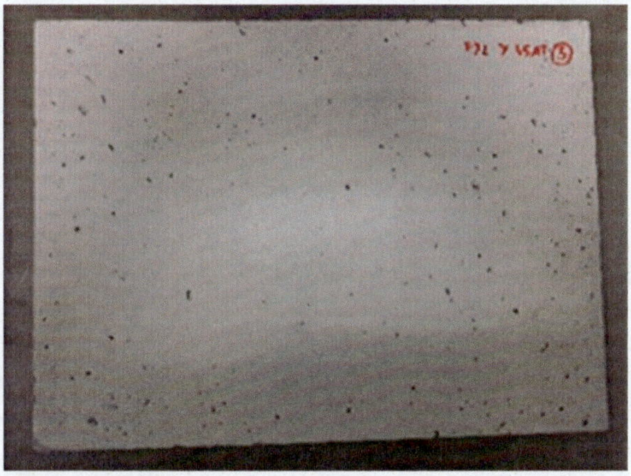

Fig. 8 Surface hardness test PYL Y1.5AT

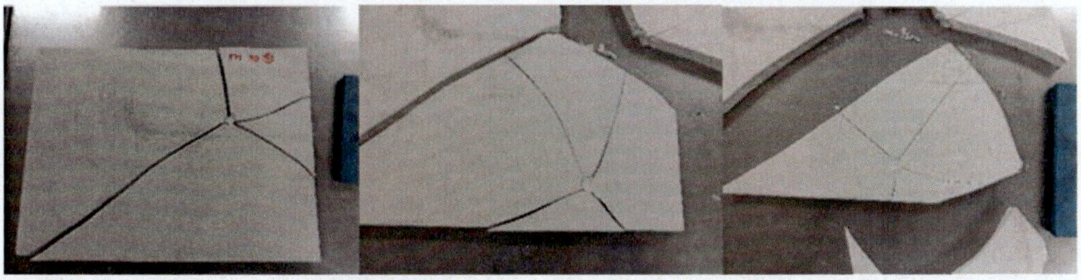

Fig. 9 Surface hardness test PYL Y0

reference plasterboard Re (Fig. 9), allowing the prefabricated material with PU (AT) to resist higher forces without breaking.

3.5 Thermal Performance

Table 6 shows the results of the test.

The test should be done with plasterboards with flat surfaces, what is impossible for gypsum plasterboards with PU (AT), as it cannot be done in the laboratory due to the nature of the material, therefore, the results are not valid. In order to calculate

Table 6 Thermal performance results

Naming	Thermal conductivity (W/m*K)	Thermal resistance (m^2*K/W)
TEST PYL Y1.5AT	0.388	0.0387
EQUATION PYL Y1.5AT	0.151	0.0993
PYL Y0	0.25	0.06

the thermal conductivity, Eq. (2) is used. This equation was determined in a previous research [8]. As it is explained in the research, the equation can be used for mixtures with polyurethane mixed with other residues

$$y = 0.0002x - 0.0256 \quad (2)$$

According to the standard EN ISO 10456:2012 [18], the thermal conductivity of gypsum plasterboards with a bulk density of 900 (kg/m^3) is 0.25 (W/m*K).

The addition of PU (AT) to the matrix produces a decrease of 39.6% on the thermal conductivity relative to the reference material. This is due to two main factors, the thermal properties of the residue and the porosity of the plasterboards. As it was proven in previous studies [7], the values obtained for polyurethane plasterboards are similar to other construction materials used for thermal insulation.

3.6 *Non-combustibility*

The reference test tubes are produced with gypsum A1, 1% fibers and 0.5% super-fluidifier additive. Table 7 shows the results for the test.

The reference test tubes have an Euroclass classification A1, meanwhile, the test tube with PU (AT) would not get, at first, the Euroclass classification A2 due to the duration of sustained flaming, it should be $\leq$20 s according to the standard. In order to be sure that the final product with PU (AT) will not get classified as A2, more test should be done EN ISO 1716:2011 [19] and EN 13823:2012 [20]. According to CTE-DB-SI [21], section 4, for ceiling and wall lining, the Euroclass classification must be B or C.

The decrease in the non-combustibility result is because the polyurethane introduced in the matrix is mixed with combustible materials.

Table 7 Non-combustibility results

Naming	Duration of sustained flaming (s)	Temperature rise of furnace (°C)	Loss of mass (%)
Y1.5AT	433	25.2	30.63
Y0	0	9.5	21.01

4 Conclusions

The bulk density of the gypsum plasterboard with PU (AT) is lower due to two aspects, the real density of the residue is lower than the gypsum one, and the residue absorbs more water during the mixing process, but it evaporates during the hardening, leading to an increase on the porosity.

The breaking load decreases, and even the plasterboard does not break, cracks are opened. This behavior is due to the elastic properties of the residue and the fiberglass.

The increase of the porosity produced by the lower density leads to a higher absorption of water (3.78% higher).

As it happens in the breaking load, the surface hardness decreases when the residue is added to the matrix, but the plasterboard does not break because of the elastic nature of the PU (AT).

Thermal resistance is higher in gypsum plasterboards with PU (AT) as a result of the insulating properties of the residue and the porosity of the plasterboard.

The addition of PU (AT) does not allow the final product to obtain an Euroclass classification A2 at first, but the results show that it is likely that it could obtain an Euroclass classification A2 or B, with which it will fulfill the requirements of the standards for its use in interior lining.

This new gypsum plasterboards with PU (AT) are more sustainable as they require less natural resources and prevent the polyurethane waste from ending up in landfills or burnt, reducing the CO_2 emissions, what presents an alternative to the development of new construction materials.

Acknowledgements This research was funded by LIFE PROGRAMME. EUROPEAN COMMISSION, grant number "LIFE 16 ENV/ES/000254". This study was carried out within the framework of the LIFE-REPOLYUSE Recovery of polyurethane for reuse in eco-efficient materials. LIFE 16 ENV/ES/000254 Project. LIFE 2016. Environment Life Programme. European Commission.

Conflicts of Interest The authors declare no conflict of interest. The funders had no role in the design of the study; in the collection, analyses, or interpretation of data; in the writing of the manuscript, or in the decision to publish the results.

References

1. Plastics—the facts 2019. An analysis of European plastics production, demand and waste data. Available online: https://www.plasticseurope.org/en/resources/publications/1804-plastics-facts-2019
2. Gómez-Rojo, R., Alameda, L., Rodríguez, A., Calderón, V., Gutiérrez-González, S.: Characterization of polyurethane foam waste for reuse in eco-efficient building materials. Polymers (2019). Available online: https://www.mdpi.com/2073-4360/11/2/359
3. Directive 2010/75/EC of the European Parliament and of the Council of 24 November 2010 on industrial emissions (integrated pollution prevention and control). Available online: https://eur-lex.europa.eu/legal-content/en/TXT/?uri=CELEX:32010L0075. Accessed on 20 Dec 2019

4. Directive 2008/98/EC of the European Parliament and of the Council of 19 November 2008 on Waste. Available online: https://eur-lex.europa.eu/legal-content/EN/TXT/?uri=CELEX:02008L0098-20180705. Accessed 20 Dec 2019
5. Melo, M.O.B.C., Da. Silva, L.B., Coutinho, A.S., Sousa, V., Perazzo, N.: Energy efficiency in building installations using thermal insulating materials in northeast Brazil. Energy Build. **47**, 35–43 (2012). https://doi.org/10.1016/j.enbuild.2011.11.021
6. Madariaga, F.J.G., Macia, J.L.: Mezclas de residuos de poliestireno expandido (EPS) conglomerados con yeso o escayola para su uso en la construcción. http://informesdelaconstruccion.revistas.csic.es/index.php/informesdelaconstruccion/article/view/589/671
7. Gutiérrez-González, S., Gadea, J., Rodríguez, A., Junco, C., Calderón, V.: Lightweight plaster materials with enhanced thermal properties made with polyurethane foam wastes. Constr. Build. Mater. **20**(28), 653–658. https://www.sciencedirect.com/science/article/pii/S0950061811006040?via%3Dihub
8. Alameda Cuenca-Romero, L.: Tesis Doctoral: Placas de yeso laminado aligeradas con residuos poliméricos reforzadas con fibras de polipropileno. Fabricación y puesta en obra. Universidad de Burgos (2014)
9. Alameda, L., Calderón, V., Junco, C., Rodríguez, A., Gadea, J., Gutiérrez-González, S.: Characterization of gypsum plasterboard with polyurethane foam waste reinforced with polypropylene fibers. Mater. Constr. **66**(324), e100 (2016). https://doi.org/10.3989/mc.2016.06015
10. Data sheet gypsum Iberyola. Placo Saint-Gobain. https://www.placo.es/products/escayola/escayola-iberyolar
11. EN 13279-1:2009. Gypsum binders and gypsum plasters—part 1: definitions and requirements. Available online: https://www.une.org/encuentra-tu-norma/busca-tu-norma/norma?c=N0043416. Accessed on 03 Dec 2019
12. EN ISO 2555:2018 Plastics—resins in the liquid state or as emulsions or dispersions—determination of apparent viscosity using a single cylinder type rotational viscometer method. Available online: https://www.une.org/encuentra-tu-norma/busca-tu-norma/norma?c=N0061259. Accessed on 14 Jan 2020
13. EN 520:2005+A1:2010: Gypsum plasterboards—definitions, requirements and test methods. Available online: https://www.une.org/encuentra-tu-norma/busca-tu-norma/norma?c=N0044718. Accessed on 03 Dec 2019
14. EN 12667:2002: Thermal performance of building materials and products. Determination of thermal resistance by means of guarded hot plate and heated flow meter methods. Products of high and medium thermal resistance. Available online: https://www.une.org/encuentra-tu-norma/busca-tu-norma/norma/?c=N0027459. Accessed 20 Dec 2019
15. EN 13501-1:2019: Fire classification of construction products and building elements—part 1: classification using data from reaction to fire tests. Available online: https://www.une.org/encuentra-tu-norma/busca-tu-norma/norma/?c=N0062154. Accessed on 09 Jan 2020
16. EN ISO 1182.2011: Reaction to fire tests for products—non-combustibility test. Available online: https://www.une.org/encuentra-tu-norma/busca-tu-norma/norma/?c=N0047993. Accessed on 10 Jan 2020
17. EN 13238:2010: Reaction to fire tests for building products—conditioning procedures and general rules for selection of substrates. Available online: https://www.une.org/encuentra-tu-norma/busca-tu-norma/norma/?c=N0047479. Accessed on 10 Jan 2020
18. EN ISO 10456:2012: Building materials and products—hygrothermal properties—tabulated design values and procedures for determining declares and design thermal values. Available online: https://www.une.org/encuentra-tu-norma/busca-tu-norma/norma?c=N0049362. Accessed on 21 Jan 2020
19. EN ISO 1716:2011: Reaction to fire tests for products—determination of the gross heat of combustion (calorific value). Available online: https://www.une.org/encuentra-tu-norma/busca-tu-norma/norma/?c=N0047995. Accessed on 10 Jan 2020
20. EN 13823:2012+A1:2016: Reaction to fire tests for building products—building products excluding floorings exposed to the thermal attack by a single burning item. Available online: https://www.une.org/encuentra-tu-norma/busca-tu-norma/norma/?c=N0056657. Accessed on 10 Jan 2020

21. CTE-DB-SI Código Técnico de la Edificación: Documento básico. Seguridad en caso de incendio. Available online: https://www.codigotecnico.org/images/stories/pdf/seguridadIncendio/DBSI.pdf. Accessed on 10 Jan 2020

Treated Municipal Solid Waste (Biomass) Based Concrete Properties—Part II: Experimental Program

Massoud Sofi, Lino Maia, Junli Liu, Ylias Sabri, Annie Zhou, Tawab Frahmand, and Priyan Mendis

Abstract Municipal Solid Waste (MSW) management is a worldwide problem growing proportionally the earth human population. The practice of incinerating garbage has ceased in some parts of the world because of air contamination and other public health issues. Environmental impact of landfilling is ever increasing. There is clearly a need to adopt cost-effective alternatives to treat MSW. This paper is a part of a major work that considers MSW based biomass as a partial replacement of sand in concrete. The product is an exciting and eco-friendly alternative for the building industry. Here, in this paper, compressive and flexural tests are conducted on samples containing 5, 10 and 15% replacement of sand by biomass. Results are presented and discussed in a view to include biomass in concrete intended for certain type applications in the construction industry such as temporary works.

Keywords Municipal solid waste · Biomass · Concrete · Mechanical properties

1 Research Objectives

The findings reported in the present paper belongs to a major works reported in two Parts. The Part I is an overview of the state of the art and the literature review of the topic [1]. The Part II is reported in this paper and is related to the experimental

M. Sofi (✉) · J. Liu · A. Zhou · P. Mendis
Department of Infrastructure Engineering, University of Melbourne, Melbourne, Australia
e-mail: massoud@unimelb.edu.au

L. Maia
CONSTRUCT-LABEST, Faculty of Engineering (FEUP), University of Porto, Porto, Portugal

Faculty of Exact Sciences and Engineering, University of Madeira, Campus Universitário da Penteada, 9020-105 Funchal, Portugal

Y. Sabri
School of Science, Centre for Advanced Materials and Industrial Chemistry, College of Science, Engineering and Health, RMIT University, Melbourne, VIC 3001, Australia

T. Frahmand
Bioelektra Australia Pty Ltd, Melbourne, Australia

V. M. C. F. Cunha et al. (eds.), *Proceedings of the 3rd RILEM Spring Convention and Conference (RSCC 2020)*, RILEM Bookseries 35,
https://doi.org/10.1007/978-3-030-76543-9_26

program and the corresponding findings. The research aims at reducing the amount of municipal waste produced by human activities goes into landfill. Using treated municipal waste by ART technology in the construction industry, specifically in concrete production, has been proposed. The optimum mix design by replacing parts of the material in the traditional concrete (cement, aggregates and water) will be investigated in this research. Besides, the mechanical properties such as compressive strength and flexural strength of the biomass concrete will also be tested in this research.

2 Experimental Programs

2.1 Methodology

After a comprehensive literature review, there is only limited literature that discussed the usage of treated municipal solid waste (TMSW) or organic biomass (OB) in the concrete that was produced by the new ART technology. The report has only addressed the microstructure, chemical component and limited discussion of the mechanical properties of the biomass concrete. Therefore, several laboratory tests were decided to determine the mechanical properties (i.e. flexural strength test and compressive strength test) of the biomass concrete and discover the optimum mixture of the concrete. Besides, a few pieces of research share similar research objectives but put emphasis on different types of biomass ash concrete; thus, the results of the experiment and the literature review are expected to be similar.

According to the literature review that has been conducted, most of the research focuses on the replacement of cement by biomass ashes. However, the new material is different from the material that has been studied previously. It has distinct characteristics and is not appropriate to replace cement in this case. Therefore, the partial proportion of sand will be replaced by the biomass ashes (TMSW) proposed in this topic.

The experiment will be testing three different mixture of concrete, namely 5, 10 and 15% replacement of sand to TMSW; whereas, the studies for 20% and higher replacement rate of concrete can be further studied in the future. The previous research shows the high-water absorbent characteristics of the organic biomass, a constant high water-to-binder ratio was used for all experiments. Additionally, several studies suggest that biomass concrete has low early strength. Therefore, hydrated lime should be added into the mixture to promote early strength in biomass concrete [2].

2.2 Materials

The materials to be used in the experiment are Ordinary Portland Cement (OPC), dry river sand, 7 and 14 mm coarse aggregate, water, TMSW (provided by Bioelektra, residue from the ART technology), and hydrated lime (Table 1).

2.3 Procedures

Specimens Preparation. There are three steps in preparing the test specimens, which are mixing, cast and curing.

Mixing. The concrete samples were mixed according to the concrete mix prepared before mixing. During the experiment, the measurement of the material was recorded to the accuracy of 0.2% for cement, water, sand, coarse aggregate, biomass ash and hydrated lime. A small amount of water was added into 7 and 14 mm coarse aggregate, whereas water of 1% weight of the sand was added into the sand. This helps promote the reactivity between the cement paste and the binder.

All the measurement was recorded and report according to AS 1012.2. Below are the mixing procedure after weighing all materials:

(a) Put sand and coarse aggregates into the mixer and had them mixed for 3 min.
(b) Added cement and hydrated lime into the mixer and continued mixing for 3 min.
(c) Supplemented water into the mixture and kept mixing for 2 min.
(d) Supplemented biomass ashes into the mixture and kept mixing for 2 min additionally.

Casting. The shape and diameter for each test specimen are specified in relevant standards. For compressive strength, the test specimen shall satisfy the following requirement: (1) the diameter of the test specimen is 100 mm in diameter; (2) the shape of the test specimen is a right cylinder, and the height is twice the diameter of the test specimen, which in this case, 200 mm (AS 1012.8.1). Also, the test specimens were prepared according to AS 1012.8.1. Whereas, for the flexure test specimen, it is a rectangular beam with a cross-section of 100 × 100 mm and height of 350 mm. The test specimens were prepared according to AS 1012.8.2. All test specimens were compacted using vibration for removing air bubbles. For each type of mixture undergoing different curing duration, three specimens were prepared for both compressive and flexural strength test.

Curing. Finally, all test specimens were stored in room temperature for air curing due to the characteristics of municipal waste. The biomass in the concrete is water absorbing, therefore water curing could be inappropriate as the mechanical properties of the biomass concrete would be heavily influenced.

Table 1 Mix proportions of the concrete specimens

No	Concrete ID	*W*/*B*	Cement	Hydrated lime	TMSW	Dry river sand	Water	Agg. (14 mm)	Agg. (7 mm)
			(kg/m^3)						
1	BM5	0.48	274	140	34.5	655.5	199	1133	16
2	BM10	0.48	274	140	69	621	199	1133	16
3	BM15	0.48	274	140	103.5	586.5	199	1133	16

Table 2 The dimension of test specimens as per Australian Standards

Test	Cross-section	Height/length	Shape	Standards
Compressive strength test	100	200	Cylinder	AS 1012.8.1
Flexure strength test	100 × 100	350	Cuboid	AS 1012.8.2

Testing

Compressive Strength Test. The test specimens for compressive strength test were casted according to the dimension given in Table 2. Since the expected concrete strength is low, ranging from 10 to 80 MPa, the restrained natural rubber capping system should be adopted. The preparation before testing followed clause 6.3.2 AS 1012.9 and the procedures described in clause 8 AS 1012.9 were applied to the experiment. The ultimate compressive strength and the failure mode of the concrete were recorded.

Flexural Strength Test. The test specimens for flexure test were prepared with the dimension mentioned in Table 2. The procedures of conducting the flexure test were specified in clause 6 AS 1012.11. Set-up of the flexural strength test is shown in Fig. 1. Equation (1) defines modulus of rupture that shall be recorded in the final report:

$$f_{cf} = PL(1000)/BD^2 \quad (1)$$

where, f_{cf} = modulus of rupture (MPa)

P = maximum applied force indicated by the testing machine (kN)

L = span length (mm)

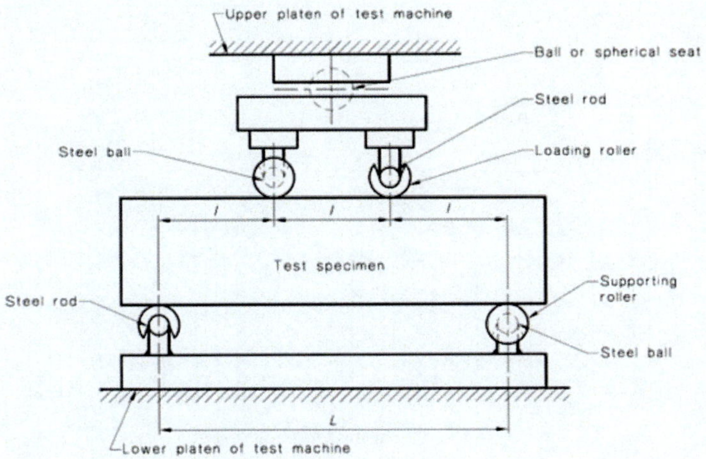

Fig. 1 Setup of the flexural strength test (AS 1012.11)

B = average width of the specimen at the section of failure (mm)

D = average depth of specimen at the section of failure (mm).

3 Results and Analysis

Both compressive and flexural strength of specimens were investigated and recorded. All results obtained from the tests are the average result of three identical concrete mix design at the same curing days.

3.1 Compressive Strength

The compressive strength of BM10 is higher than BM5 and BM15 in 3-day, 7-day, 14-day and 28-day curing, as presented in Table 3 and Figs. 2, 3 and 4. This is perhaps because BM5 has too much void in the concrete as the water escaped during mixing and curing stage, leading to partial cement paste not effectively reacting with water. Other reason could be there is not enough municipal waste (aggregate) for the mix design, where it escaped during the mixing stage.

Table 3 Average compression strength of BM5, BM10 and BM15

Concrete mix	Average compression strength (MPa)			
	3-day	7-day	14-day	28-day
BM5	12.88	14.42	16.36	19.16
BM10	19.22	22.00	24.79	29.60
BM15	17.38	21.14	22.78	27.70

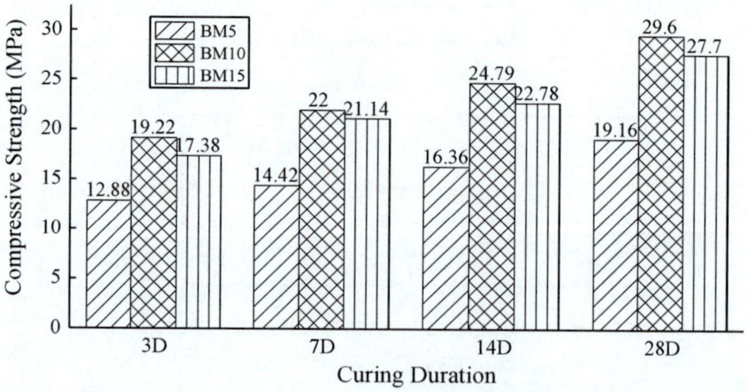

Fig. 2 Compression strength results based on curing duration

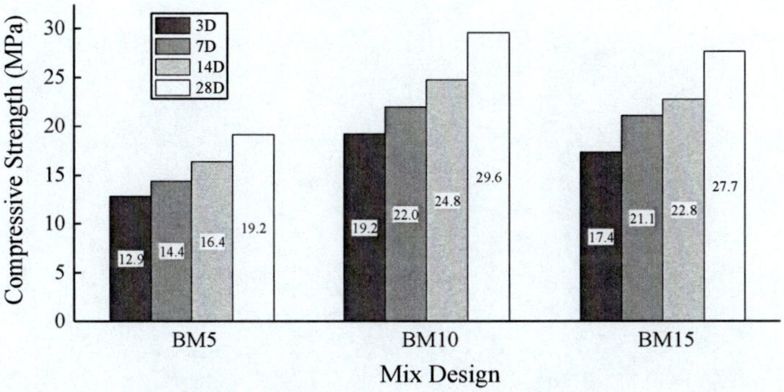

Fig. 3 Compression strength results based on mix design

Fig. 4 Compression strength development of BM5, BM10 and BM15

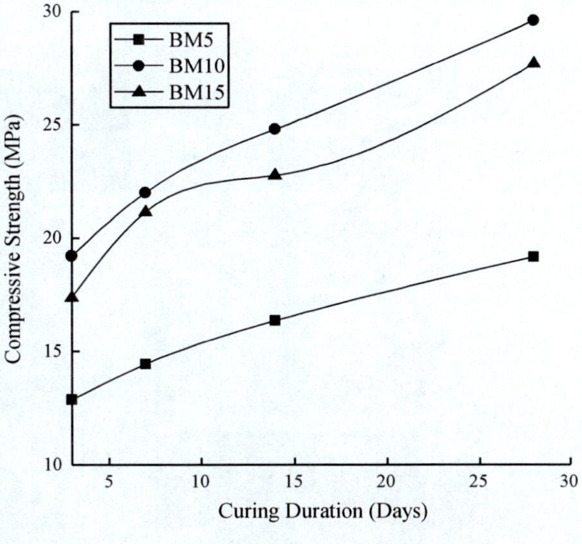

On the other side, there is excessive amount of municipal waste in BM15, which lumps together and does not mix well with the cement paste. Consequently, the compressive load could not be transferred evenly to the whole concrete. This is caused by the lower compressive strength in the municipal waste lumps, which is shown in Fig. 6d. The reaction between water and cement paste in stages increases the compressive strength of the biomass concrete proportionately, which is shown in Figs. 3, 4. Figures 5, 6 and 7 shows the observed failure modes.

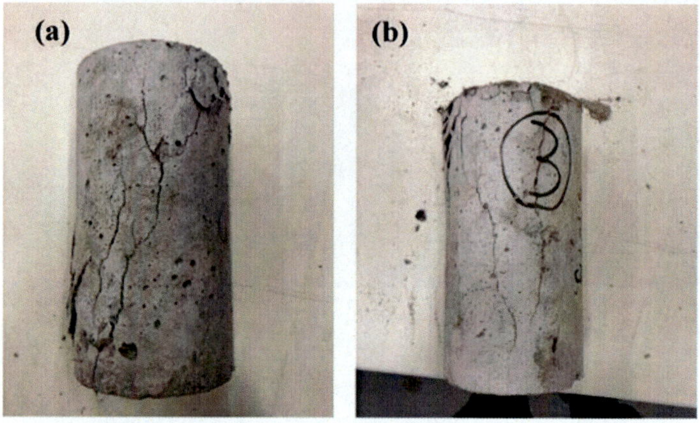

Fig. 5 **a** Shearing failure mode; **b** splitting failure mode

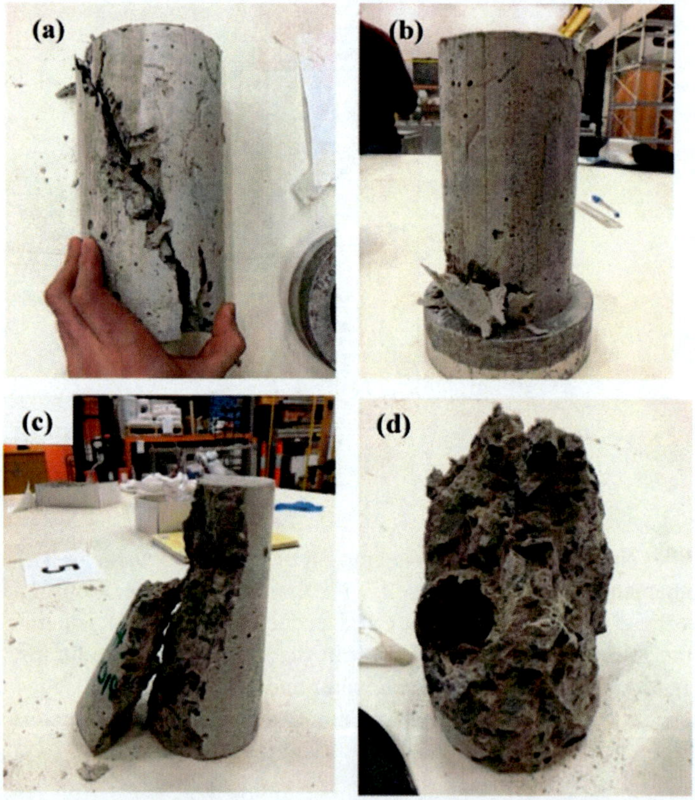

Fig. 6 **a** Shearing failure at 45° angle, **b** splitting failure at the bottom edge of the biomass concrete, **c** the mixture of shearing and splitting failure, and **d** municipal waste lump in biomass concrete

Fig. 7 **a** Compression test and **b** surface of the BM5 after compression test

3.2 *Flexural Strength*

The flexural strength of BM10 is slightly higher than BM5, and BM15 is 3-days testing; however, they converge into almost same flexural strength towards 7-day, 14-day and 28-day testing, which is shown in Figs. 8 and 9. From the experimental results, it indicates that BM15 is slightly higher than BM5 and BM10 by 0.3 and 0.1 MPa, but overall, they have similar flexural strength at 28-days curing. Figure 10a shows the failing pattern and the overall setup of the flexural test.

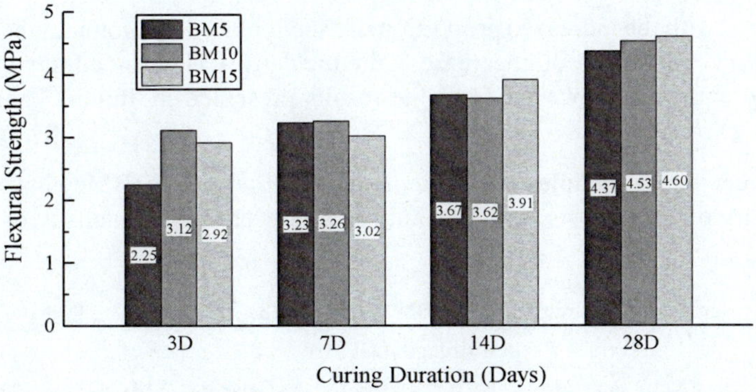

Fig. 8 Flexural strength test results based on curing duration

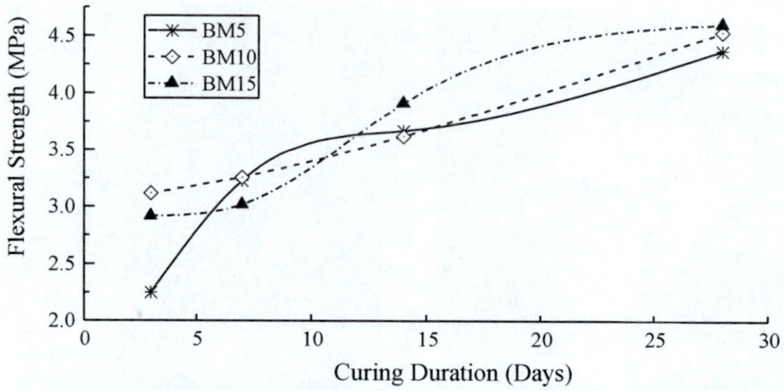

Fig. 9 Flexural strength development of BM5, BM10 and BM15

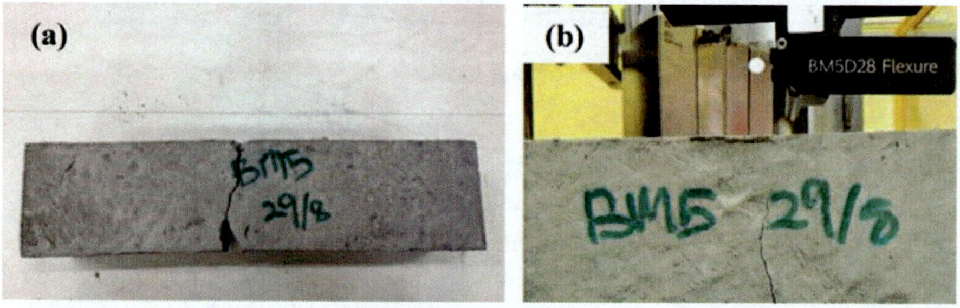

Fig. 10 **a** Flexural strength test failing pattern, and **b** flexural test for BM5 at day 28

4 Discussions

It is expected with the increased proportion of sand replaced by biomass, the strength and the cured concrete would decrease while the density is designed to be identical to ordinary concrete. However, from the results presented in Tables 3 and 4, it is found that:

(1) The density of samples with a replacement rate of 5% (BM5) have a mean density of 2100 kg/m^3 which is noticeably lower than the density of samples

Table 4 Average flexural strength of BM5, BM10 and BM15

Concrete sample	Average flexural strength (MPa)			
	3-days	7-days	14-days	28-days
BM5	2.25	3.23	3.67	4.37
BM10	3.12	3.26	3.62	4.53
BM15	2.92	3.02	3.91	4.60

with replacement rates of 10% (BM10) and 15% (BM15), having a mean density of 2324 kg/m^3 and 2312 kg/m^3 respectively.

(2) The standard deviation of density ascends with an increased rate of replacement.

(3) The compressive strength of samples with 10% sand replacement outperforms the compressive strength of samples with 5 and 10% replacement.

(4) Samples with 15% biomass replacement have more flexural strength comparing to other samples.

(5) The development of both compressive and flexural strength shows a linear pattern after three days, which indicates the strength may subject to further development.

4.1 Characteristics of Biomass

Biomass is a mix of organic and inorganic waste with inconsistent chemical and physical properties. Therefore, the higher replacement level could lead to increasing variation in the concrete strength, as reflected by the standard deviation of the strengths. It also absorbs water which reduces workability and hydration activity resulting in lower strengths. Additionally, due to its fibre-like structure, the flexural strength increases with an addition of biomass, which can still be observed after concrete hardening (Fig. 10b). During the bending testing, the fibre-like components in the biomass can bridge the cracks. In other words, the presence of fiber-like components prevents opening of the microcrack elsewhere in the matrix [3]. However, the water-absorbing feature of biomass hinders the effective reaction to take place which bonds particles together.

4.2 Lower Density and Strength for BM5

As biomass is added to concrete to replace sand, the density of the reference concrete is reduced. The biomass absorbs the water in the mortar before it has time to hydrate. Insufficient water present to participate in the chemical reaction with cement causes lower strengths of BM5. BM10 and BM15. The higher the biomass content, the lower the strength.

4.3 Comparison Between BM10 and BM15

It is found that the compression strength of BM10 is slightly higher than the compression strength of BM15 at different stages. In terms of flexural strength, BM15 outperforms BM10 by a small amount, but the difference can be ignored at the 28-day stage. Although the theoretical optimum strength will be closer to 10% replacement rate, more usage of biomass is recommended. Such low strength concrete can be used in non-structural members or members subject to minimal load; thus, the aim is to increase the usage of biomass instead of achieving higher strength. A 5% more replacement rate can mean a significant amount in mass production of biomass concrete as an effective way of waste treatment.

4.4 Recommendation on Future Research

It is recommended to perform testing with 10%, 15% or even higher replacement rate of biomass and seek an area of application. Modulus of elasticity, splitting tensile strength are also anticipated to be investigated for different levels of replacement. Additionally, it is also worth attempting wet curing to avoid water escape and maximize hydration reactions. Finally, it is recommended to test the strengths in longer terms, e.g. 40, 60, 120 days etc. As compared to sand, biomass tend to retain more moisture immediately after mixing; the retained water may participate in the reaction over a longer period. The extended development is reflected by the linear shape of the plot and shows a trend to develop for a more extended period continuously. Thus, it is worth testing the strengths of samples cured for a longer period.

5 Conclusion

Followings are the conclusion drawn based on the results obtained from the research:

(1) The compressive strength of the biomass concrete at 10% replacement of sand is higher than 5 and 15% replacement rate.
(2) The distribution of the load throughout the concrete is not even based on the failure mode observed from the compression test.
(3) The flexural strength of the biomass concrete with 15% replacing level is slightly higher than that with 5 and 10% sand replacement.

The conclusion of the studies indicates the feasibility of mixing municipal waste into concrete mix design which will help in reducing waste production and prevent more waste goes into landfill.

The results presented indicate that although the strength reduced with increase of biomass contents, flexural strength remains in the same range as the reference mix. More experimental work needs to be conducted to increase the confidence level in the results.

Acknowledgements This research is financially supported by the Melbourne Research Scholarship offered by the University of Melbourne and the Australian Research Council's Discovery Early Career Researcher Grant (DE170100165, DE 2017 R1). This work is financially supported by: Base Funding—UIDB/04708/2020 of the CONSTRUCT—Instituto de I&D em Estruturas e Construes—funded by national funds through the FCT/MCTES (PIDDAC). This work is funded by national funds through FCT—Fundao para a Cincia e a Tecnologia, I.P., under the Scientific Employment Stimulus—Institutional Call—CEECINST/00049/2018.

References

1. Sofi, M., Maia, L., Liu, J., Sabri, ., Zhou, A., Frahmand, T., Mendis, P.: Treated municipal solid waste (biomass) based concrete properties—part I: state of the art. Springer in RILEM Bookseries of the 3rd RILEM SPRING Convention (2020)
2. George, S., Sofi, A.J.M.T.P.: Enhancement of fly ash concrete by hydrated lime and steel fibres. Mater. Today **4**(9), 9807–9811 (2017)
3. Soltan, D.G., Li, V.C.: A self-reinforced cementitious composite for building-scale 3D printing. Cem. Concr. Compos. **90**, 1–13 (2018)

Treated Municipal Solid Waste (Biomass) Based Concrete Properties—Part I: State of the Art

Massoud Sofi, Lino Maia, Junli Liu, Ylias Sabri, Zhiyuan Zhou, Tawab Frahmand, and Priyan Mendis

Abstract Municipal Solid Waste (MSW) management is a worldwide problem growing with the increase of global human population. The practice of incinerating garbage has ceased in some parts of the world because of air contamination and other public health issues. Environmental impact of landfilling is ever increasing. There is clearly a need to adopt cost-effective alternatives to treat MSW. This paper is a part of a major work that considers MSW based biomass as a partial replacement of sand in concrete. The product of the global work is an exciting and eco-friendly alternative for the building industry, especially concrete intended for certain types of applications in the construction industry such as temporary works. Here, in this paper, an overview of the state of the art on the topic is presented.

Keywords Municipal solid waste · Biomass · Concrete · Mechanical properties

M. Sofi (✉) · J. Liu · Z. Zhou · P. Mendis
Department of Infrastructure Engineering, University of Melbourne, Melbourne, Australia
e-mail: massoud@unimelb.edu.au

L. Maia
CONSTRUCT-LABEST, Faculty of Engineering (FEUP), University of Porto, Porto, Portugal

Faculty of Exact Sciences and Engineering, University of Madeira, Campus Universitário da Penteada, 9020-105 Funchal, Portugal

Y. Sabri
School of Science, Centre for Advanced Materials and Industrial Chemistry, College of Science, Engineering and Health, RMIT University, Melbourne, VIC 3001, Australia

T. Frahmand
Bioelektra Australia Pty Ltd, Melbourne, Australia

V. M. C. F. Cunha et al. (eds.), *Proceedings of the 3rd RILEM Spring Convention and Conference (RSCC 2020)*, RILEM Bookseries 35,
https://doi.org/10.1007/978-3-030-76543-9_27

1 Introduction

1.1 Technology of Waste Treatment

Waste generation is an inevitable by-product of human activities. On a global scale, approximately 3.5 million tonnes of municipal solid waste (MSW) are being produced daily [1]. In the year of 2010–2011, Australia alone has a waste production of 2.2 tonnes per capita. Although 60% of this generated waste was either recycled or recovered, there is still 40% of the remaining sent for landfill disposal. Over time, the degradation of the waste resulting in the landfills to emit leachate and gases to the environment and waterways. It is estimated that landfills have contributed approximately 20% of the global greenhouse gases (methane and carbon dioxide) emissions and leachings of toxic chemicals, such as mercury, arsenic, beryllium, boron, cadmium, lead, thallium and hydrocarbon compounds, into the environment [1], causing harmful threats to human health, environment and ecological system. Furthermore, landfills cause nuisance, such as flies, odours, smoke and noise, while the increase in vermin surrounding landfills becomes an issue with other adverse health effects, such as congenital disabilities, respiratory illnesses and even cancer [2].

However, as the technology for waste management improves, there are many methods to treat MSW effectively and environmentally friendly. For example, Bioelektra Group has invented the Advanced Recycling Technology (ART) of MSW for waste treatment. It operates unsorted MSW using the innovative RotoSTERIL mechanical heat treatment technology. This revolutionary method is based on a new incredibly useful technological process which uses the autoclave roto sterile BCG 7000, a machine that has been designed to treat unsorted MSW in the sterilization process.

The waste is physically and chemically processed. The waste is also sterilized to eliminate odours. Besides, the MSW collected from the household will be transported to the 'reception hall', which is the only place that the workers contact the MSW. This design of the waste management facility has significantly reduced the growth of bacteria and pests that brings harmful threats to the human and surrounding environment. Materials such as metals, plastic and glasses are separated during different stages of separation using advanced technology (i.e. magnetic separator, laser detection optical sorter etc.) leaving 30% of the entire processed waste as biodegradable fraction also called biomass.

Finally, all the separated fractions are collected and transferred to be recycled recovered and reused, leaving no waste to be landfill, promoting the circular economy and sustainable development [3]. Bioelektra Group has a pilot plant in Rozanski in west Poland which has been operational for nine months, and it has proven the reliability and practicality of the ART technology on recycling MSW.

1.2 Pollution by Construction Material Production and Green Alternatives

The destruction of the environment always tags along with the development of the city and technology. These human activities have resulted in many major natural severe disasters such as wildfires, tsunami, flooding and drought event, depletion of the ozone layer, rising of sea level, global warming etc.

The construction and mining industries have contributed a significant proportion to these events. The activity of producing and transporting construction material has emitted a high amount of carbon dioxide (CO_2). Amongst all the construction material, the highest CO_2 producing ingredient is cement [4]. Cement production in China was 1.39 billion metric tons in 2008 [5], accounting for 50% of the world's output [6]. Large quantities of air pollutants are emitted from cement production, including SO_2, NO_X, CO, and PM, and therefore cement industry has been identified as a primary source of pollution in China. Cement production is expected to reach 3.5 billion metric tons by the end of 2050 if its consumption remains constant globally [7].

The fundamental utilization of Ordinary Portland Cement (OPC) is to make concrete and mortar. The reason why cement is popular and widely used over other construction material is due to its ability to hold the structure together and is relatively cheap compared to other construction material such as steel and timber. To promote sustainable and green construction in the industry, the behaviour of including agricultural waste and industrial waste into concrete or mortar have been studied and implemented in the field. There are several benefits associated with it, which provides for enhanced mechanical properties, energy-efficient and cost-effective.

1.3 Research Objectives

This research aims at reducing the amount of municipal waste produced by human activities goes into landfill. Using treated municipal waste by ART technology in the construction industry, specifically in concrete production, has been questioned. The work is reported in two Parts. Part I is the present paper where an overview of the state of the art on the topic is presented. The Part II is related to the experimental program and the corresponding findings.

2 State of the Art

2.1 *Utilization of Organic Biomass Ashes in Australian Construction Industry*

The use of organic and inorganic wastes has always been a featured topic and has attracted more attention in recent years [8]. One of the major drivers is rapid population growth and urbanisation, which result in ascending waste treatment demand [9]. Currently, the most commonly used methods for waste treatment are landfill and waste-to-energy incineration, which has been a long-term environmental concern [10]. Common issues with landfill include contamination of land and underground water [11] due to leaching and the high cost of maintenance associated with it [12], which makes it a non-economic and risky way of waste treatment.

Although waste-to-energy (WTE) is widely adopted to transform waste into energy [13], Connett [14] argued that it is not a suitable way of treating waste, considering sustainability as the key issue in the twenty-first century. Besides, air pollution due to combustion [15] and its low efficiency [16] prompt new ways of waste treatment need to be investigated.

Another driver is that cement production is one of the significant sources of greenhouse gas emission including sulfur dioxide, nitrogen oxides, carbon monoxide, particle matters and carbon dioxide [7, 17] and it generates 5–7% of all anthropogenic carbon dioxide. Furthermore, there are potential hazards affecting workers' health as it is being handled during transportation and occupation [18].

Moreover, China's new policy of tightening controls on importing foreign waste has pressurised first-world countries like Australia relying on exporting for waste treatment. Additionally, it is suggested some other major importers like Malaysia and Thailand have also announced a tightening of their waste importing policies while others were taking actions forcing Australia to look for new ways to treat or export approximately 1.29 million tonnes of waste [19]. Despite the political issues associated with these changes, it is crucial for the world, especially for countries like Australia, relying on exporting for waste treatment.

2.2 *Organic Biomass Ash from ART*

The material in this research will be focusing on is MSW treated by Vortex-oscillation technology. MSW quite differ from agro-waste and is widely available from every household. Agro-waste consists of animal waste (e.g. manure, animal carcasses), crop waste (e.g. cornstalk, sugarcane bagasse, pruning), food processing waste and hazardous and toxic waste (e.g. pesticides, herbicides, insecticides) [20]. In contrast, MSW comprises of paper, plastic, cardboard, metals, glass, rags, kitchen waste, food waste, rubber, ash and fine etc.

The treatment of MSW through Vortex-oscillation technology involves sequential step and involves sterilization of the waste material. Later organic part is then separated from the inorganic component and further treated as aforementioned. The final product ready for mixing is low-weight, not reactive and having a fibre-like microstructure [8]. This method minimises treatment cost and environmental impact while also having a positive social impression. According to Sofi et al. [8], the sustainability to use treated MSW (TMSW) produced by the ART technology as partial replacement of construction material is expected and demonstrated in the future. According to Bioelektra, the TMSW created is sterile and free of contaminants [3]. Besides, it has high energetic value, low humidity and other physical characteristics which make it an ideal alternative replacement of the construction material for green energy production.

3 Previously Investigated Materials

The increasing need for creating eco-friendly construction materials as sustainable solutions to both waste treatment and building material production inspires professionals with diverse backgrounds. A wide range of materials has been studied by scholars from different parts of the world. Among these studies and researches, agro-waste like rice husk [9, 21] and its ash [18], bagasse [21], oil palm shell [22], corn [10], oyster shell [9], ground coffee waste [23], coconut shell [24] and plant ashes [25] and industrial waste like wood ash [11], treated waste foundry sand [12], wood fibre [26], biomass boiler ash and green liquor dregs from paper from waste paper industry [27] are investigated. Fibres material is mostly used to replace parts of the aggregates in the concrete while most of the ashes were used to replace cement in the mixture.

Even though the wide range of waste materials have been studied, the use of municipal waste in the fabrication of concrete has yet to be further investigated. It has been partially explored by Syarif et al. [21] but only as a component of the "recycling organic waste" which is treated by burning to a temperature of 1450 °C before it is ready for mixing as part of concrete.

3.1 Characteristics of Biomass Concrete

Mechanical Properties

Organic waste materials, mainly biomass ash from various agriculture waste act differently when participating as part of the cementitious mixes depending on their chemical characteristics and the amount being added.

According to Martínez-Lage et al. [27], concrete with 10% paper ashes replacement of cement has higher compressive strength at 7 and 28 days of maturity compared to the OPC concrete. Besides, this statement is further supported by a literature review [28] stating that there is an increase in mechanical strength when the substitution rate of cement to wood fly ash is at 10%. However, there is a result showing that the higher substitute rate of cement in the concrete, the lower the strength of the concrete [8].

A replacement of 0–20% of either binder or aggregates in cement is most widely experimented by researchers, whereas some tested higher replacement rate of 40% [29]. Generally, additional waste material would reduce the compressive strength and flexural strength as supplementary waste material would dilute the cement within the mix and absorb water, reducing pozzolanic reaction rate [30].

Most studies have tested a replacement rate of 20% of the cement and the result for this proportion level of waste replacement provides an adequate strength of the concrete [9, 11, 12]. However, Sofi et al. [8] tested the vortex-treated biomass and the sample containing 20% was still wet and did not bind together. Therefore, they concluded that the maximum dosage of vortex-treated biomass should not exceed 15%. The vortex-treated biomass is also the material being studied in this research. The conclusion from the previous research helps to identify a proper range of dosage for the replacement material.

Workability

Workability as another critical feature of concrete mix is also affected by adding waste replacements, especially biomass ashes. The smaller particle size of biomass ashes gives it a larger surface area to react with other materials, which leads to the increased water demand of the concrete mix and reduces consistency to maintain required slump value [15]. Certain materials such as bamboo, sisal and oyster shell also have higher water absorption ability to cause increased water demand [10]. It is commonly known that the higher amount of water in the concrete mix will primarily reduce the strength of hardened concrete; this makes workability become one of the challenges of using biomass materials as a replacement in concretes.

However, some researchers have been trying to study concrete with biomass component (mainly agro-waste) [28, 30]. They utilize superplasticizer to improve the workability of the concrete mix and metakaolin to enhance the strength of the hardened concrete [28, 30]. The researches show that the addition of superplasticizer into the mixture has successfully increased the workability of the concrete mix. Nevertheless, from these literature reviews, there is no equation or ratio to determine the dosage of superplasticizers. Therefore, it can be concluded that the dosage of superplasticizers is dependent on the targeted slump of each research. In this research, superplasticizers will also be incorporated into the concrete mixture to enhance concrete performance.

3.2 *Factors Affecting Concrete Strength*

Aggregates

Aggregates account for about 60–80% of the concrete volume and are approximately 70–80% of its weight. Therefore, it is imperative to add the right aggregates into the concrete mix to obtain satisfactory strength. Aggregates will determine the concrete's elastic, thermal properties and dimensional stability of the concrete. Aggregates' features such as toughness, shape, size, density, soundness and specific gravity will also determine the concrete's mechanical properties and workability [31]. According to Haque et al. [32], the compressive strength of the concrete increases with larger aggregates. Vilane and Sabelo [33] stated that increase of aggregate size will improve the workability of the concrete. However, Ogundipe et al. [31] claimed that a particular aggregate size in a specific concrete mix gives optimal compressive strength and using larger aggregate will decrease the strength of the concrete. Also, finer aggregates are recommended in the concrete mix to make it more resistant to flexural stresses.

Curing Method

Different curing methods will affect the concrete strength. According to Araldi et al. [34], humidity and temperature will affect the behavior of concrete. There are several curing methods such as sprinkle of water, membrane curing and steam curing etc. The most common practice for concrete curing found in lots of researches is standard curing in which the samples are submerged into water at room temperature at 23 °C. Li et al. found that steam curing has satisfactory performance on the removal strength of concrete with large amount of mineral admixture like FA and GGBS [35].

Water-to-Cement Ratio (w/c)

For pure OPC cement with certain amount of cement content, the lower w/c ratio contributes to higher compressive strength [36]. This is due to the decreased distances between cement particles when the w/c ratio is lower [37]. However, the use of very low w/c ratio could adversely affect the workability [36]. According to the experiment conducted by Felekoğlu et al. [38] on fresh self- compacting concrete, the optimum water to cement (w/c) ratio ranges from 0.48 to 0.6. Besides, Elinwa and Mahmood [39] utilized w/c ratio of 0.565 in the wood ash concrete. Moreover, due to the existence of carbon which will promote the water absorbability in the concrete will result in reduced workability of the cement paste. Similarly for the biomass concrete, the biomass as fiber material could absorb water and reduce workability of concrete. Therefore, superplasticizers shall be used in the experiment to achieve targeted workability.

4 Conclusions

Based on the literature review and on the previously investigated materials, it could be concluded that incorporating municipal solid waste like biomass in concrete can produce more sustainable concrete. It helps in reducing waste production and preventing more waste to go into landfill. The biomass concrete should be air-cured as the municipal waste in the concrete will absorb more water and become weaker if it is cured underwater. It is crucial to check such possibility in an experimental program, which is reported in the Part II having the title 'Treated Municipal Solid Waste (Biomass) based Concrete Properties – Part II: Experimental program' [40].

Acknowledgements This research is financially supported by the Melbourne Research Scholarship offered by the University of Melbourne and the Australian Research Council's Discovery Early Career Researcher Grant (DE170100165, DE 2017 R1). This work is financially supported by: Base Funding—UIDB/04708/2020 of the CONSTRUCT—Instituto de I&D em Estruturas e Construções—funded by national funds through the FCT/MCTES (PIDDAC). This work is funded by national funds through FCT—Fundação para a Ciência e a Tecnologia, I.P., under the Scientific Employment Stimulus—Institutional Call—CEECINST/00049/2018.

References

1. Mian, M.M., Zeng, X.L., Bin Nasry, A.A., Al-Hamadani, S.M.Z.F.: Municipal solid waste management in China: a comparative analysis. J. Mater. Cycles Waste Manag. **19**(3), 1127–1135 (2017)
2. Zhang, D., Huang, G., Xu, Y., Gong, Q.: Waste-to-energy in china: key challenges and opportunities. Energies **8**(12), 14182–14196 (2015)
3. Bioelektra Group: Bioelektra Group Homepage (2019). Available: http://bioelektra.com/en/
4. Petkar, S.S.: Environmental Impact of Construction Materials and Practices (2014). Available: https://www.researchgate.net/publication/290427381_Environmental_Impact_Of_Construction_Materials_And_Practices?channel=doi&linkId=56973eca08aea2d74374bf64&showFulltext=true
5. Hu, H., Kavan, P.: Energy consumption and carbon dioxide emissions of China's non-metallic mineral products industry: present state, prospects and policy analysis. Sustainability **6**(11), 8012–8028 (2014)
6. Salazar, K., Kimball, S.M.: Mineral Commodity Summaries, 2009. Government Printing Office, Washington (2009)
7. Lei, Y., Zhang, Q.A., Nielsen, C., He, K.B.: An inventory of primary air pollutants and CO_2 emissions from cement production in China, 1990–2020. Atmos. Environ. **45**(1), 147–154 (2011)
8. Sofi, M., Sabri, Y., Zhou, Z., Mendis, P.: Transforming municipal solid waste into construction materials. Sustainability **11**(9), 2661 (2019)
9. Prusty, J.K., Patro, S.K., Basarkar, S.S.: Concrete using agro-waste as fine aggregate for sustainable built environment—a review. Int. J. Sustain. Built Environ. **5**(2), 312–333 (2016)
10. Mo, K.H., Alengaram, U.J., Jumaat, M.Z., Yap, S.P., Lee, S.C.: Green concrete partially comprised of farming waste residues: a review. J. Clean. Prod. **117**, 122–138 (2016)
11. Udoeyo, F.F., Inyang, H., Young, D.T., Oparadu, E.E.: Potential of wood waste ash as an additive in concrete. J. Mater. Civ. Eng. **18**(4), 605–611 (2006)

12. Kaur, G., Siddique, R., Rajor, A.: Properties of concrete containing fungal treated waste foundry sand. Constr. Build. Mater. **29**, 82–87 (2012)
13. Ryu, C., Shin, D.: Combined heat and power from municipal solid waste: current status and issues in South Korea. Energies **6**(1), 45–57 (2013)
14. Connett, P.: Municipal waste incineration: a poor solution for the first century. Presented at the 4th Annual International Management Conference, Amsterdam (1998)
15. Ban, C.C., Ramli, M.: The implementation of wood waste ash as a partial cement replacement material in the production of structural grade concrete and mortar: an overview. Resour. Conserv. Recycl. **55**(7), 669–685 (2011)
16. Zhang, D.L., Huang, G.Q., Xu, Y.M., Gong, Q.H.: Waste-to-energy in China: key challenges and opportunities. Energies **8**(12), 14182–14196 (2015)
17. Bildirici, M.E.: Cement production, environmental pollution, and economic growth: evidence from China and USA. Clean Technol. Environ. Policy **21**(4), 783–793 (2019)
18. Vishwakarma, V., Ramachandran, D.: Green concrete mix using solid waste and nanoparticles as alternatives—a review. Constr. Build. Mater. **162**, 96–103 (2018)
19. OECD: OECD Environmental Performance Reviews: Australia 2019. OECD Publishing (2019)
20. Agamuthu, P.: Challenges and opportunities in agro-waste management: an Asian perspective. In: Inaugural Meeting of First Regional 3R Forum in Asia, pp. 11–12 (2009)
21. Syarif, M., Sampebulu, V., Tjaronge, M.W., Nasruddin: Characteristic of compressive and tensile strength using the organic cement compare with portland cement. Case Stud. Constr. Mater. **9** (2018)
22. Mannan, M.A., Ganapathy, C.: Concrete from an agricultural waste-oil palm shell (OPS). Build. Environ. **39**(4), 441–448 (2004)
23. Kua, T.A., Arulrajah, A., Horpibulsuk, S., Du, Y.J., Shen, S.L.: Strength assessment of spent coffee grounds-geopolymer cement utilizing slag and fly ash precursors. Constr. Build. Mater. **115**, 565–575 (2016)
24. Ganiron, Jr., T.U.: Sustainable management of waste coconut shells as aggregates in concrete mixture. J. Eng. Sci. Technol. Rev. **6**(5) (2013)
25. Gonzalez-Kunz, R.N., Pineda, P., Bras, A., Morillas, L.: Plant biomass ashes in cement-based building materials. Feasibility as eco-efficient structural mortars and grouts. Sustain. Cities Soc. **31**, 151–172 (2017)
26. Xu, R.S., He, T.S., Da, Y.Q., Liu, Y., Li, J.Q., Chen, C.: Utilizing wood fiber produced with wood waste to reinforce autoclaved aerated concrete. Constr. Build. Mater. **208**, 242–249 (2019)
27. Martínez-Lage, I., et al.: Concretes and mortars with waste paper industry: biomass ash and dregs. J. Environ. Manag. **181**, 863–873 (2016)
28. Teixeira, E.R., Camoes, A., Branco, F.G.: Valorisation of wood fly ash on concrete. Resour. Conserv. Recycl. **145**, 292–310 (2019)
29. Shah, P.A., Mehta, J.G., PathariyaSaraswati, C., RanaJaykrushna, A., Patel, A.N.: Sugarcane baggase ash and pozzocrete as an techno-economical solution in design mix concrete. Indian J. Appl. Res. **4**(5) (2014)
30. Zeidabadi, Z.A., Bakhtiari, S., Abbaslou, H., Ghanizadeh, A.R.: Synthesis, characterization and evaluation of biochar from agricultural waste biomass for use in building materials. Constr. Build. Mater. **181**, 301–308 (2018)
31. Ogundipe, O.M., Olanike, A.O., Nnochiri, E.S., Ale, P.O.: Effects of coarse aggregate size on the compressive strength of concrete. Civ. Eng. J. Tehran **4**(4), 836–842 (2018)
32. Haque, M., Tuhin, I., Farid, M.S.S.: Effect of aggregate size distribution on concrete compressive strength. SUST J. Sci. Technol. **19**(5), 35–39 (2012)
33. Vilane, B.R.T., Sabelo, N.: The effect of aggregate size on the compressive strength of concrete. J. Agric. Sci. Eng. **2**(6), 66–69 (2016)
34. Araldi, P., Balestra, C.E.T., Savaris, G.: Influence of multiple methods and curing temperatures on the concrete compressive strength. J. Eng. Proj. Prod. Manag. **9**(2), 66–73 (2019)
35. Li, M., Wang, Q., Yang, J.: Influence of steam curing method on the performance of concrete containing a large portion of mineral admixtures. Adv. Mater. Sci. Eng. **2017**, 1–11 (2017)
36. Neville, A.M.: Properties of Concrete, 5th edn. Pearson Education Limited, England (2011)

37. Bentz, D.P., Aitcin, P.C.: The hidden meaning of water-cement ratio. Concr. Int. **30**(5), 51–54 (2008)
38. Felekoglu, B., Turkel, S., Baradan, B.: Effect of water/cement ratio on the fresh and hardened properties of self-compacting concrete. Build. Environ. **42**(4), 1795–1802 (2007)
39. Elinwa, A.U., Mahmood, Y.A.: Ash from timber waste as cement replacement material. Cem. Concr. Compos. **24**(2), 219–222 (2002)
40. Sofi, M., Maia, L., Liu, J., Sabri, Y., Zhou, A., Frahmand, T., Mendis, P.: Treated Municipal Solid Waste (Biomass) Based Concrete Properties—Part II: Experimental Program. In: Springer in RILEM Bookseries of the 3rd RILEM SPRING Convention (2020)

Recovery of Phosphorous from Sewage Sludge Ash Prior to Utilization as Secondary Resource in Concrete and Bricks

Lisbeth M. Ottosen, Gunvor M. Kirkelund, and Pernille E. Jensen

Abstract Sewage sludge ash (SSA), which today is residual waste, can be separated into two resources, phosphorous and a particulate material for use in production of concrete or brick, by electrodialytic separation (EDS). Three SSAs from different sewage sludge mono-incineration plants were included in this investigation. Overall they had similar characteristics, but still the differences meant that the EDS process needs optimization for each ash type. Under the same experimental conditions, 80% P was recovered from two of the SSAs whereas only 65% was recovered from the third SSA. After EDS, the investigation points at a decrease in Ca and P may be beneficial if using the SSA-EDS in concrete. The investigation also showed that the investigated SSAs had high Fe contents, which may be problematic if used in brick production. In conclusion, the investigation points at a potential for SSA to be considered as secondary resource in the construction materials after EDS.

Keywords Sewage sludge ash · Phosphorous · Supplementary cementitious material · Fired clay bricks

1 Introduction

Ashes from incineration of sewage sludge (SSA) are often residual waste, i.e. removed from the overall material cycle. The ashes might be secondary resources. Research is carried out on utilization of SSA as supplementary cementitious material in concrete [1–4] or clay replacement in fired clay bricks [5–8]. However, attention should be payed to the high content of phosphorous (P), as P is a finite, essential resource, which is in list of critical raw materials for the EU [9]. The P concentration in SSA is typically 5–10 wt% [1], and even higher concentrations have been found, 11 wt% P [10], 11.9 wt% P [11] and 12.3 wt% P [12]. The P content in phosphate rock is generally reported as P_2O_5, and the economic grade varies from 25 to 37% P_2O_5 [13], corresponding to 11–16 wt% P. The P content in SSAs are thus generally lower;

L. M. Ottosen (✉) · G. M. Kirkelund · P. E. Jensen
Department of Civil Engineering, Technical University of Denmark, Building 118, 2800 Lyngby, Denmark
e-mail: LO@byg.dtu.dk

V. M. C. F. Cunha et al. (eds.), *Proceedings of the 3rd RILEM Spring Convention and Conference (RSCC 2020)*, RILEM Bookseries 35,
https://doi.org/10.1007/978-3-030-76543-9_28

however, some SSAs have concentrations in the same range. Since the P concentration is high in SSA, P recovery should be considered prior to use in construction materials, otherwise P is lost for recovery. The high P content in SSA has reached awareness in some EU countries and legislation supporting/requesting P recovery, e.g. in Germany, where the sewage sludge ordinance includes requirements for P recovery from SSA [14]. Research and development are progressing to implement P recovery from SSA. The methods under development are based thermochemical treatment or acid extraction, the latter being reported most extensively in literature.

Research on recovery of P from SSA and following use of the SSA in construction materials is scarce. Donatello et al. [15] used sulfuric acid to wash out P and tested the pozzolanic activity of the SSA. The acid washing decreased the pozzolanic activity according to the SAI and Frattini tests. Kappel et al. [16] used EDS (described in Sect. 1.1) for recovery of P prior to use in mortar. They reported a loss in compressive strength from 60 to 55 MPa when 20% cement was replaced with SSA after EDS treatment (SSA-EDS) and milling. They noted that the workability of mortar with SSA-EDS decreased. Ottosen et al. [8] reported a screening with SSA-EDS as clay replacement in bricks (10–60% replacement). The apparent density decreased and the porosity increased with increasing clay substitution. When firing the brick material with 60% clay replacement at 1050 °C a significant glassy phase formed. The results generally showed that 60% replacement of clay was probably too much, but with 10% replacement, the properties were changed only slightly compared to the reference. The mentioned investigations points at potential uses for the SSA after P recovery, but the potential needs unfolding prior to a full evaluation of the options.

Evaluating the use potential for use of SSA secondary raw material includes knowledge on amounts and variety in SSA characteristics over time and between plants. Worldwide, 1.7 mio tons SSA are produced annually and the amount is increasing [17]. At a local, Danish scale, the two large (out of three) sewage sludge incinerators have an annual production of about 20,000 t SSA (containing 1000–2000 t P). Annually, 13,500 t P is applied as commercial fertilizer in Denmark [18], i.e. 8–14% of this P could be recovered from SSA. Denmark uses 4–5 mio m^3 concrete annually [19] corresponding to 1–1.5 mio tons cement. After acid recovery of P, about half of the SSA is left [10], i.e. 10,000 tons in the Danish case. Thus, SSA after P recovery can replace only about 0.2% cement in average. However, the SSA has coloring potential for the concrete [2] and brick [8], which gives the construction materials with SSA new interesting aesthetical potentials, possibly adding value to these materials. To scale the use of SSA as secondary resource it is beneficial if the SSA characteristics varies only little over time and between plants so the P recovery processes and the mix design for construction materials need minimum of adjustments. In this investigation, SSA from three different mono-incineration plants were collected and treated by EDS. The aim is to investigate if the differences between the three ashes are in a range, where they can be considered being the same raw material.

1.1 *Electrodialytic Separation for Recovery of Phosphorous*

EDS recover P and remove heavy metals simultaneously in a two compartment electrodialytic cell [10] (Fig. 1). The compartments are separated by a cation exchange membrane, which only allows cations to pass. The SSA is suspended in water in the anode compartment. When the electric DC current is applied, the anolyte with SSA suspension is acidified due to electrolysis of H_2O at the anode ($H_2O \rightarrow 2H^+ + \frac{1}{2}O_2$ (g) + $2e^-$). During the acidification, heavy metals and P are extracted from the SSA, but whereas the cationic heavy metals are transported by electromigration over the cation exchange membrane and concentrate in the cathode compartment, the extracted P remains in the filtrate of the SSA suspension as negatively charged ions or neutral molecules. Hereby simultaneous extraction and separation is obtained. At the end of the treatment, the recovered P is in the filtrate of the anolyte and the solubilized heavy metals in the catholyte. Filtration separates SSA-EDS and the filtrate with the recovered P. Previous research has shown 80–90% recovery by EDS [10, 20].

2 Materials and Methods

The work includes SSAs from three different mono-incineration plants for sewage sludge, where the sludge was incinerated in a fluidized bed combustor. The sewage sludge incinerated originated from treatment of urban wastewater. The three ashes are: SSA-A. From Germany, SSA-B From Spildevandscenter Avedøre, BIOFOS, Denmark, and SSA-C from Lynetefællesskabet, BIOFOS, Denmark.

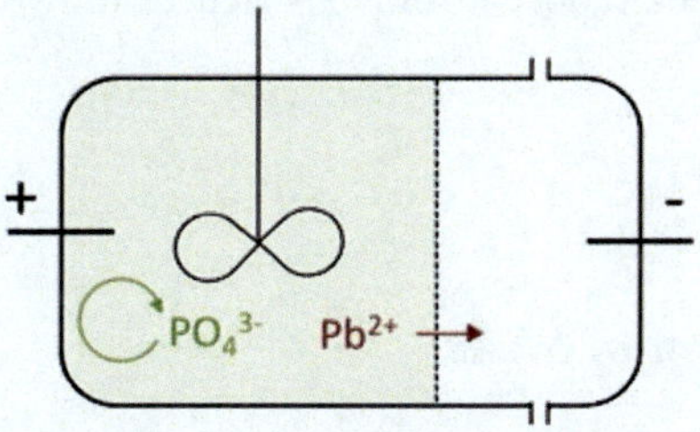

Fig. 1 Principle of EDS cell. During treatment, the ash is suspended in water in the anode compartment where P concentrates in the filtrate, and the heavy metals electromigrate into the catholyte passing the cation exchange membrane (marked as dotted line)

2.1 Ash Characterization

Concentrations of P, Al, Ca, Fe, K, Mg, Na, P, S, Cu, Pb and Zn were measured after pre-treatment (acid digestion) in accordance to Danish Standard DS259: 1.0 g ash and 20.0 ml (1:1) HNO_3 was heated at 200 kPa (120 °C) for 30 min. Filtration through 0.45 μm filter. Concentrations were measured with ICP-OES in the filtrate. SSA pH was measured in suspension: 10.0 g ash in 25 ml distilled water. After 1 h agitation, pH was measured directly in the suspension with Radiometer electrodes. Loss on ignition (LoI) was found after 30 min at 550 °C. Water content was measured as weight loss after 24 h at 105 °C. Three to five replicates of each of these analyses were made. Solubility in water was evaluated as weight loss after washing 50.0 g SSA in 500 ml distilled water three times.

2.2 Electrodialytic Separation Experiments

The EDS laboratory cell following Fig. 1 was used. The compartments were separated by a cation exchange membrane (CEM) from IONICS. The electrodes were made of platinum coated titanium wire (diameter 3 mm) and the length of the electrodes inside the cell was approximately 4 cm. The power supply was Hewlett Packard E3612A. One EDS experiment was made with each of the three SSAs. The experimental conditions were similar. The SSA was suspended in tap water. The liquid:solid ratio of 14 was obtained by suspending 25 g SSA in 350 ml water. The SSA was kept suspended by an overhead stirrer (RW11 basic from IKA). In the cathode compartment 500 mL 0.01 M $NaNO_3$ adjusted to 2 with HNO_3 was circulated. The experiments were conducted at a constant current of 50 mA applied to the electrodes. The duration of all experiments was 1 week (chosen from experiences in [8]).

3 Results

3.1 Characterization of SSAs

Some characteristic of the three SSAs are in Table 1.

In general, the characteristics for the three SSAs were quite similar. The three ashes all had a distinct rusty-red color. The solubility of the SSAs was between 1.4 and 2.0% showing that the ashes contained a minor soluble fraction. They all had a low water content, which means that chemical reactions during storing have been limited. The SSAs were alkaline meaning that hydroxides were leached or formed when suspending the SSAs in water. The major differences between the SSAs (in Table 1) were: (I) The LoI was very low for SSA-A and SSA-B (0.2 and 0.3%) revealing a good incineration. LoI for SSA-C was higher (3%), (II) The Fe

Table 1 Selected characteristics of the three investigated SSAs: pH, conductivity, LoI, water content and solubility and concentrations (found by acid digestion)

	SSA-A	SSA-B	SSA-C
pH	11.4 ± 0.2	9.8 ± 0.01	10.2 ± 0.05
LoI 550 °C (%)	0.3 ± 0.005	0.6 ± 0.02	3.0 ± 0.12
Water content (%)	0.2 ± 0.04	0.3 ± 0.2	0.2 ± 0.05
Water solubility (%)	1.4	1.9	2.0
Al (g/kg)	17.5 ± 0.8	21.3 ± 0.3	13.8 ± 0.5
Ca (g/kg)	119 ± 4.5	122 ± 2.0	139 ± 3.6
Fe (g/kg)	93.4 ± 3.6	68.3 ± 1.6	41.5 ± 0.2
K (g/kg)	6.7 ± 0.3	5.4 ± 0.05	4.3 ± 0.03
Mg (g/kg)	9.0 ± 0.3	15.4 ± 0.3	15.4 ± 1.8
Na (g/kg)	2.5 ± 0.07	2.7 ± 0.03	2.1 ± 0.05
P (g/kg)	92.9 ± 4.2	91.8 ± 1.5	56.8 ± 2.3
S (g/kg)	7.9 ± 0.2	6.9 ± 0.03	n.m.
Cu (mg/kg)	2090 ± 90	490 ± 3	500 ± 2
Pb (mg/kg)	97 ± 0.7	158 ± 1	103 ± 5
Zn (mg/kg)	2490 ± 127	2160 ± 34	2840 ± 13

concentration was only half in SSA-C compared to SSA-A, (III) The P concentration was low (56.8 g/kg) in SSA-C compared to the two other SSAs (about 90 g/kg) and (IV) The Cu concentration was about 4 times higher in SSA-A than in the two other SSAs.

3.2 *Recovery of Phosphorous and Separation of Heavy Metals by EDS*

Table 2 shows the P recovery and mass loss from the three EDS experiments. Figure 2 shows the distribution of P, Cu, Pb and Zb at the end of the EDS experiments.

Mass balances are given as the mass of an element in different parts of the EDS cell at the end of the experiment over the mass of the element initially in the SSA. The mass balances for the different elements in the three EDS experiments were between 85 and 115% (data not shown), which are acceptable. The exceptions from

Table 2 pH of SSA and mass loss by the end of the 7 day EDS experiments

	EDS-A	EDS-B	EDS-C
Mass loss (g)	9.8	11.6	12.2
Mass loss (%)	39	46	49
Recovered P (%)	65	81	82

Percentage of recovered P (in anolyte filtrate) and removed Cu, Pb and Zn (in cathode compartment)

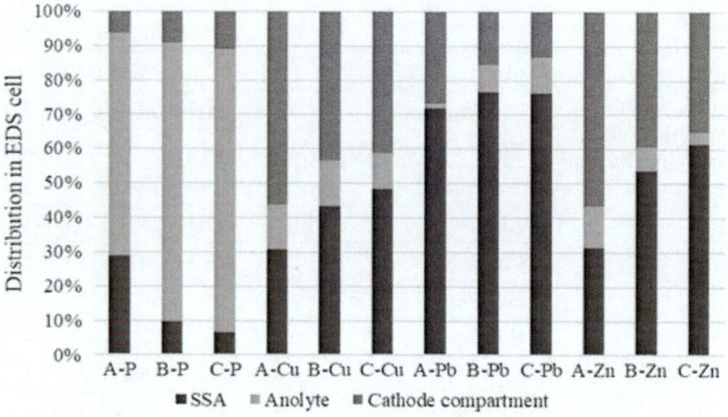

Fig. 2 Distribution of P, Cu, Pb and Zn in the electrodialytic cell at the end of the three EDS experiments (SSA = solid ash residue, anolyte = filtrate, cathode compartment = at cathode + in membrane + in catholyte)

this are Pb in experiment SSA-A (mass balance 145%) and Fe in all experiments (68–75%).

The P recovery was very similar in the EDS experiments B and C, where about 80% P was in the anolyte filtrate by the end of the experiments (Fig. 2). In EDS-A on the contrary, only 65% P was recovered into the anolyte and 29% remained in the SSA. The Cu, Pb and Zn released from the SSA during EDS were removed into the catholyte, and thus the separation of P and heavy metals was successful. In all three SSAs, 70–75% Pb was still in the SSA after EDS, which was the highest percentage of the three heavy metals in every of the experiments. Few percentages of more Zn than Cu was recovered in every experiment, and the order in percentages remaining in the SSA was the same for Cu and Zn: SSA-C > SSA-B > SSA-C.

3.3 *Characterization of SSA-EDSs*

Characteristics of the three SSA-EDSs are in Table 3. The SSAs were acidified from the original alkaline to acidic (3.4–4.2), and the acidification had resulted in a SSA mass loss of between 39 and 49% (Table 1). There is no correlation between final pH after EDS and mass loss. The major differences in the concentrations are: (I) Significantly higher Ca and P concentrations in SSA-EDS-A than in the two other SSAs, (II) The Fe concentration was about half in SSA-C compared to the two other SSAs, (III) A much higher Cu concentration in SSA-EDS-A, which was also the case before EDS (Table 1), and (IV) the Zn concentrations were different in all three ashes.

Table 3 Selected characteristics of SSA-EDSs

	SSA-EDS-A	SSA-EDS-B	SSA-EDS-C
pH	3.4	4.2	3.6
Al (g/kg)	7.4 ± 0.1	12.3 ± 0.5	9.1 ± 0.8
Ca (g/kg)	23.2 ± 0.4	6.7 ± 0.4	4.5 ± 0.4
Fe (g/kg)	95.3 ± 1.2	91.5 ± 3.8	46.1 ± 2.1
K (g/kg)	2.9 ± 0.04	3.8 ± 0.2	3.4 ± 0.3
Mg (g/kg)	3.2 ± 0.05	4.0 ± 0.2	4.6 ± 0.6
P (g/kg)	40.1 ± 0.6	16.6 ± 1.0	7.4 ± 0.5
S (g/kg)	2.0 ± 0.05	0.8 ± 0.05	n.m.
Cu (mg/kg)	1210 ± 0.1	460 ± 0.4	430 ± 2.8
Pb (mg/kg)	162 ± 0.4	191 ± 0.1	160 ± 2.2
Zn (mg/kg)	1040 ± 0.2	1630 ± 0.8	2890 ± 12

Concentrations are based on acid digestion

4 Discussion

The EDS recovery of P was successful (in relation to the 80% recovery target [14]) from experiments EDS-B and EDS-C, whereas only 65% P was recovered from experiment EDS-A. The P concentrations were very similar in SSA-A and SSA-B (Table 1). The experimental conditions were the same, and thus the acid production from the electrolysis at the anode was the same. The pH reached the lowest level in SSA-EDS-A (Table 3). Thus, the buffering capacity for SSA-A was less than for the two other SSAs, which corresponds well to SSA-A containing the largest the fraction not being dissolved during EDS. The difference in P extraction must be found in differences in the P speciation in the SSAs. The P speciation was not determined in the present study, but the EDS results clearly points at the importance of an increased understanding of the P speciation. Whitlockite (calcium magnesium phosphate mineral) has previously been determined in SSA [21], but otherwise little has been reported on P speciation in SSAs. The knowledge gap also includes the link between the firing characteristics at the different incinerators and incineration parameters, which are known to play a major role in the ash characteristics for another type of biomass ash, namely wood ash [22].

The distinct rusty-red color revealed a high iron oxide content. The Fe concentration in Tables 1 and 3 were measured after acid digestion, and shows the acid soluble concentration of Fe rather than the total concentration. In [23] it was found for two SSAs that the concentration of Fe was two times higher when measured as total concentration by XRF than by ICP after acid digestion after DS259 (as in the present investigation). The Al, Ca, K, Mg, P and S concentrations found with XRF were generally in good agreement with the concentrations after acid digestion according to DS259. Figure 3 shows the acid soluble concentrations of Al, Ca, Fe K, Mg and P in the investigated SSAs and SSA-EDSs found after DS259 (Tables 1 and 3).

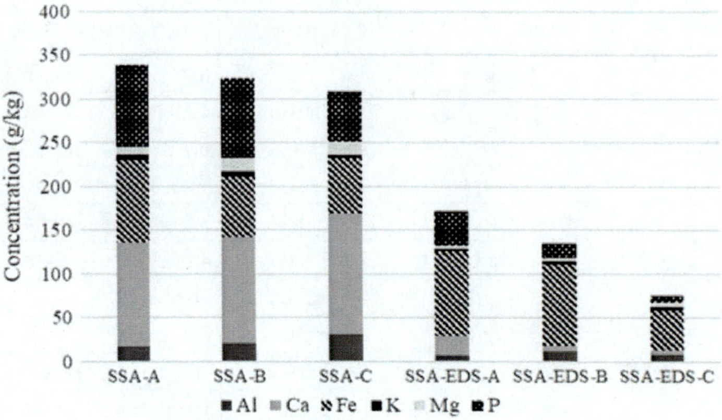

Fig. 3 Acid soluble concentrations [g/kg] of Al, Ca, Fe, Mg, P and S in the three ashes before and after EDS

Based on XRD analysis, Kappel et al. [16] identified quartz, feldspar and hematite in SSA and SSA-EDS (in SSA samples from the same plant as SSA-B). Cheeseman et al. [21] reported quarts, whitlockite and hematite in an SSA from UK, and Anderson [24] quartz, calcite, hematite, anhydride, feldspar and glassy phase ~70% in another SSA from UK. Thus, quartz and hematite were identified in these reported cases. The quartz may originate from the fluidized bed. The high concentration of hematite is due to the use of iron salts to precipitate P in the wastewater treatment facility, which is the most commonly used chemical. Other facilities use Al as major salt and this strongly influences the chemical properties of the SSA [5, 23].

The most significant decrease during EDS was in the Ca concentration followed by P (Fig. 3). The total Fe concentration must be expected higher than the acid soluble fraction in Fig. 3 (as mentioned in the paragraph above), but as seen the acid soluble fraction increased, however, taking the solubilized fraction during EDS into account, Fe has overall been removed from the SSAs. The elements in Fig. 3 do not account for the major mass of the SSAs. As quarts was reported present in SSAs [11, 21, 24], and since quarts sand (the part originating from the fluidized bed) is insoluble during EDS, Si must be expected prevailing in the investigated SSA and SSAs as well.

In relation to concrete, EN450-1 (2012) opens for the use of co-combustion ashes with coal and sewage sludge. The investigated ashes are though from mono-incineration of SSA. However, comparing the concentrations of Ca, Mg, P and S (calculated to oxides) to the requirements described in EN450-1 (2012) might point on limitations in the use, if the upper concentrations in the requirements are exceeded. Requirements are that the reactive Ca shall not exceed 10% CaO per mass, the total MgO content shall not exceed 4%, the total content of P_2O_5 not 5% and the SO_3 content shall not exceed 3%. The methods prescribed used in EN450-1 (2012) are not the same as used in the present investigation, but the comparison of the measured concentrations and the prescribed limiting values points at some issues, where further investigations are needed. These points are marked in orange in Table 4, and as seen

Table 4 Contents of Ca, Mg, P and S in the SSAs and SSA-EDSs compared to limiting values in EN450-1

	SSA-A	SSA-B	SSA-C	SSA-EDS-A	SSA-EDS-B	SSA-EDS-C
CaO	16,7	17,1	19,3	3,2	0,9	0,6
MgO	1,5	2,6	2,6	0,5	0,7	0,8
P_2O_5	21,3	21,0	13,0	9,2	3,8	1,7
SO_3	2,0	1,7	n.a.	0,5	0,2	n.a.

The colors indicate concentrations needing more investigation (orange) or concentrations, which seems non-problematic (green). It must be stressed that these are indications since the analysis used are not the same as prescribed in EN450-1

it is the content of CaO and P_2O_5 in all three SSAs before EDS, and P_2O_5 in SSA-A after EDS. Table 4 indicates that the problematic issues with high concentrations of CaO (which can potentially be reactive) and P_2O_5 were solved during EDS, however, a thorough investigation is needed to conclude.

In relation to brick production, the hematite content was considered responsible for formation of a glassy-like phase when clay/SSA-EDS was fired at 1050 °C, a phase strongly influenced the properties of the brick material [8]. Even though the total concentration of Fe was not determined in the investigated SSAs and SSA-EDSs, the Fe concentration was very high in all cases as the acid soluble (DS259) concentrations were between 4% (Table 1) and 9.5% (Table 3). This is thus a general point for further investigation in relation to use of SSA-EDS in brick manufacturing.

5 Conclusion

SSA is a potential resource for the critical raw material P, and some countries have legislation supporting/requesting P recovery. In Denmark, about 7–14% of the imported commercial fertilizer could be replaced with P extracted from Danish SSAs. EDS is a method under development for P recovery. The current investigation showed, that method need optimization for the actual SSA, since sufficient P was recovered from only two out of the three SSAs, when using the same experimental parameters. After P recovery, about 50–60% of the SSA remains after EDS, and this fraction was evaluated as raw material for concrete or brick on basis of the chemical composition. The SSAs all contained a high content of Fe (both prior to and after EDS), and this content should be investigated in relation to use in bricks, as iron oxide is known to form a glassy like phase in bricks, changing the properties. For use in concrete, the high concentration of Ca (which can be problematic if reactive) in SSA was decreased during EDS to a low, and likely non-problematic level. The same goes for P. The investigation points at EDS having potential to separate SSA into two resources, P and material for production of construction materials. It also points at P speciation and mineralogy/composition of the treated SSA to be major focus points in future research.

References

1. Cyr, M., Coutand, M., Clastres, P.: Technological and environmental behavior of sewage sludge ash (SSA) in cement-based materials. Cem. Concr. Res. **37**, 1278–1289 (2007)
2. Kappel, A., Ottosen, L.M., Kirkelund, G.M.: Colour, compressive strength and workability of mortars with an iron rich sewage sludge ash. Constr. Build. Mater. **157**, 1199–1205 (2017)
3. Krejcirikova, B., Ottosen, L.M., Kirkelund, G.M., Rode, C., Peuhkuri, R.H.: Characterization of sewage sludge ash and its effect on moisture physics of mortar. J. Build. Eng. **21**, 396–403 (2019)
4. Rutkowska, G., Wichowski, P., Franus, M., Mendryk, M., Fronczyk, J.: Modification of ordinary concrete using fly ash from combustion of municipal sewage sludge. Materials **13**(487), 1–17 (2020)
5. Wiebusch, B., Seyfried, C.F.: Utilization of sewage sludge ashes in the brick and tile industry. Water Sci. Technol **36**(11), 251–258 (1997)
6. Lin, D.F., Weng, C.H.: Use of sewage sludge ash as brick material. J. Environ. Eng. **127**(10), 922–927 (2001)
7. Anderson, M., Skerratt, R.G.: Variability study of incinerated sewage sludge ash in relation to future use in ceramic brick manufacture. Br. Ceram. Trans. **102**(3), 109–203 (2003)
8. Ottosen, L.M., Bertelsen, I.M.G., Jensen, P.E., Kirkelund, G.M.: Sewage sludge ash as resource for phosphorous and material for clay brick manufacturing. Constr Build Mater 249, 118684 (2020)
9. COM: 490 On the 2017 List of Critical Raw Materials for the EU (2017)
10. Ottosen, L.M., Jensen, P.E., Kirkelund, G.M.: Phosphorous recovery from sewage sludge ash suspended in water in a two-compartment electrodialytic cell. Waste Manage. **51**, 142–148 (2016)
11. Krüger, O., Adam, C.: Recovery potential of German sewage sludge ash. Waste Manage. **45**, 400–406 (2015)
12. Ebbers, B., Ottosen, L.M., Jensen, P.E.: Comparison of two different electrodialytic cells for separation of phosphorus and heavy metals from sewage sludge ash. Chemosphere **125**, 122–129 (2015)
13. Gupta, D.K., Chatterjee, S., Datta, S., Veer, V., Walther, C.: Review. Role of phosphate fertilizers in heavy metal uptake and detoxification of toxic metals. Chemosphere **108**, 134–144 (2014)
14. Verordnung zur neuordnung der klärschlammverwertung. Bundesgesetzblatt Jahrgang Teil I Nr. 65 (2017)
15. Donatello, S., Freeman-Pask, A., Tyrer, M., Cheeseman, C.R.: Effect of milling and acid washing on the pozzolanic activity of incinerator sewage sludge ash. Cem. Concr. Compos. **32**, 54–61 (2010)
16. Kappel, A., Viader, R.P., Kowalski, K.P., Kirkelund, G.M., Ottosen, L.M.: Utilization of electrodialytically treated sewage sludge ash in mortar. Waste Biomass Valor. **9**(12), 2503–2515 (2018)
17. Donatello, S., Cheeseman, C.R.: Recycling and recovery routes for incinerated sewage sludge ash (ISSA): a review. Waste Manage. **33**(11), 2328–2340 (2013)
18. Poulsen, H.D., Møller, H.B., Klinglmair, M., Thomsen, M.: En fosforvidenssyntese. Fosfor i dansk landbrug—ressource og miljøudfordring (in Danish). Report from Aarhus Universitet, DCE (2019)
19. Dansk Beton: Bæredygtig beton initiativ. Roadmap mod 2030, Halvering af CO2-udledningen fra betonbyggeri (in Danish), Copenhagen, Denmark (2019)

20. Guedes, P., Couto, N., Ottosen, L.M., Ribeiro, A.B.: Phosphorus recovery from sewage sludge ash through an electrodialytic process. Waste Manage. **34**, 886–892 (2014)
21. Cheeseman, C.R., Sollars, C.J., McEntee, S.: Properties, microstructure and leaching of sintered sewage sludge ash. Resour. Conserv. Recycl. **40**, 13–25 (2003)
22. Sigvardsen, N.M., Kirkelund, G.M., Jensen, P.E., Geiker, M.R., Ottosen, L.M.: Impact of production parameters on physiochemical characteristics of wood ash for possible utilisation in cement-based materials. Resour. Conserv. Recycl. **145**, 230–240 (2019)
23. Ottosen, L.M., Kirkelund, G.M., Jensen, P.E.: Extracting phosphorous from incinerated sewage sludge ash rich in iron or aluminum. Chemosphere **91**(7), 963–969 (2013)
24. Anderson, M.: Encouraging prospects for recycling incinerated sewage sludge ash (ISSA) into clay-based building products. J. Chem. Technol. Biotechnol. **77**, 352–360 (2002)